STUDY GUIDE

Steven E. Bassett
Southeast Community College

THE HUMAN BODY
in Health and Disease

Martini • Bartholomew • Welch

PRENTICE HALL, Upper Saddle River, NJ 07458

Acquisitions Editor: Halee Dinsey
Project Manager: Don O'Neal
Special Projects Manager: Barbara A. Murray
Production Editor: Meaghan Forbes
Supplement Cover Manager: Paul Gourhan
Supplement Cover Designer: PM Workshop Inc.
Manufacturing Buyer: Dawn Murrin
Cover Image: ©Image Bank, Terje Rakke

Upper Saddle River, NJ 07458

Printed in the United States of America

10 9 8 7 6 5 4 3 2 1

ISBN 0-13-017266-9

Prentice-Hall International (UK) Limited, London
Prentice-Hall of Australia Pty. Limited, Sydney
Prentice-Hall Canada, Inc., Toronto
Prentice-Hall Hispanoamericana, S.A., Mexico
Prentice-Hall of India Private Limited, New Delhi
Pearson Education Asia Pte. Ltd., Singapore
Prentice-Hall of Japan, Inc., Tokyo
Editora Prentice-Hall do Brazil, Ltda., Rio de Janeiro

Contents

CHAPTER 1

An Introduction to Anatomy and Physiology

Anatomy and physiology are part of our daily life. Almost all of us are interested in our own body, how it looks, how it's put together, and what makes it work. In addition to our natural curiosity about the human body, we are fascinated by the complexity of our organs and systems and intrigued by the diverse processes that can disrupt their functions. We spend a significant amount of time and money trying to ensure that our bodies look good and that they are functioning at optimal efficiency.

The term anatomy is derived from a Greek word that means to "cut open," or "dissect." **Anatomy** is the study of the **structure** of the human body. **Physiology** is the science that attempts to explain the physical and chemical processes occurring in the body. Anatomy and physiology are interrelated subjects that provide the foundation for the medical sciences, health-related sports and fitness programs, and clinical applications.

Chapter 1 provides the framework for understanding some basic concepts and learning terminology that will get you started on a successful and worthwhile journey through the human body. By the end of the chapter you should be familiar with the language of anatomy and physiology, basic anatomical concepts, and the physiological processes that make life possible.

Chapter Objectives:

1. Define the terms anatomy and physiology.
2. Identify the major levels of organization in humans and other living organisms.
3. Explain the importance of homeostasis.
4. Describe how positive and negative feedback are involved in homeostatic regulation.
5. Use anatomical terms to describe body regions, body sections, and relative positions.
6. Identify the major body cavities and their subdivisions.
7. Distinguish between visceral and parietal portions of serous membranes.
8. Describe the four common techniques used in a physical examination.

PART I: OBJECTIVE-BASED QUESTIONS

OBJECTIVE 1: Define the terms **anatomy** and **physiology** (textbook p. 3).

After studying p. 3 in the text, you should be able to answer the following questions related to the study of anatomy and physiology.

_____ 1. The study of the body structures is known as _____.

a. anatomy
b. biology
c. homeostasis
d. physiology

_____ 2. The study of how the body functions is known as _____.

a. anatomy
b. biology
c. homeostasis
d. physiology

_____ 3. The word _____ means "to cut open."

a. anatomy
b. biology
c. homeostasis
d. physiology

_____ 4. The study of the chemical processes that take place to keep us alive is known as _____.

a. anatomy
b. biology
c. homeostasis
d. physiology

_____ 5. The study of how bones are connected is called _____; the study of the function of bones is called _____.

a. anatomy, physiology
b. physiology, anatomy
c. anatomy, biology
d. anatomy, homeostasis

_____ 6. The scientific study of living organisms, including humans, is called _____.

a. anatomy
b. biology
c. homeostasis
d. physiology

OBJECTIVE 2: Identify the major levels of organization in humans and other living organisms (textbook pp. 3–5).

After studying pp. 3–5 in the text, you should be able to identify the various levels of organization that exist within all living organisms.

_____ 1. The text discusses six levels of organization. The level of organization ascends from the simplest to the most complex level. Based on that concept, which of the following is the correct sequence of organization?

a. cells–molecules–tissues–organs–organ systems–organism
b. molecules–cells–tissues–organs–organ systems–organism
c. molecules–organs–tissues–cells–organ systems–organism
d. molecules–tissues–cells–organ systems–organs–organism

_____ 2. Which of the following is considered to be the smallest living unit in the body?

a. atom
b. cell
c. molecule
d. tissue

_____ 3. Which of the following statements is true?

a. If enough cells malfunction, eventually the organ will malfunction.
b. If an organ malfunctions, eventually the cells will malfunction.
c. Both a and b are true.
d. Neither a nor b is true.

_____ 4. Within each level of organization there are sublevels. Which of the following is the correct sequence of sublevels and levels, ascending from the simplest to the most complex?

a. atoms–molecules–organelles–cells–tissues
b. atoms–molecules–organelles–tissues–cells
c. cells–atoms–molecules–organelles–tissues
d. molecules–organelles–atoms–cells–tissues

_____ 5. Because the skin is made of epithelial tissue, neural tissue, and connective tissue, it is considered to be _____.

a. a tissue
b. an organ
c. an organ system
d. none of the above

6. Identify the various levels of organization on the diagram in Figure 1–1.

Figure 1–1 Levels of Organization

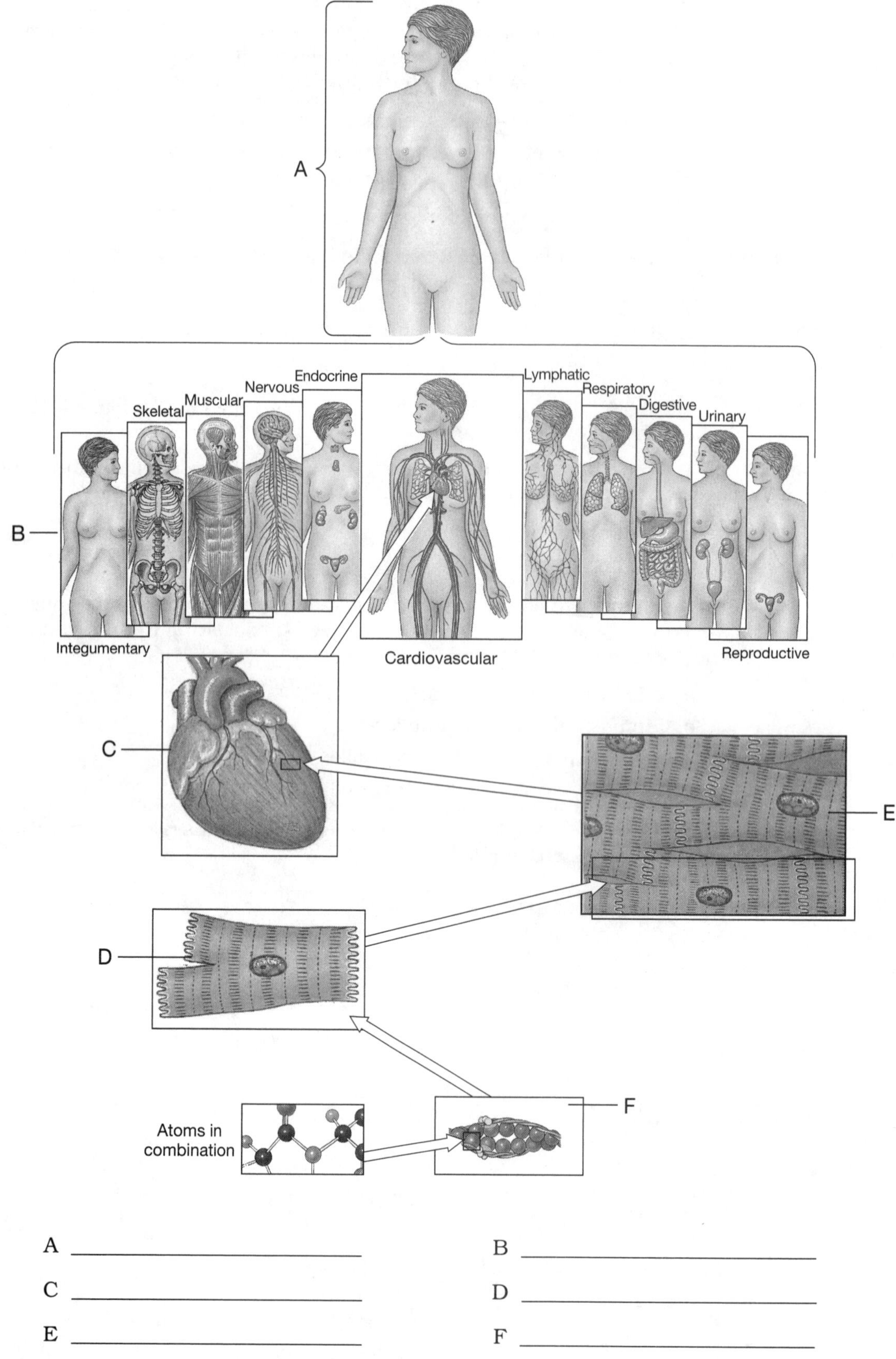

A ______________________ B ______________________

C ______________________ D ______________________

E ______________________ F ______________________

OBJECTIVE 3: Explain the importance of homeostasis (textbook p. 5).

After studying p. 5 in your text, you should have a working understanding of the word *homeostasis*. The concept of homeostasis will recur throughout the text.

_____ 1. The term homeostasis refers to _____.

a. a constantly changing environment
b. an internal environment that does not change very much
c. an unchanging internal environment
d. the study of physiology

_____ 2. A person in homeostasis is _____.

a. healthy
b. immune to diseases
c. sick
d. fluctuating between being healthy and being sick

_____ 3. Which of the following statements is true?

a. When our organs begin to malfunction, the mechanisms that control homeostasis will begin to fail. We will then become sick.
b. When the mechanisms that control homeostasis begin to fail, our organs will malfunction. We will then become sick.
c. When we get sick, our organs will begin to malfunction. The mechanisms that control homeostasis will then fail.
d. When we get sick, the mechanisms that control homeostasis begin to fail. Our organs will then begin to malfunction.

_____ 4. Which of the following sentences uses the word homeostasis correctly?

a. A person may need to go to the doctor's office if he or she is out of homeostasis.
b. If a person is not in homeostasis, he or she is sick.
c. Doctors strive to put us back into homeostasis if we are sick.
d. All the above use the word homeostasis correctly.

5. The mechanisms that control homeostasis are activated in response to a _____________.

OBJECTIVE 4: Describe how positive and negative feedback are involved in homeostatic regulation (textbook pp. 5–6).

After studying pp. 5–6 in the text, you should be able to describe how positive and negative feedback mechanisms are involved in homeostasis.

_____ 1. The mechanisms that control homeostasis are typically activated by a stimulus. A stimulus is _____.

a. a nerve impulse
b. anything that creates an impulse
c. anything that produces a response
d. both b and c

_____ 2. The temperature of our house is under homeostatic control. This control is analogous to the body's homeostatic control because _____.

a. as the temperature changes, the thermostat is turned on or off, just as the mechanisms in our body respond to stimuli
b. the control of our body temperature is known as thermoregulation, which is related to the temperature of the house
c. the temperature of the house fluctuates very little
d. the thermostat acts as a stimulus, since it produces a response

_____ 3. You can think of a positive feedback mechanism as one that causes a response in a constant direction, whereas a negative feedback mechanism causes fluctuating responses. Based on that information, which of the following represents a positive feedback mechanism?

a. the clotting of blood
b. the dilation and constriction of blood vessels
c. the fluctuation of body temperature
d. the response of the eye's pupil to light

4. The rising and falling of body temperature is a negative feedback response. Explain why, then, the rise of body temperature (as in the case of a fever) is a positive feedback response.

5. Which feedback mechanism provides long-term control?

6. On the diagram in Figure 1–2, identify which letters represent the stimulus and which letters represent the response.

Figure 1–2 Feedback Mechanisms and Temperature Control

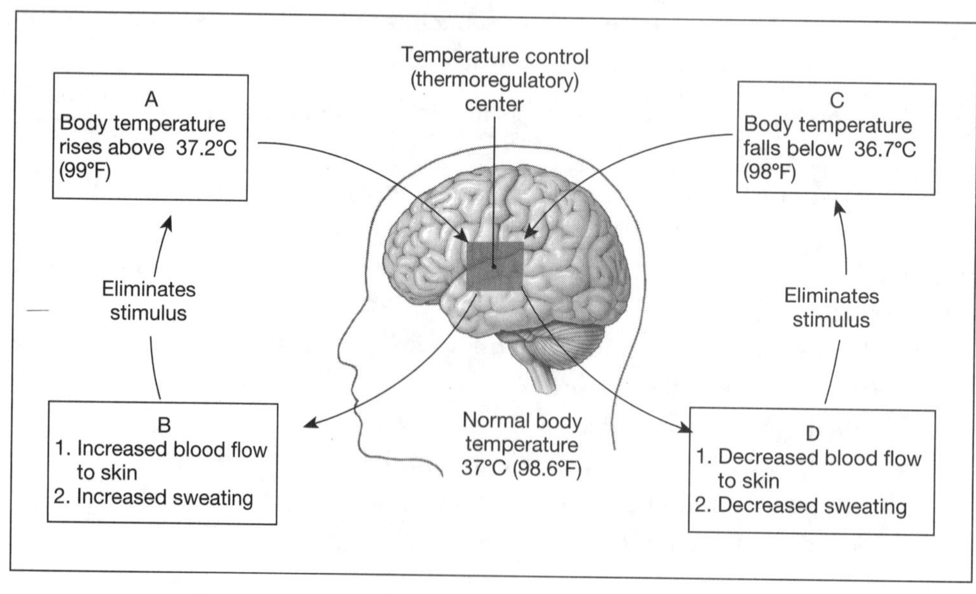

A. Stimulus _____ and _____

B. Response _____ and _____

C. Does this diagram represent a negative feedback mechanism or a positive feedback mechanism? ____________________

OBJECTIVE 5: Use anatomical terms to describe body regions, body sections, and relative positions (textbook pp. 7–12).

After studying pp. 7–12 in the text, you should understand and be able to use the anatomical terms described. It is important to learn these terms, as they will be used throughout the text.

_____ 1. The reproductive organs are located in which body region?

a. abdominal
b. epigastric
c. hypogastric
d. pelvic

_____ 2. The abdominal quadrant that contains the majority of the stomach is called the _____.

a. right upper quadrant
b. left upper quadrant
c. left lower quadrant
d. none of the above

_____ 3. Which of the following best describes the location of the navel relative to the nipple area?

a. below
b. below and to the right
c. lateral and superior
d. medial and inferior

4. The elbow is (proximal or distal) to the shoulder.

5. Describe the standard reference position of the body known as the anatomical position of the body.

6. Identify the sectional plane that describes this sectioned finger. *Hint:* You will need to determine if the fingernail is an anterior or a posterior structure in the anatomical position. The dotted arrow represents the line of dissection through the finger.

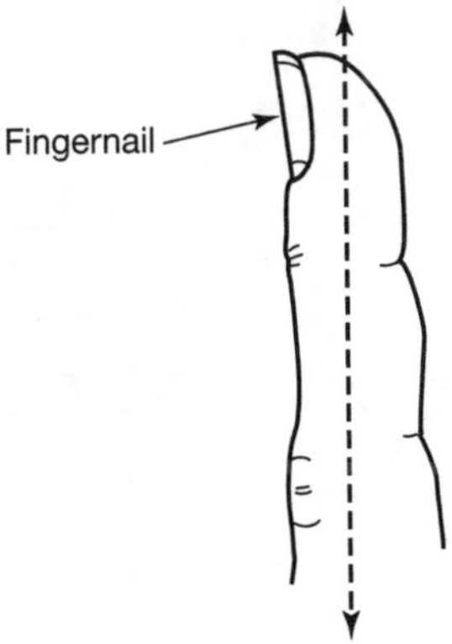

7. Identify the anatomical landmarks on the models in Figure 1–3.

Figure 1–3 Anatomical Landmarks

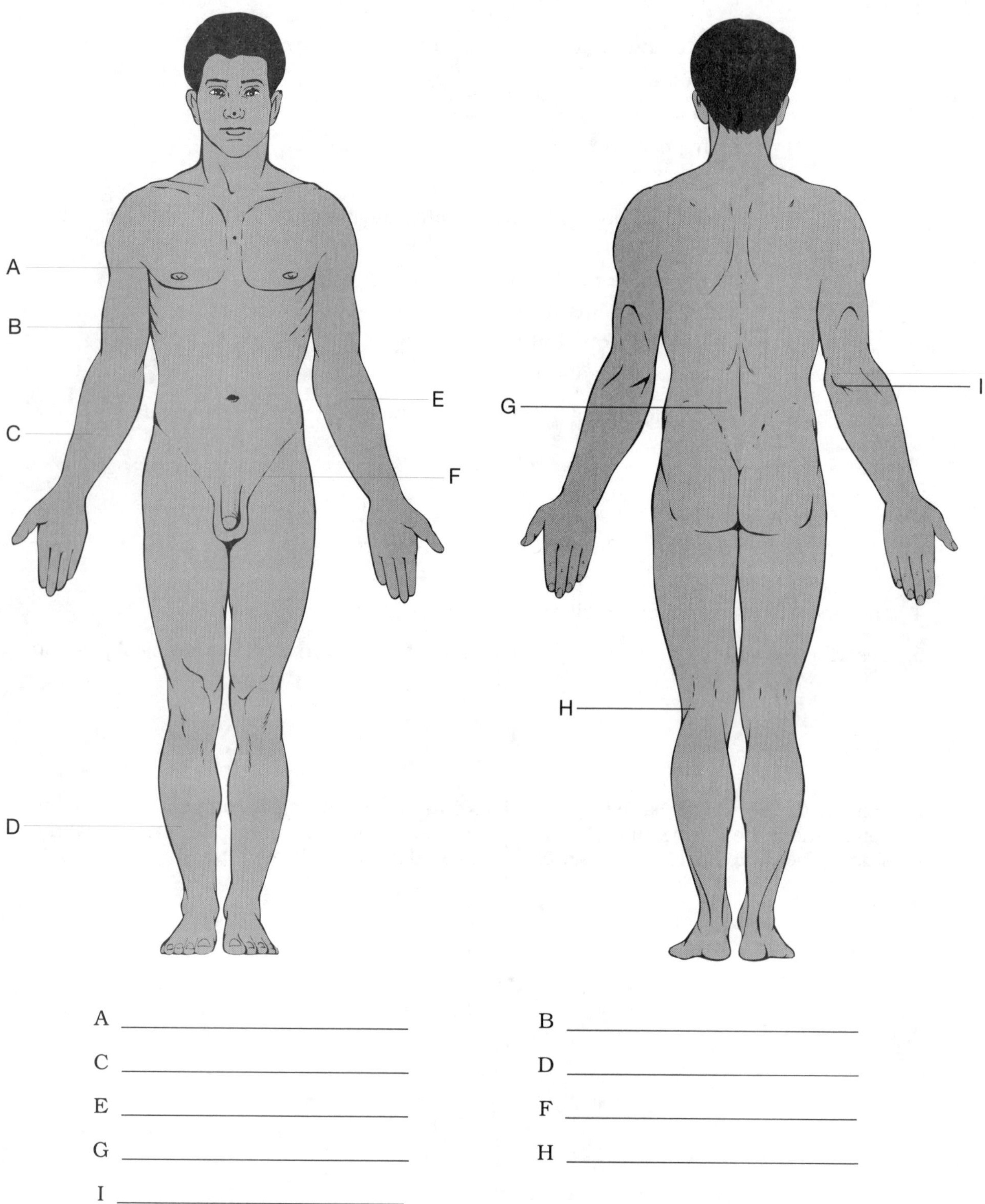

A ______________________ B ______________________

C ______________________ D ______________________

E ______________________ F ______________________

G ______________________ H ______________________

I ______________________

8. Answer the following questions in reference to Figure 1–4.
   a. The stomach is found in which quadrant?
   b. The liver is found in which quadrant?
   c. The spleen is found in which quadrant?
   d. The appendix is found in which quadrant?

Figure 1–4 Abdominal Quadrants

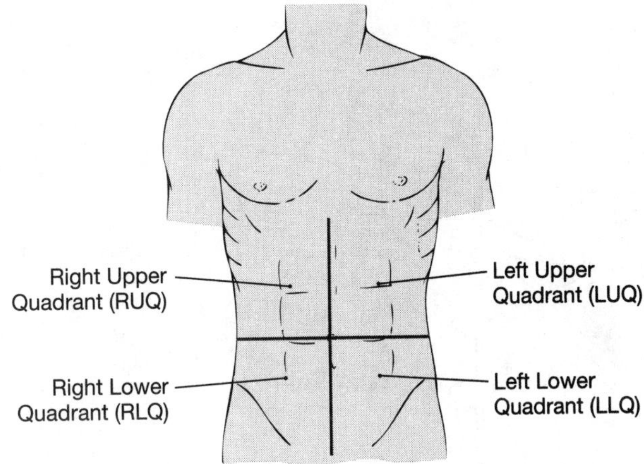

9. Identify the various abdominopelvic regions in Figure 1–5.

Figure 1–5 Abdominopelvic Regions

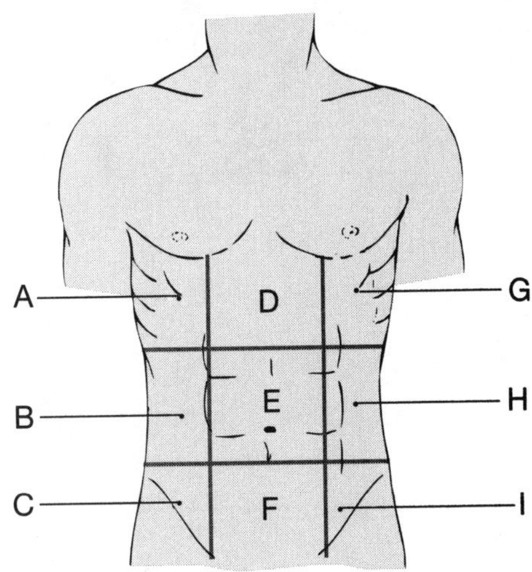

A ______________________ B ______________________

C ______________________ D ______________________

E ______________________ F ______________________

G ______________________ H ______________________

I ______________________

10. In Figure 1–6 identify which letter represents the frontal plane, the sagittal plane, and the transverse plane.

Figure 1–6 Body Planes

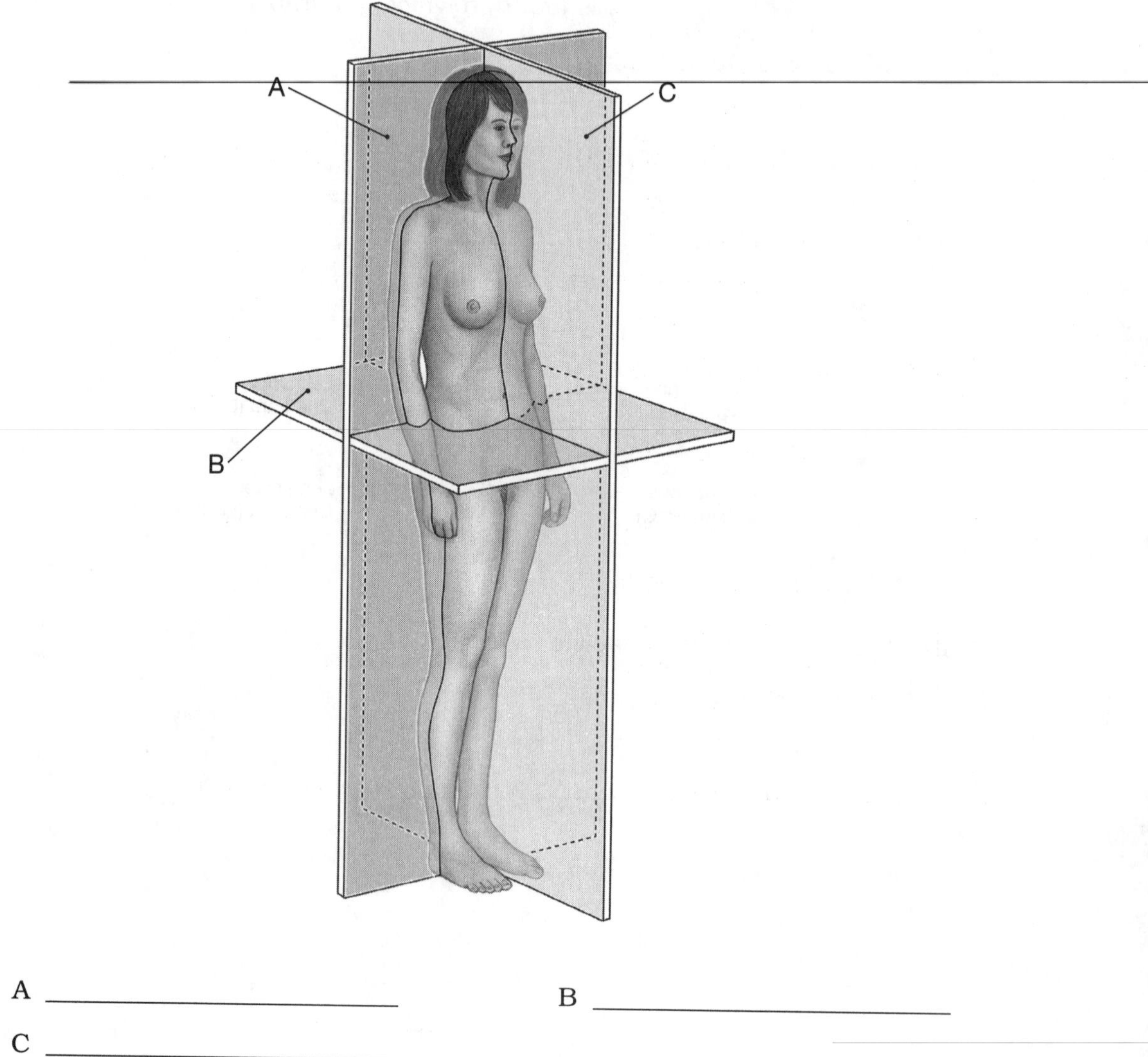

A ______________________ B ______________________

C ______________________

OBJECTIVE 6: Identify the major body cavities and their subdivisions (textbook pp. 12–14).

Study Figure 1-8 in your textbook carefully, and use it to answer the following questions.

_____ 1. Which of the following is located in the dorsal body cavity?

a. brain
b. diaphragm
c. kidneys
d. lungs

_____ 2. Which of the following is located in the ventral body cavity?

a. diaphragm
b. urinary bladder
c. visceral membranes
d. all the above

_____ 3. The mediastinum is located in the _____.

a. abdominal cavity
b. pleural cavities
c. sternum area
d. ventral cavity

_____ 4. Which of the following is not a membrane of the body cavities?

a. mediastinum
b. pericardium
c. peritoneum
d. pleura

5. What separates the abdominopelvic cavity from the pleural cavity?

6. Identify and label the various body cavities in Figure 1–7.

Figure 1–7 Body Cavities

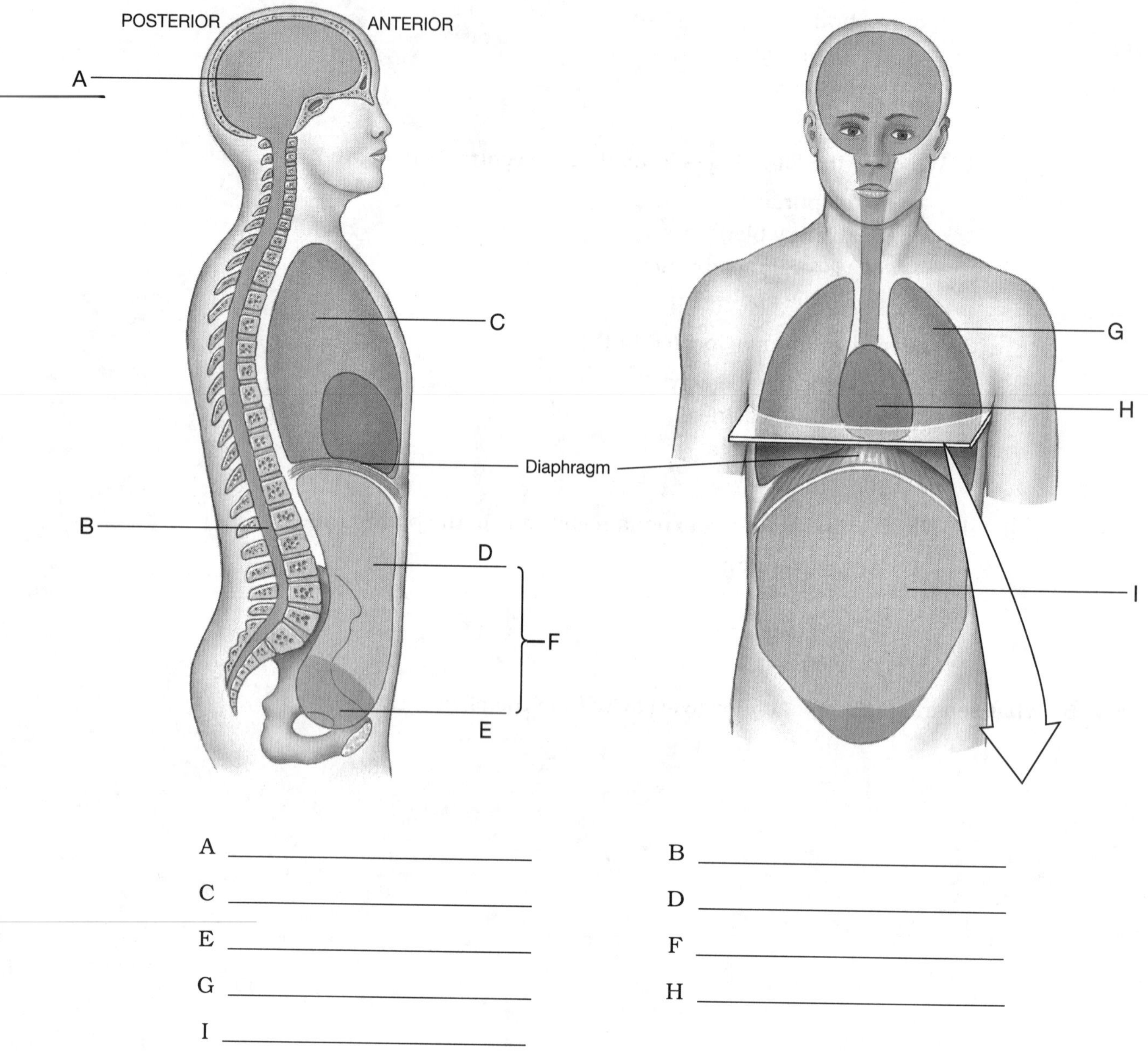

A ____________________ B ____________________

C ____________________ D ____________________

E ____________________ F ____________________

G ____________________ H ____________________

I ____________________

OBJECTIVE 7: Distinguish between visceral and parietal portions of serous membranes (textbook pp. 12–14).

After studying pp. 12–14 in the text, you should be able to describe the distinguishing features of the visceral and parietal portions of the serous membranes that line the thoracic and abdominopelvic cavities.

_____ 1. The visceral pleural membrane covers the outer _____.

a. surface of the heart
b. surface of the lungs
c. surface of the peritoneal cavity
d. surface of the thoracic cavity

_____ 2. The parietal pericardial membrane forms the _____.

a. lining of the heart
b. lining of the lungs
c. lining of the pericardial cavity
d. lining of the peritoneal cavity

3. Use one of these words, visceral or parietal, to complete this statement: The membrane that covers the surface of the intestines is the ______________________ membrane.

4. Use one of these words, visceral or parietal, to complete this statement: The membrane that lines the body cavity is the ______________________ membrane.

5. Use one of these words, visceral or parietal, to complete this statement: The membrane that is adjacent to the organ it is covering is called the ______________________ membrane.

6. Based on your answers to the previous questions, describe the difference between a visceral membrane and a parietal membrane.

OBJECTIVE 8: Describe the four common techniques used in a physical examination (textbook p. 14).

After studying p. 14 in the text, you should be able to describe and explain how the four physical examination techniques are used to assess a patient's health.

_____ 1. Which of the following techniques are physicians using when they use their fingers, placed "under" the lower jaw, to feel for swollen glands?

a. auscultation
b. inspection
c. palpation
d. percussion

_____ 2. Which of the following techniques are physicians using when they place two fingers on the patient's back and, using fingers on the other hand, thump the fingers on the back?

a. auscultation
b. inspection
c. palpation
d. percussion

_____ 3. Which of the following techniques are physicians using when they use a stethoscope to listen for the proper closing of the heart valves?

a. auscultation
b. inspection
c. palpation
d. percussion

_____ 4. At the medical clinic, a physician determines that you have a pulse rate of 68 beats per minute by using the _____ technique.

a. auscultation
b. inspection
c. palpation
d. percussion

_____ 5. Information derived from the four physical examination techniques can be recorded as _____ of the patient.

a. homeostasis
b. normal values
c. vital signs
d. All the above are correct.

PART II: CHAPTER-COMPREHENSIVE EXERCISES

Match the terms in column A with the definitions or phrases in column B.

MATCHING I

	(A)	(B)
_____	1. Abdominal cavity	A. The study of the structure of the body
_____	2. Anatomy	B. The basic living unit of life
_____	3. Auscultation	C. A group of organized tissues working together
_____	4. Cell	D. A state of well-being
_____	5. Diaphragm	E. A mechanism that returns the body to homeostasis after minor fluctuations from the norm
_____	6. Dorsal cavity	F. A hormone that is involved in a positive feedback mechanism during labor
_____	7. Frontal plane	G. Divides body into anterior and posterior sections
_____	8. Homeostasis	H. Cavity containing the brain and spinal cord
_____	9. Negative feedback	I. Separates the thoracic cavity from the abdominopelvic cavity
_____	10. Organ	J. Cavity containing the intestines
_____	11. Oxytocin	K. The use of a stethoscope to assess the body's condition

MATCHING II

	(A)	(B)
_____	1. Palpation	A. The study of "how living things work"
_____	2. Pelvic cavity	B. A group of organized cells working together
_____	3. Physiology	C. Out of homeostasis
_____	4. Positive feedback	D. A mechanism that is perpetuated by more stimuli
_____	5. Sagittal plane	E. Divides body into left and right sections
_____	6. Sick	F. Divides body into superior and inferior sections
_____	7. Thoracic cavity	G. Cavity containing the heart and lungs
_____	8. Tissue	H. Membrane covers the outside of the organ
_____	9. Transverse plane	I. Membrane lines the cavity itself
_____	10. Parietal	J. Contains the urinary bladder
_____	11. Visceral	K. Feeling the body for abnormal tissue growth

MATCHING III

	(A)	(B)
_____	1. Antebrachial	A. Refers to the upper arm
_____	2. Antecubital	B. Refers to the anterior side of the elbow
_____	3. Axilla	C. Refers to the posterior side of the knee
_____	4. Brachial	D. Refers to the lower arm
_____	5. Carpal	E. Refers to the wrist
_____	6. Cervical	F. Refers to the ankle
_____	7. Crural	G. Refers to the neck region
_____	8. Cubital	H. Refers to the armpit
_____	9. Popliteal	I. Refers to the elbow region
_____	10. Sural	J. Refers to the anterior lower leg
_____	11. Tarsal	K. Refers to the posterior lower leg

CONCEPT MAP I

This concept map will help you organize the relationships of the various body cavities. Use the following terms to complete the map by filling in the boxes identified by the circled numbers, 1–9.

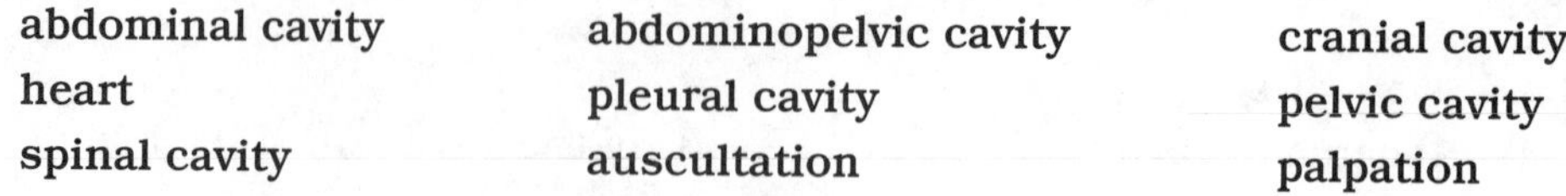

abdominal cavity	**abdominopelvic cavity**	**cranial cavity**
heart	**pleural cavity**	**pelvic cavity**
spinal cavity	**auscultation**	**palpation**

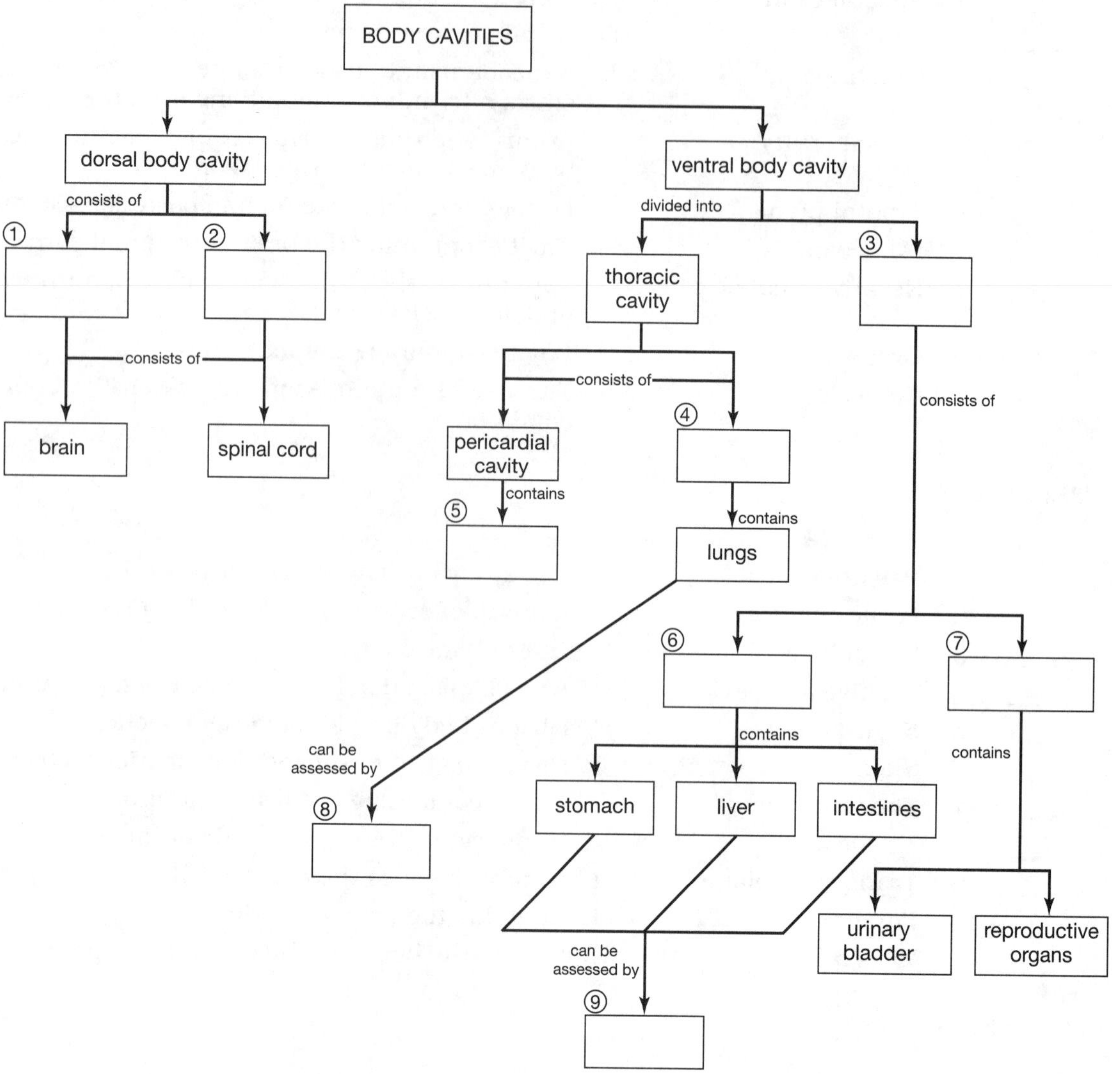

BODY CAVITIES

dorsal body cavity

consists of

① ②

consists of

brain

spinal cord

ventral body cavity

divided into

thoracic cavity

③

consists of

consists of

pericardial cavity

④

contains

⑤

contains

lungs

⑥

⑦

can be assessed by

⑧

contains

contains

stomach

liver

intestines

urinary bladder

reproductive organs

can be assessed by

⑨

CONCEPT MAP II

This concept map summarizes and organizes the levels of organization described in Chapter 1. Use the following terms to complete the map by filling in the boxes identified by the circled numbers, 1–4.

molecules **multicellular organisms** **tissues**
unicellular organisms

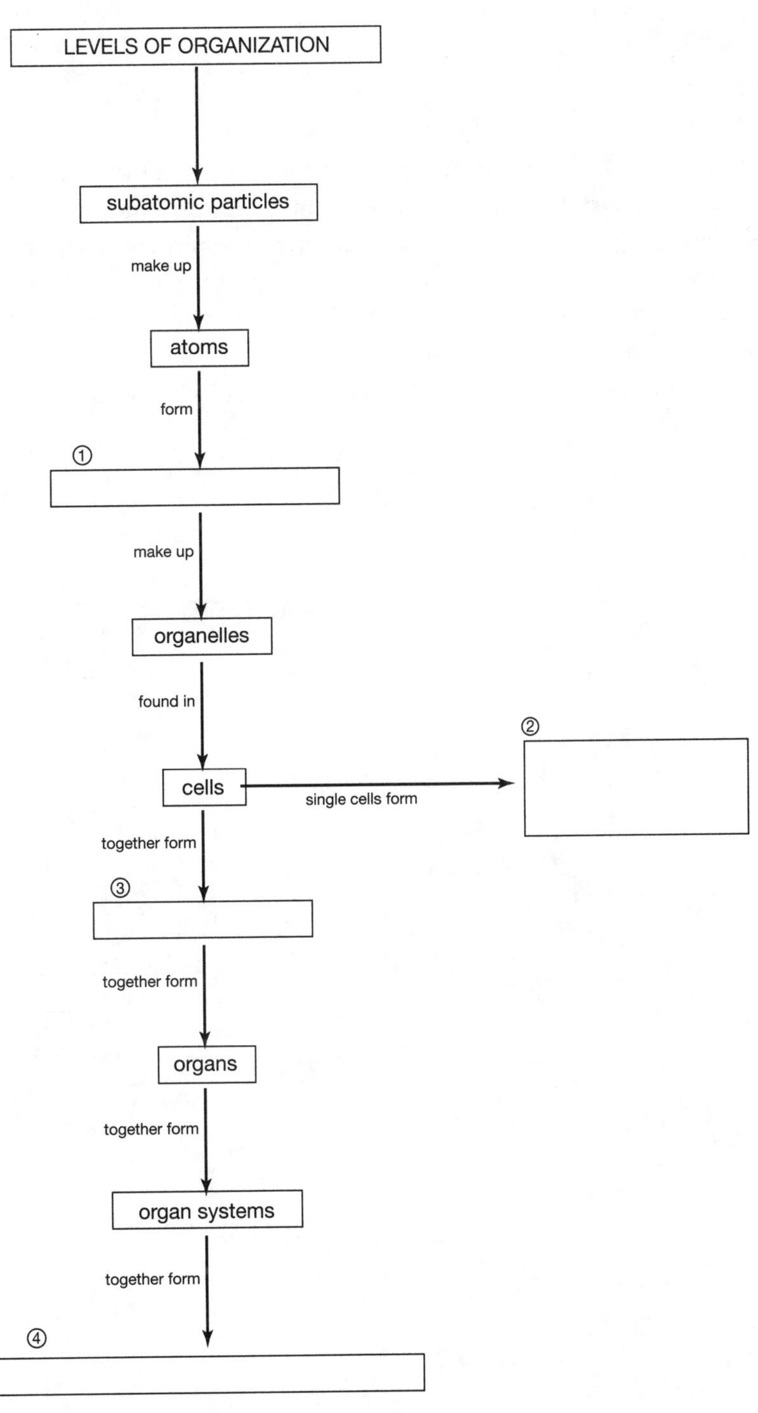

CROSSWORD PUZZLE

The following crossword puzzle reviews the material in Chapter 1. To complete the puzzle you have to know the answers to the clues given, and must be able to spell the terms correctly.

ACROSS

3. The depression located posterior to the knee is known as the _____ region.
9. When the body is healthy, it is said to be in _____.
11. The homeostatic mechanism that is constantly fluctuating is called _____ feedback.

DOWN

1. The study of how the body works
2. A technique to determine the presence of abnormal tissue mass
4. A technique to determine the density of a certain tissue
5. Many of the scientific words we use are derived from the Greek and the _____ languages.
6. The abdominal region located superior to the umbilicus is the _____.
7. The wrist bones are collectively called the _____.
8. The body region typically referred to as the elbow is anatomically called the _____ region.
10. The visceral _____ membranes surround the lungs.

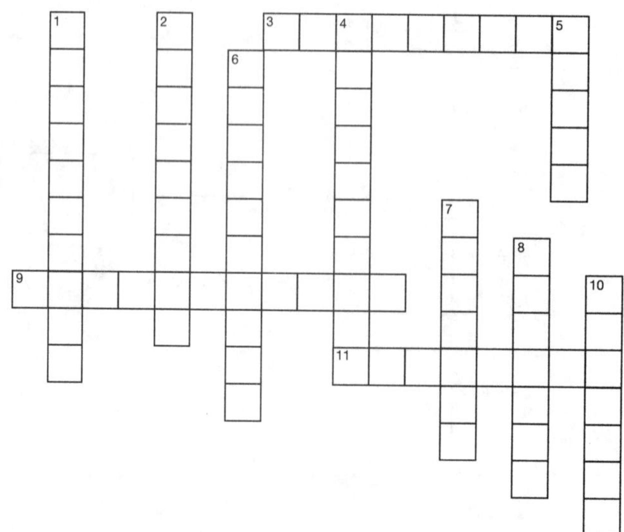

FILL-IN-THE-BLANK NARRATIVE

Use the following terms to fill in the blanks of this summary of Chapter 1.

confusion	**homeostasis**	**hypogastric**
Latin	**negative**	**organ**
popliteal	**positive**	**tissues**
urinary bladder	**stimuli**	**auscultation**
percussion	**density**	

Anatomists and physiologists use terms derived from Greek and (1) __________ to describe the structures and functions of the body. Scientists use these terms, rather than common, everyday words to avoid (2) __________ and to describe precisely what they mean. For example, rather than referring to the back of the knee, which can be interpreted as either the posterior side of the patella or the bend at the back of the knee, anatomists refer to the (3) __________. This term precisely locates the area.

Use of proper terminology also helps doctors pinpoint pain in the body and the organs that may be involved. Pain in the (4) __________ region suggests that the organ affected may be the lower intestines or perhaps the (5) __________. If the patient has a pain in the epigastric or hypochondriac regions, the physician may be able to use the (6) __________ technique to listen for the sounds made by air moving in and out of the lungs. The physician may also use the (7) __________ technique to determine if there is a fluid buildup in the lungs, which would change the (8) __________ of the lungs.

One of the ways the human body is characterized is by levels of organization. Because of this division into levels, if an (9) __________ fails, it will affect the entire organ system of which it is a part. An organ can fail if the (10) __________ are not working together properly. When all the systems are functioning properly, the body is said to be in (11) __________. This steady-state condition is due to the operations of the (12) __________ and (13) __________ feedback systems. The latter system is constantly making the proper adjustments in response to the changing (14) __________ in our environment.

CLINICAL CONCEPT

The following clinical concept applies the information presented in Chapter 1. Following the application is a set of questions to help you understand the concept.

Nurse Abby walks up to Doctor Bart and begins to describe to the doctor the condition of the patient in room 304. Nurse Abby says the patient is complaining of pain in the tummy area. Doctor Bart asks the nurse to be more specific such as, where in the "tummy" area is the pain? Is the pain on the right side or the left side of the "tummy?" Nurse Abby asks the doctor if he means on her right or on the patient's right. Doctor Bart is starting to get a bit upset.

Head nurse Hanna discovers the difficulty Doctor Bart is having with Nurse Abby. Head nurse Hanna goes into room 304 and walks up to Doctor Bart and explains that the patient in room 304 is complaining of pain in the lower right quadrant, specifically in the right inguinal region. Doctor Bart palpates the region and then orders surgery for the patient.

1. What is the anatomical position for the human body?

2. When we are referring to the right side of the body, are we referring to our right side or to the patient's right side?

3. Why do you suppose the doctor ordered surgery for the patient?

4. What does it mean to "palpate"?

5. Why was Doctor Bart having difficulty understanding Nurse Abby?

THIS CONCLUDES CHAPTER 1 EXERCISES

CHAPTER 2

Chemistry and the Human Body

All living and nonliving things are composed of various combinations of basic units of matter called **atoms.** The unique characteristics of each object are determined by the combinations and interactions of its constituent atoms.

The study of anatomy and physiology begins with the most basic principles of **chemistry,** the branch of science that deals with atoms and their interactions. To be successful in your study of anatomy and physiology, you will need to know how simple chemical components combine to form more complex forms that make up the cells, tissues, and organs of the human body.

The exercises in this chapter examine how the chemical processes need to continue throughout life in an orderly and timely sequence if homeostasis is to be maintained. The activities are designed to measure your knowledge of chemical principles and to help you apply these principles to the structure and functions of the cells, tissues, organs, and organ systems in the human body.

Chapter Objectives:

- Describe the basic structure of an atom.
- Describe the different ways in which atoms combine to form molecules and compounds.
- Distinguish between decomposition and synthesis chemical reactions.
- Distinguish between organic and inorganic compounds.
- Explain how the chemical properties of water are important to the functioning of the human body.
- Describe the pH scale and the role of buffers.
- Describe various functions of inorganic compounds.
- Discuss the structure and functions of carbohydrates, lipids, proteins, nucleic acids, and high-energy compounds.
- Describe the role of enzymes in metabolism.
- Describe the use of radioisotopes in visualizing organs and the treatment of diseases.

PART I: OBJECTIVE-BASED QUESTIONS

OBJECTIVE 1: Describe the basic structure of an atom (textbook pp. 19–21).

After studying pp.19–21 in the text, you should be able to describe the general makeup of an atom.

_____ 1. An atom consists of a nucleus surrounded by an electron shell. Which of the following subatomic particles is not found in the nucleus of an atom?

a. electrons
b. neutrons
c. protons
d. both a and b

_____ 2. In order for an atom to be electrically neutral, the atom has to have _____.

a. an equal number of electrons and neutrons
b. an equal number of electrons and protons
c. an equal number of neutrons and protons
d. equal numbers of neutrons, protons, and electrons

_____ 3. Which of the following uniquely identifies an atom?

a. the number of electrons
b. the number of neutrons
c. the number of protons
d. the number of protons plus the number of neutrons

_____ 4. The atomic weight (mass number) of an atom can be determined by _____.

a. adding the number of neutrons and the number of electrons
b. adding the number of neutrons and the number of protons
c. adding the number of protons, electrons, and the number of neutrons
d. adding the number of protons and electrons

_____ 5. Two isotopes of an element–sodium, for example–differ from each other in their _____.

a. number of neutrons and electron number
b. number of neutrons and mass number
c. number of neutrons and proton number
d. number of protons and mass number

6. Identify the number of neutrons, protons, and electrons and their locations in the atom shown in Figure 2–1.

    a. Number of neutrons _____ Location: (nucleus or electron shell?)
    b. Number of protons _____ Location: (nucleus or electron shell?)
    c. Number of electrons _____ Location: (nucleus or electron shell?)

Figure 2–1 Atomic Structure

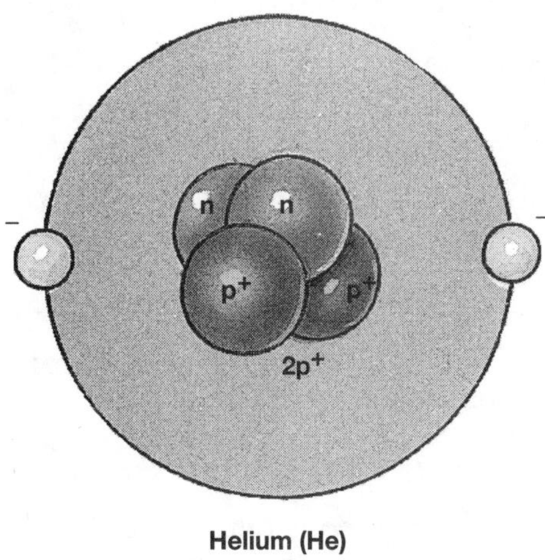

Helium (He)

7. Draw an atom of lithium showing the number of protons, electrons, and neutrons it contains.

8. Examine Figure 2–2. Identify the difference between the hydrogen atom and its isotopes, going from diagram a to c. Draw a similar diagram of boron (^{11}B) and its (^{12}B) isotope.

Figure 2–2 Hydrogen Atoms

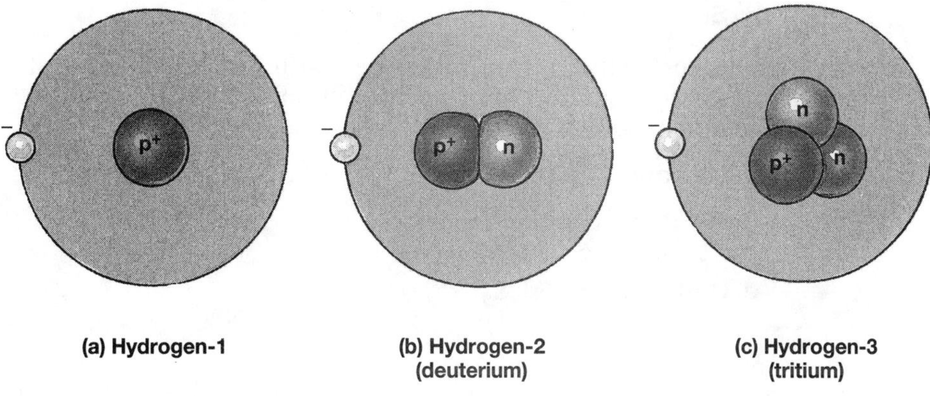

(a) Hydrogen-1 **(b) Hydrogen-2 (deuterium)** **(c) Hydrogen-3 (tritium)**

OBJECTIVE 2: Describe the different ways in which atoms combine to form molecules and compounds (textbook pp. 22–23).

After studying pp. 22–23 in the text, you should be able to distinguish between the two main types of bonds: ionic and covalent. You should understand the information in Table 2-1 pertaining to the periodic table.

Table 2-1 The Periodic Table and Ionic Charges

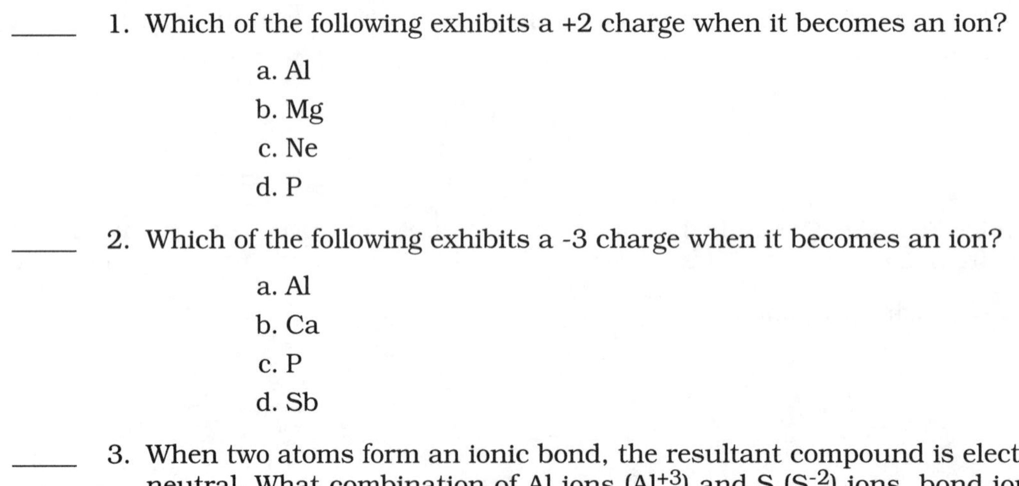

+1	+2		+3	+4	-3	-2	-1	0
H								He
Li	Be	Transition elements	B	C	N	O	F	Ne
Na	Mg		Al	Si	P	S	Cl	Ar
K	Ca		Ga	Ge	As	Se	Br	Kr

The table shows that:
When H becomes an ion, it exhibits a charge of +1.
When Al becomes an ion, it exhibits a charge of +3.
When Br becomes an ion, it exhibits a charge of -1.

These are general rules to follow when considering bonding between ions. You will learn the exceptions to these generalized rules in chemistry courses. Use the information in Table 2–1 to answer the following questions.

_____ 1. Which of the following exhibits a +2 charge when it becomes an ion?

a. Al
b. Mg
c. Ne
d. P

_____ 2. Which of the following exhibits a -3 charge when it becomes an ion?

a. Al
b. Ca
c. P
d. Sb

_____ 3. When two atoms form an ionic bond, the resultant compound is electrically neutral. What combination of Al ions (Al^{+3}) and S (S^{-2}) ions bond ionically to produce a neutral compound?

a. 1 Al and 1 S (AlS)
b. 2 Al and 2 S (Al_2S_2)
c. 2 Al and 3 S (Al_2S_3)
d. 3 Al and 2 S (Al_3S_2)

_____ 4. Covalent bonds involve sharing of electrons between atoms. *Typically,* we find that H (hydrogen) forms one bond; O (oxygen) forms two bonds; N (nitrogen) forms three bonds; and C (carbon) forms four bonds. We represent covalent bonds by drawing lines between the atoms, where each line represents one bond. Based on this information, which of the following is a correct structural formula for CO_2?

a. O-C-O b. O=C=O c. C-O-O d. C=O=O

_____ 5. Which one of the following molecules is drawn correctly as far as showing the proper number of covalent bonds?

a. H=H b. O-O c. N≡N d.

```
   H
   |
H—N—H
   |
   C—H
   ||
   O
```

6. Use the periodic table to complete Table 2–2 by filling in boxes A through K.

Table 2–2 Examples of Atoms, Ions, and Isotopes

Neutral atom	Number of protons	Number of electrons	Number of neutrons	Mass number
Li	3	3	4	D=
Ca	A=	20	20	40
Al	13	B=	14	27
O	8	8	C=	16
Ion	**Number of protons**	**Number of electrons**	**Symbol**	
Li	3	F=	Li^+	
Ca	20	18	G=	
Al	E=	10	Al^{+3}	
O	8	10	O^{-2}	
Isotope	**Number of protons**	**Number of neutrons**	**Mass number**	
H=	3	5	8	
^{42}Ca	20	J=	42	
^{29}Al	13	16	K=	
^{18}O	I=	10	18	

OBJECTIVE 3: Distinguish between decomposition and synthesis chemical reactions (textbook pp. 23–24).

After studying pp. 23–24 in the text, you should be able to describe the differences between decomposition and synthesis reactions, and to name the two processes that constitute metabolism.

_____ 1. The sum of all the chemical reactions in the body that keep us alive is known as _____.

a. equilibrium
b. homeostasis
c. metabolism
d. molecular reactions

_____ 2. Which of the following is a decomposition reaction?

a. Hydrogen + oxygen → water
b. NaCl → Na^+ + Cl^-
c. NH_3 + H^{+1} → NH_4^+
d. Pepsinogen + hydrochloric acid → pepsin

_____ 3. Which of the following is a synthesis reaction?

a. $C_6H_{12}O_6 \rightarrow C_3H_6O_3 + C_3H_6O_3$
b. Carbon + oxygen → CO_2
c. X + Z → XZ
d. Both b and c are correct.

_____ 4. Which of the following represents a decomposition reaction?

a. A + B ↔ AB
b. AB + CD → AD + BC
c. AB → A + B
d. C + D → CD

5. When the rate of the synthesis reaction exactly balances the rate of the decomposition reaction, the result is ______________________.

OBJECTIVE 4: Distinguish between organic and inorganic compounds (textbook p. 24–31).

After studying p. 24–31 in the text, you should be able to classify the chemical substances associated with the human body as either organic or inorganic compounds.

_____ 1. Which of the following is an inorganic molecule?

a. CH_4
b. CO_2
c. glucose
d. None of the above is inorganic.

_____ 2. Use the terms *organic* or *inorganic* to answer the following questions.

a. Carbon dioxide is a/an ______________________ molecule.
b. Water is a/an ______________________ molecule.
c. Glucose is a/an ______________________ molecule.
d. ______________________ molecules tend be very large molecules.

3. What element distinguishes organic compounds from inorganic compounds?

____

4. Organic compounds contain carbon, yet carbon dioxide is considered to be inorganic. Why is carbon dioxide placed in the inorganic compound category?

OBJECTIVE 5: Explain how the chemical properties of water are important to the functioning of the human body (textbook pp. 24–25).

After studying pp. 24–25 in the text, you should be able to explain why water is vital to the body.

_____ 1. In a solution, water is the _____.

a. solvent
b. solute
c. organic molecule
d. all the above

_____ 2. The material that is dissolved in the solution is the _____.

a. ion
b. solute
c. solution
d. solvent

_____ 3. Which of the following special properties of water are essential to survival?

a. Most of our chemical reactions occur in water.
b. Water has a high heat capacity.
c. Water helps distribute body heat.
d. All the above.

_____ 4. One of water's unique characteristics is that it prevents rapid changes in body temperature. This characteristic is due to water's _____.

a. ability to dissociate
b. ability to dissolve many different kinds of molecules
c. high heat capacity
d. neutral pH

5. Under normal circumstances, what fraction of the weight of an individual is accounted for by water?

OBJECTIVE 6: Describe the pH scale and the role of buffers (textbook pp. 25–27).

After studying pp. 25–27 in the text, you should be able to explain the pH scale and describe how buffers work to maintain homeostasis.

_____ 1. Which of the following pH values denotes the most acidic solution?

a. 3
b. 5
c. 7
d. 9

_____ 2. The following pH values represent alkaline solutions. Even though the solutions are alkaline, which one is more acidic than the others?

a. 9
b. 11
c. 12
d. 14

_____ 3. How many times more acidic is a solution with a pH value of 7 than one with a pH value of 9?

a. 2

b. 10

c. 20

d. 100

_____ 4. A compound that helps maintain a constant pH is _____.

a. a base

b. a buffer

c. a hydroxide ion

d. an alkaline chemical

_____ 5. pH is a measure of the concentration of _____ in solution.

a. acid

b. buffers

c. hydrogen ions

d. hydroxide ions

6. Which solution is the most acidic? (Remember, the pH of a solution is a measure of the concentration of hydrogen ions in the solution.)

a. a solution with 0.0001 mole of H^+/L

b. a solution with 0.000001 mole of H^+/L

c. a solution with 0.00000001 mole of H^+/L

d. a solution with 0.0000000001 mole of H^+/L

7. Homeostasis is regulated primarily by chemical reactions in our body fluids. If the _____ of the fluids changes, chemical reactions will be altered, and homeostasis will be disrupted.

8. In Figure 2–3, identify by letter the region or point on the pH scale that corresponds to the acidic region of the pH scale, the alkaline region of the pH scale, and approximate pH values for blood, urine, water, and stomach acid.

Figure 2–3 pH and Hydrogen Ion Concentration

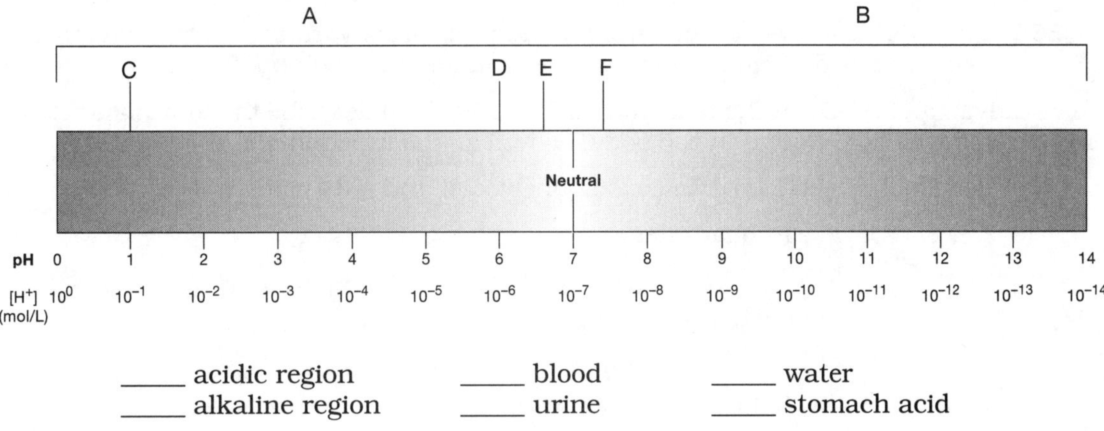

_____ acidic region _____ blood _____ water
_____ alkaline region _____ urine _____ stomach acid

OBJECTIVE 7: Describe the various functions of inorganic compounds (textbook pp. 24–27).

After studying pp. 24–27 in the text, you should be able to describe the physiological roles of hydrogen ions, water, and salts.

_____ 1. Which of the following dissociate to produce hydrogen ions in solution?

a. acids
b. bases
c. buffers
d. salts

_____ 2. If we have excess hydrogen ions in the stomach fluid, its pH will be _____ than normal. To correct this imbalance, we need to remove the excess hydrogen ions via the action of a _____.

a. higher, base
b. higher, buffer
c. lower, base
d. lower, buffer

_____ 3. Antacids are buffers because they _____.

a. increase hydrogen ion concentration
b. increase the base concentration to neutralize the acid
c. remove excess hydrogen ions
d. stimulate the buffers in our fluid systems

_____ 4. Most chemical reactions in the body occur in _____.

a. solution
b. a solvent with ions in it
c. water
d. all the above

_____ 5. Which of the following statements is true?

a. Electrolytes are charged particles.
b. Electrolytes dissociate into ions.
c. Electrolytes will form covalent bonds.
d. Ions can become electrolytes.

OBJECTIVE 8: Discuss the structure and functions of carbohydrates, lipids, proteins, nucleic acids, and high-energy compounds (textbook pp. 27–31).

After studying pp. 27–31 in the text, you should be able to describe the four major classes of organic molecules found in the body, as well as the function of ATP.

_____ 1. Which of the following consist of glycerol and fatty acids?

a. carbohydrates
b. fats
c. lipids
d. steroids

_____ 2. Amino acids bonded together form _____.

a. carbohydrates
b. lipids
c. nucleic acids
d. proteins

_____ 3. Which of the following is (are) associated with carbohydrates?

a. DNA
b. fatty acids
c. glycogen
d. steroids

_____ 4. A steroid molecule is an example of a _____.

a. carbohydrate
b. fat
c. high-energy compound
d. lipid

_____ 5. Which of the following describes a phospholipid?

a. It consists of 1 glycerol, 2 fatty acids, and 1 phosphate.
b. It consists of 1 glycerol and 3 fatty acids.
c. It consists of 1 glycerol, 2 fatty acids, and 1 carbohydrate.
d. It consists of 1 glycogen, 2 fatty acids, and 1 phosphate.

_____ 6. Which of the following is a high-energy compound?

a. ATP
b. carbohydrate
c. DNA
d. protein

7. Complete the portion of the DNA structure shown in Figure 2–4 by identifying the bases (guanine, cytosine, adenine, or thymine) that correspond to the blank boxes labeled A–E.

Figure 2–4 Nucleic Acids: DNA

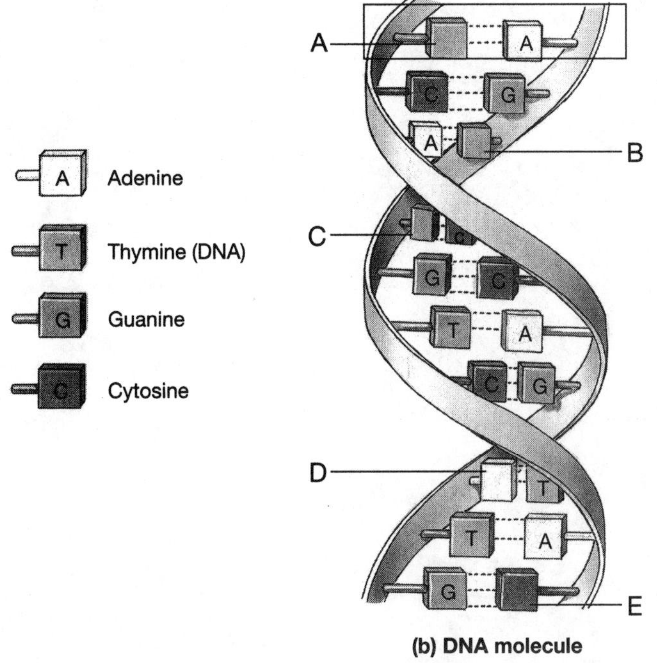

(b) DNA molecule

A ____________________ B ____________________

C ____________________ D ____________________

E ____________________

8. Table 2–3 will help you categorize the various classes of organic molecules. Place an X in the appropriate column for the terms in the first column.

Table 2–3 Classification of Various Organic Molecules

	Carbohydrate	Lipid	Protein	Nucleic acid	High-energy compound
Amino Acid					
ATP					
Cholesterol					
Cytosine					
Disaccharide					
DNA					
Fatty acid					
Glucose					
Glycerol					
Glycogen					
Guanine					
Monosaccharide					
Nucleotide					
Phospholipid					
Polysaccharide					
RNA					
Starch					

OBJECTIVE 9: Describe the role of enzymes in metabolism (textbook pp. 28–30).

After studying pp. 28–30 in the text, you should be able to discuss the nature and function of enzymes in chemical reactions in the body, and how they function in an effort to maintain homeostasis.

_____ 1. Which of the following is affected by an enzyme?

a. the concentration of the substrate
b. the concentration of the reactants
c. the pH of a solution
d. the rate of a reaction

_____ 2. In the reaction A + B → C, which letter(s) represent the substrates in a reaction involving an enzyme?

a. A and B
b. A and C
c. B and C
d. C only

_____ 3. Enzymes are what type of organic molecule?

a. carbohydrate
b. high-energy compound
c. nucleic acid
d. protein

_____ 4. In a chemical reaction such as A + B → C, if you were to run out of _____, the reaction would stop.

a. buffers
b. enzymes
c. products
d. substrate

5. High body temperatures, as in the case of a fever, will ______________________ the enzymes, rendering them nonfunctional.

OBJECTIVE 10: Describe the use of radioisotopes in visualizing organs and the treatment of diseases (textbook pp. 32–34).

After studying pp. 32–34 in the text, you should be able to explain what radioisotopes are and how they are used. You should also be able to explain the significance of half-life.

_____ 1. Cobalt-58 has a half-life of 71 days. This means that _____.

a. Cobalt-58 will become cobalt-29 (half of 58) in 71 days
b. Cobalt-58 will decompose in 35-1/2 days. This is half of 71 days
c. Cobalt-58 will decompose in 71 days
d. half of the original amount of radioactive material will decompose in 71 days

_____ 2. Radioactive iodine is used to assess the activity of the thyroid gland because _____.

a. iodine acts as a chemical messenger for thyroid activity
b. radioactive iodine has a short half-life
c. the thyroid gland secretes iodine
d. the thyroid gland uses iodine to manufacture its hormones

_____ 3. Your textbook discusses positron emission tomography. What is a positron? (You may have to do a little research to answer this question.)

a. a particle that is emitted from a proton
b. a particle with a positive charge that is emitted from an electron
c. a particle with the same mass as an electron but possessing a positive charge instead of a negative charge
d. a positively charged particle that is emitted from neutrons

_____ 4. Which of the following is the purpose for using monoclonal antibodies?

a. Since monoclonal antibodies are an antibody structure, they destroy the abnormal cells.
b. The monoclonal antibodies transfer attached radioisotopes to the abnormal cells.
c. Monoclonal antibodies are more potent than abnormal antibodies.
d. Monoclonal antibodies produce antigens, which then destroy the abnormal cells.

_____ 5. Radioactive isotopes used in nuclear medicine typically _____.

a. are extremely powerful and are therefore effective in killing cancer cells, for example
b. have a half-life of 71 days or more
c. have a relatively long half-life
d. have a relatively short half-life

PART II: CHAPTER-COMPREHENSIVE EXERCISES

Match the terms in column A with the definitions or phrases in column B.

MATCHING I

	(A)	(B)
_____	1. atom	A. Smallest unit of matter
_____	2. atomic number	B. Smallest unit of living matter
_____	3. buffer	C. Has a negative charge
_____	4. cell	D. Number of protons in an atom
_____	5. covalent	E. A type of bond not associated with ions
_____	6. decomposition	F. An atom with a charge
_____	7. disaccharide	G. The breaking down of compounds
_____	8. dissociation	H. The breaking of ionic bonds
_____	9. electron	I. Anything that resists changes in pH
_____	10. enzyme	J. Sucrose is an example
_____	11. ion	K. Acts as a catalyst to speed up chemical reactions
_____	12. monoclonal antibody	L. Can be used to target cancerous antigens

MATCHING II

	(A)	(B)
_____	1. isotopes	A. Has a positive charge
_____	2. lipids	B. Has no charge
_____	3. metabolism	C. Two atoms of the same element having different numbers of neutrons
_____	4. monosaccharide	D. The sum total of all the chemical reactions in the body
_____	5. neutron	E. The manufacture of compounds
_____	6. polysaccharide	F. The liquid portion of a solution
_____	7. positron	G. The material dissolved in solution
_____	8. proton	H. Glucose is an example
_____	9. solutes	I. Starch is an example
_____	10. solvent	J. Mostly insoluble in water
_____	11. substrate	K. The name for reactants when enzymes are involved
_____	12. synthesis	L. A particle the size of an electron but possessing a positive charge

CONCEPT MAPS

These concept maps summarize and organize some of the ideas discussed in Chapter 2. Use the following terms to complete the map by filling in the boxes identified by the circled numbers, 1–11.

atoms	**sucrose**	**carbohydrates**
proteins	**DNA**	**protons**
isotope	**steroid**	**ion**
ATP	**radioactive**	

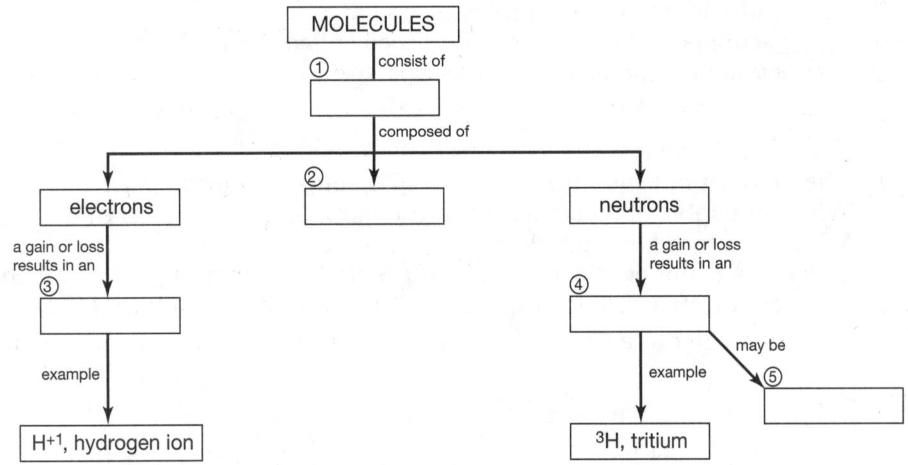

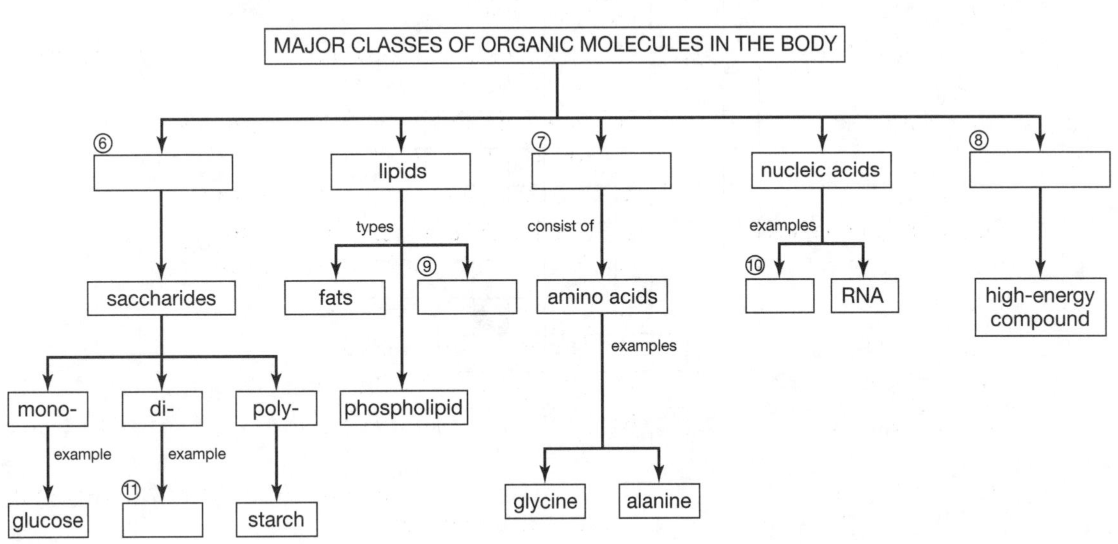

CROSSWORD PUZZLE

The following crossword puzzle reviews the material in Chapter 2. To complete the puzzle, you have to know the answers to the clues given, and you must be able to spell the terms correctly.

ACROSS

1. The greater the concentration of _____ ions in solution, the lower the pH.
2. The disaccharide sucrose consists of fructose bonded to _____.
5. Two atoms that have the same number of protons but a different number of neutrons are called _____.
6. The high-energy compound produced by the body is _____ triphosphate.
8. Fats and steroids are examples of _____.
10. A_____ consists of a chain of molecules called amino acids.
11. Monoclonal antibodies will target specific _____.

DOWN

1. The amount of time it takes for a radioisotope to decompose.
3. When phosphorus becomes an ion, it exhibits a _____ 3 charge.
4. When an atom loses an electron, it exhibits a _____ charge.
7. The mass number of an atom changes if its number of _____ changes.
9. An atom or molecule that has a positive or a negative charge is an _____.

FILL-IN-THE-BLANK NARRATIVE

Use the following terms to fill in the blanks of this summary of Chapter 2.

ATP	**buffers**	**carbohydrate**
chemical	**covalent**	**polysaccharide**
pH	**glycogen**	**homeostasis**
metabolism	**glucose**	**hydrogen ions**
protons	**drop**	**enzymes**
hormones	**tracer**	

Atoms contain subatomic particles called electrons, (1) ___________, and neutrons. When electrons of atoms are involved in (2) ___________ reactions, they create either ionic or (3) ___________ bonds. When bonding occurs, new molecules or compounds are formed.

One such molecule in the body is glucose, which is a type of (4) ___________. The liver can convert excess glucose into a(n) (5) ___________ called (6) ___________, which is stored in the liver and muscle tissues. When the body requires energy, the liver converts glycogen to (7) ___________. Glucose then undergoes a series of chemical reactions that eventually produce the high-energy molecule called (8) ___________. This conversion is one of the millions of reactions that cells undergo every day to maintain life. The sum of all these reactions is called (9) ___________.

Chemical reactions in the body must occur very rapidly. To speed up these reactions, the body produces catalysts called (10) ___________. Often, (11) ___________ are produced in chemical reactions. If the concentration of hydrogen ions becomes too high, the pH will begin to (12) ___________. If the pH changes very much, enzymes will be denatured, and chemical reactions will cease. (13) ___________ in the body prevent drastic changes in pH by removing hydrogen ions from solution, thereby stabilizing the (14) ___________ and maintaining (15) ___________ in the body.

The various glands of the body must produce (16) ___________ to maintain homeostasis. For example, if the thyroid gland is suspected of malfunctioning, the physician can monitor it by using a radioactive (17) ___________. Once the problem has been determined, the physician can take appropriate steps to help the patient regain homeostasis.

CLINICAL CONCEPTS

The following clinical concept applies the information presented in Chapter 2. Following the application is a set of questions to help you understand the concept.

Patricia called the doctor's office because she was having severe "stomach" pain. The doctor's nurse asked her to describe specifically where the pain was located, and she indicated that it was in the lower portion of her "belly." The nurse determined that the pain was coming from the right inguinal region rather than the so-called belly or stomach area, so she advised her to go to the emergency room for further evaluation.

At the hospital, the doctor palpated the right inguinal region, and Patricia yelled out in pain. The doctor suggested that it could be appendicitis or possibly a bleeding ulcer in the large intestine. Before ordering surgery to examine the appendix, the doctor decided to order a specific test designed to detect internal bleeding in the intestines. This test consists of a specially prepared solution of ^{99}Tc (technetium-99), which is an isotope of ^{98}Tc. This solution was injected into Patricia. If internal bleeding was occurring, the solution containing ^{99}Tc would be found in high concentrations in the localized area of bleeding.

After a few minutes of examination of the "pictures" taken by the gamma ray camera, the doctor found an ulcer that was bleeding. The ulcer was inflamed in the area of a nerve and thus was causing the severe pain.

1. How was the technetium-99 solution used to avoid exploratory surgery?

2. If an atom has an altered number of neutrons, many times it will become a radioactive isotope. Technetium-98 (^{98}Tc) typically has 55 neutrons. How many neutrons does ^{99}Tc have?

3. What does the number 99 represent in ^{99}Tc?

4. Why did the nurse ask Patricia to be more specific about the location of her pain?

THIS CONCLUDES CHAPTER 2 EXERCISES

CHAPTER

3

Cells: Their Structure and Function

Just as atoms are the building blocks of molecules, **cells** are the highly organized basic structural units found in all living things. In the human body, cells originate from a single fertilized egg, and additional cells are produced by the division of preexisting cells. These cells become specialized in the process called **differentiation,** forming tissues and organs that perform specific functions.

The **cell theory,** which provides the foundation for the study of anatomy and physiology, incorporates the following basic concepts:

- Cells are the building blocks of *all* living things.
- Cells are produced by the division of preexisting cells.
- Cells are the smallest units of structure and function.
- Each cell maintains homeostasis at the cellular level.
- Homeostasis at the tissue, organ, system, and organism levels reflects the combined and coordinated actions of many cells.

The exercises in Chapter 3 relating to cells will reinforce your ability to learn, synthesize, and apply the principles of cell biology to the structure and function of the human body. They will also help you understand the important role cells play in maintaining homeostasis.

Chapter Objectives:

1. List the functions of the cell membrane and the structures that perform those functions.
2. Describe the ways cells move materials across the cell membrane.
3. Describe the organelles of a typical cell and indicate their specific functions.
4. Explain the functions of the cell nucleus.
5. Summarize the process of protein synthesis.
6. Describe the process of mitosis and explain its significance.
7. Define differentiation and explain its importance.
8. Describe disorders caused by abnormal mitochondria or lysosomes.
9. Describe the development of cancer and its possible causes.

PART I: OBJECTIVE-BASED QUESTIONS

OBJECTIVE 1: List the functions of the cell membrane and the structures that perform those functions (textbook p. 38–42).

The cell membrane forms the outer boundary of the cell and provides more than just a boundary to hold in cell organelles. After studying pp. 38–42 in the text, you should be able to describe how the cell membrane controls what moves into and out of the cell.

1. A cell membrane is made mostly of ____________.

   a. phospholipid, carbohydrates, protein, and cholesterol
   b. phospholipid, carbohydrates, protein, and receptors
   c. phospholipid, carbohydrates, receptors, and vitamins
   d. phospholipid, carbohydrates, receptors, and water

____ 2. Both the external and the internal environments of a cell are mostly water. What feature of the cell membrane prevents it from dissolving?

   a. The membrane is a double-layered structure.
   b. The membrane is soluble in water.
   c. The protein molecules of the cell membrane are too large to dissolve.
   d. The tail (fatty-acid) portion is not soluble in water.

____ 3. The cell membrane is ____.

   a. freely permeable
   b. impermeable
   c. selectively permeable
   d. solid

4. Structurally, the cell membrane is a (n) ________________________.

5. The outer boundary of the intracellular material is called the ____________________.

6. Identify the cell membrane components in Figure 3–1.

Figure 3–1 The Cell Membrane

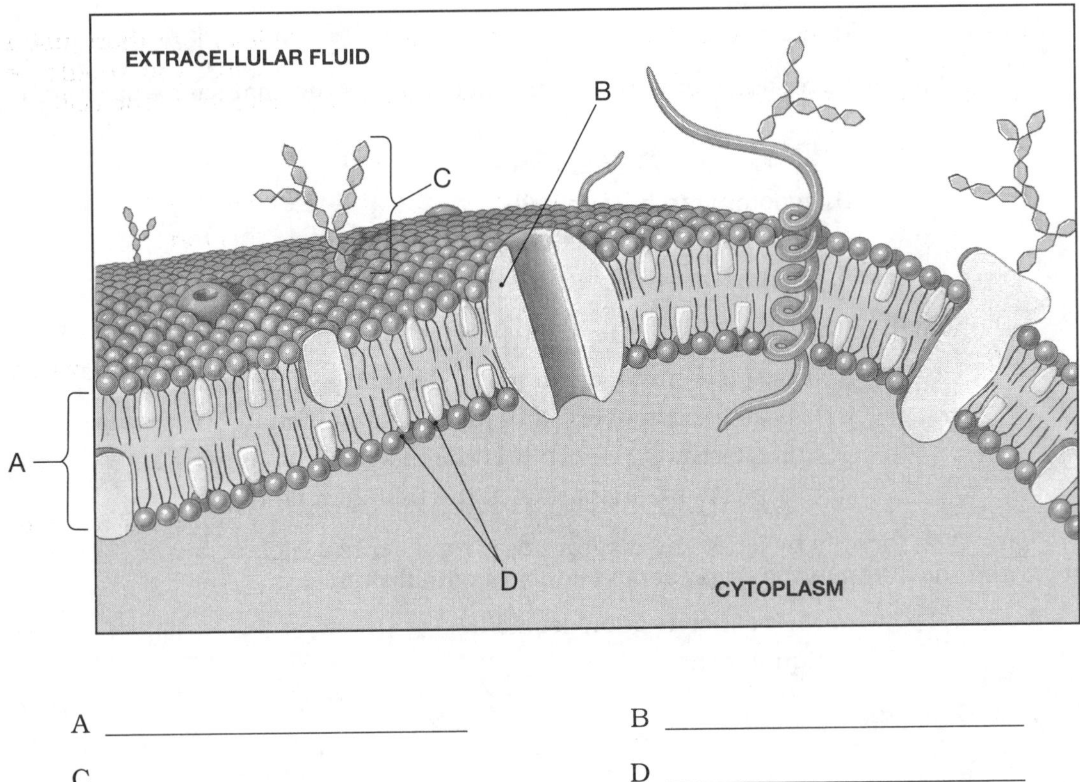

A ______________________ B ______________________

C ______________________ D ______________________

OBJECTIVE 2: Describe the ways cells move materials across the cell membrane (textbook pp. 42–46).

After studying pp. 42–46 in the text, you should be able to describe how a cell obtains nutrients and gets rid of wastes.

_____ 1. Which of the following statements is true?

a. Diffusion and osmosis are synonymous terms.
b. Diffusion is a type of osmosis.
c. Diffusion is passive transport, whereas osmosis is active transport.
d. Osmosis is a type of diffusion.

_____ 2. Which of the following describes a hypertonic solution?

a. The solution (external environment) has a greater concentration of solutes than the internal environment of the cell.
b. The solution (external environment) has a lower concentration of solutes than the internal environment of the cell.
c. The solution (external environment) has more water than the internal environment.
d. The solution (external environment) has less water than the internal environment.

_____ 3. A cell will swell with water if it is placed in a (an) _____ solution.

a. hypertonic
b. hypotonic
c. isotonic
d. normal saline

_____ 4. Many intravenous solutions are normal saline. A normal saline solution is _____.

a. hypertonic to human cells
b. hypotonic to human cells
c. isotonic to human cells
d. used because the body needs sodium chloride (salt)

_____ 5. Which of the following is true of phagocytosis?

a. It is active transport; it is the discharge of material.
b. It is active transport; it is the taking in of material.
c. It is passive transport; it is the discharge of material.
d. It is passive transport; it is the taking in of material.

6. In Figure 3–2, identify by letter the diagram that represents a high concentration gradient and the diagram that represents a state of equilibrium.

a. High concentration gradient. _____
b. Equilibrium. _____

Figure 3–2 Diffusion

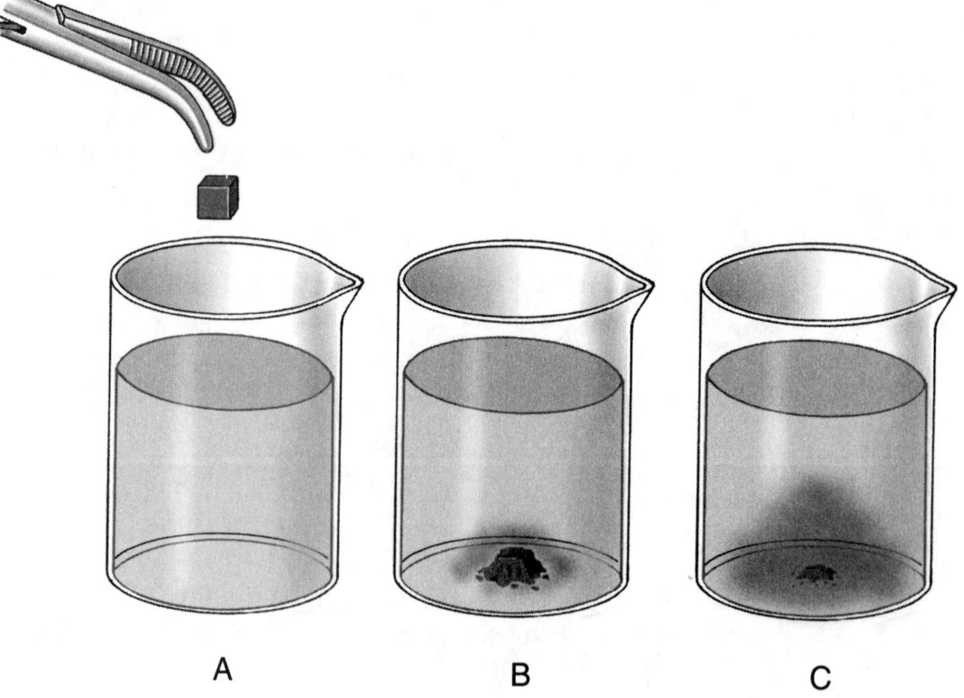

7. Each diagram in Figure 3–3 shows a cell in a container that represents the cell's environment. Which diagram represents a cell in a hypertonic solution; a hypotonic solution; an isotonic solution?

____ a. Hypertonic environment.
____ b. Hypotonic environment.
____ c. Isotonic environment.

Figure 3–3 Osmotic Environments

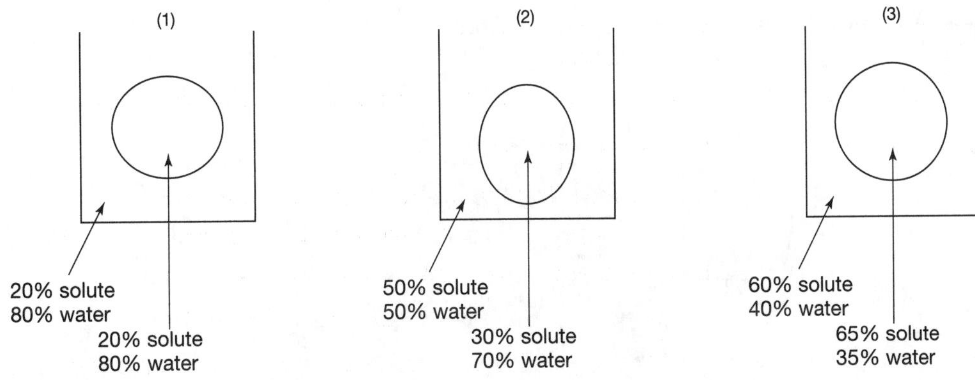

OBJECTIVE 3: Describe the organelles of a typical cell and indicate their specific functions (textbook pp. 47–48).

After studying pp. 47–48 in the text, you should be able to list the organelles in a typical cell and describe the essential function of each. Just as each organ of the body has a specific function that is necessary to keep it alive, the organelles of an individual cell have specific functions to keep the cell alive. If the cells stay alive and are healthy, the body is in homeostasis.

____ 1. The major function of ribosomes is ____.

a. to alter proteins
b. to engulf bacteria
c. to produce adenosine triphosphate
d. to produce protein

____ 2. The major function of mitochondria is ____.

a. to alter protein
b. to produce adenosine triphosphate
c. to produce carbohydrates
d. to produce protein

____ 3. The major function of lysosomes is ____.

a. to release digestive enzymes
b. to engulf material
c. to produce carbohydrates
d. to produce protein

____ 4. The major function of the Golgi apparatus is ____.

a. to alter protein
b. to engulf material
c. to produce carbohydrates
d. to produce protein

_____ 5. Which of the following statements is true regarding the functions of the endoplasmic reticulum?

a. SER synthesizes carbohydrates, and RER synthesizes protein.
b. SER synthesizes protein, and RER synthesizes carbohydrates.
c. SER has protein-producing ribosomes attached to it, and RER does not.
d. SER and RER have the same function.

6. Identify the cell organelles in Figure 3–4.

Figure 3–4 A Generalized Cell of the Human Body

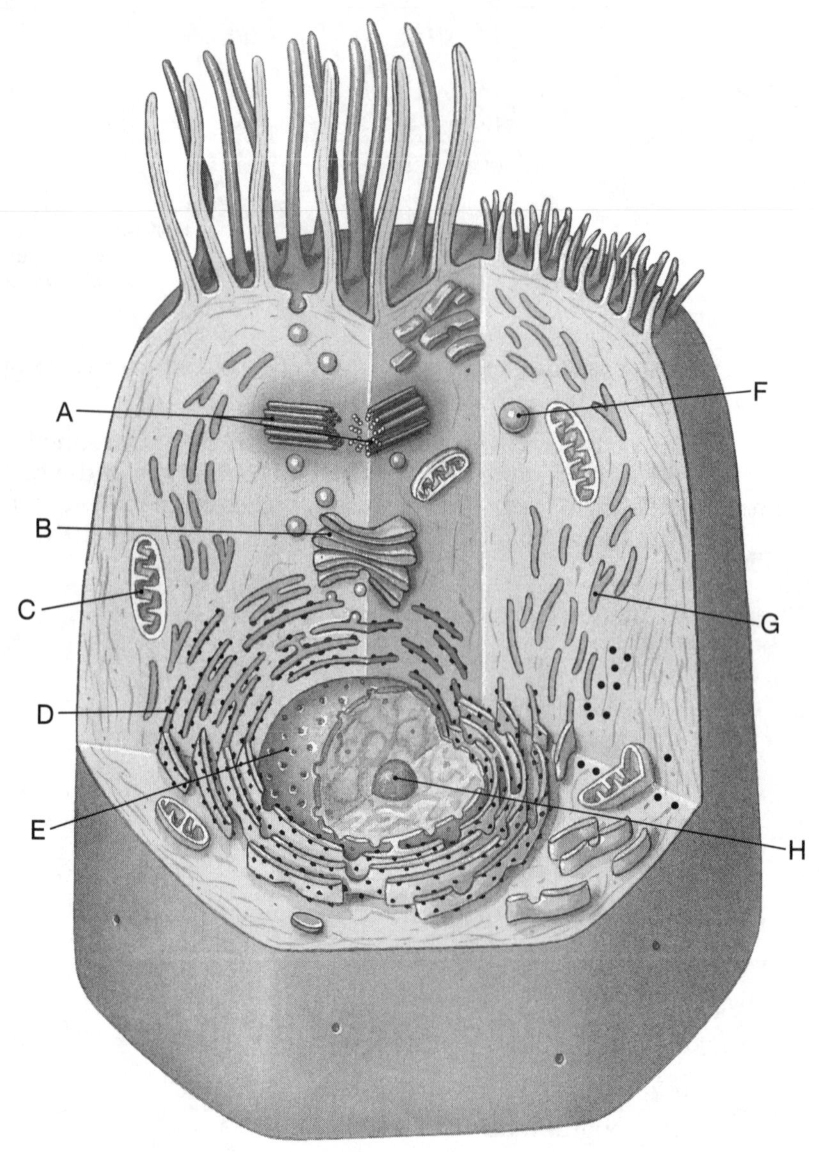

A ____________________ B ____________________

C ____________________ D ____________________

E ____________________ F ____________________

G ____________________ H ____________________

OBJECTIVE 4: Explain the functions of the cell nucleus (textbook pp. 48–49).

After studying pp. 48–49 in the text, you should be able to describe how DNA in the cell's nucleus exerts control over the cell.

_____ 1. How does the nucleus exert its control over the cell?

a. The nucleus controls the cell by regulating protein production.
b. The nucleus controls the cell via chemical communication through the nuclear pores.
c. The nucleus controls the cell via the nucleolus.
d. The nucleus has an ideal location in the cell—it is centrally located.

_____ 2. Ribosomal proteins and mRNA (messenger ribonucleic acid) are produced in the _____.

a. nucleolus
b. nucleus
c. cytoplasm
d. mitochondria

_____ 3. There are _____ chromosomes in the nucleus of a cell, and these chromosomes are made of _____.

a. 23, DNA
b. 23, protein
c. 46, DNA
d. 46, RNA

_____ 4. The nuclear pores permit the exit of _____ into the cytosol so it (they) can travel to the ribosomes.

a. RNA
b. DNA
c. chromosomes
d. nucleoli

_____ 5. Our genetic coded blueprint is located in the _____.

a. nucleolus of a cell
b. nucleus of a cell
c. proteins of a cell
d. ribosomes of a cell

6. Identify the structures in Figure 3–5.

Figure 3–5 A Generalized Cell of the Human Body

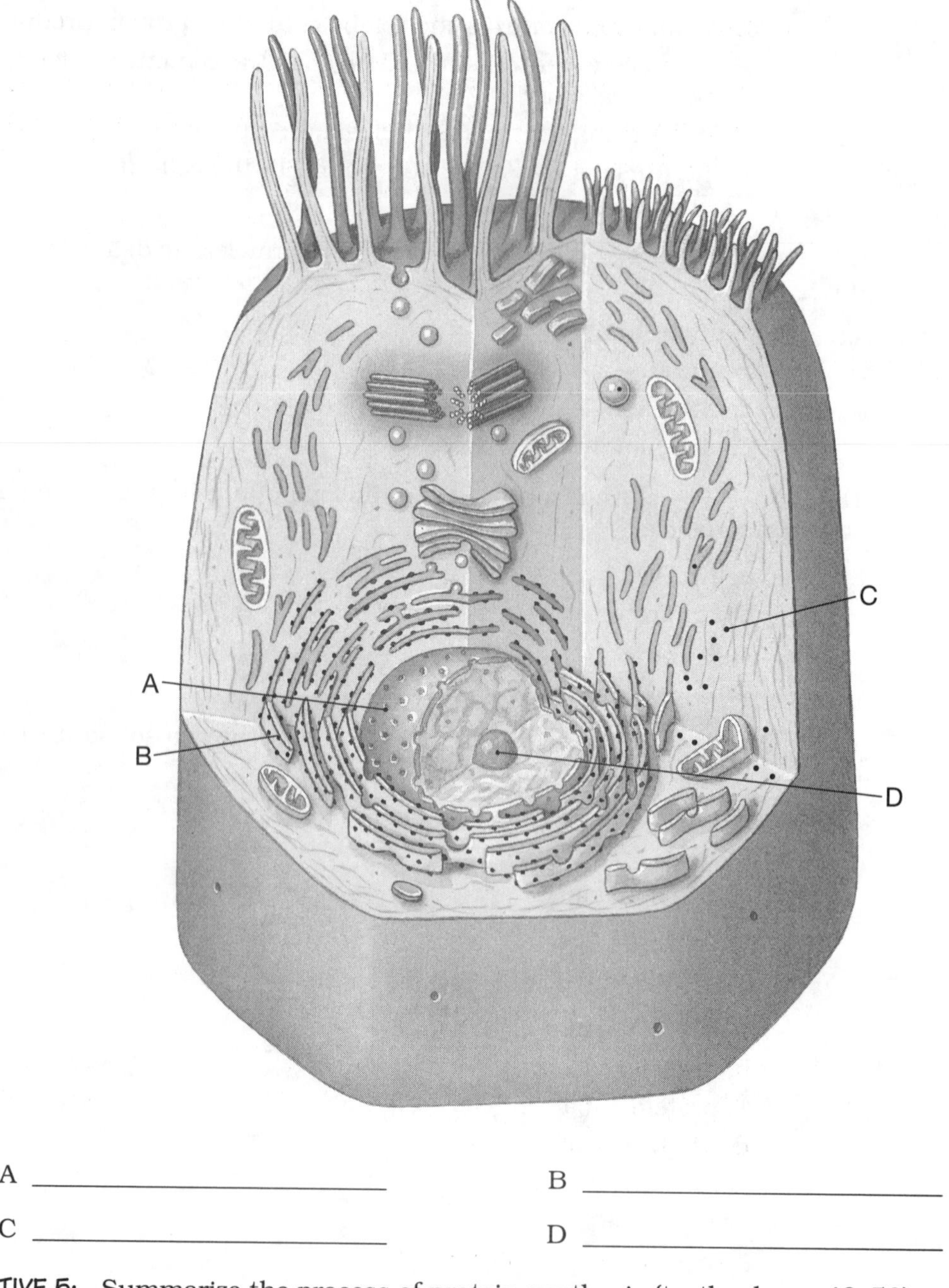

A ______________________ B ______________________

C ______________________ D ______________________

OBJECTIVE 5: Summarize the process of protein synthesis (textbook pp. 49–50).

After studying pp. 49–50 in the text, you should be able to describe how information coded in DNA in the nucleus is transmitted to the ribosomes in the cytoplasm. The ribosomes use the coded information to synthesize protein.

____ 1. Which of the following sequences of events correctly describes protein synthesis?

a. DNA transcribes the coded message to RNA; RNA (messenger RNA) exits the nuclear pores; RNA translates the coded message to ribosomes; ribosomes make protein.

b. DNA transcribes the coded message to RNA; RNA (transfer RNA) exits the nuclear pores; RNA carries the coded message to the ribosomes, where translation occurs; ribosomes make protein.

c. DNA translates the coded message to RNA; RNA exits the nuclear pores; RNA carries the coded message to the ribosomes, where transcription occurs; ribosomes make protein.

d. DNA transcribes the coded message to RNA; RNA (messenger RNA) exits the nuclear pores and translates the coded message to transfer RNA; tRNA carries the coded message to the ribosomes; ribosomes make protein.

____ 2. The DNA transmits a chemical message to RNA. This step is called ____.

a. codon
b. synthesis
c. transcription
d. translation

____ 3. The RNA that exits the nuclear pores to the cytosol on its way to the ribosomes is called ____.

a. messenger RNA
b. ribosomal RNA
c. transcription RNA
d. transfer RNA

____ 4. The process by which RNA gives the coded message to the ribosomes is called ____.

a. codon
b. synthesis
c. transcription
d. translation

____ 5. In order to make a protein according to the instructions coded in the DNA, ribosomes need amino acids. Where do the ribosomes get the amino acids?

a. from the cytosol
b. from the nucleus
c. from the SER
d. from within the ribosomes

____ 6. If the DNA triplet is TAG, the corresponding codon on the messenger RNA will be ____.

a. ACT
b. AGC
c. ATC
d. AUC

_____ 7. If the messenger RNA has the codons CCC CGG UUA, the corresponding transfer RNA anticodons will be _____.

a. GGG GCC AAU
b. CCC CGG TTA
c. GGG GCC AAT
d. CCC CGG TTU

8. A protein consists of a sequence of _________ bonded together.

9. Identify the letters in Figure 3–6 that represent the following: DNA, mRNA, tRNA, ribosomes, nuclear pore, the process of transcription, and the process of translation.

Figure 3–6 Protein Synthesis

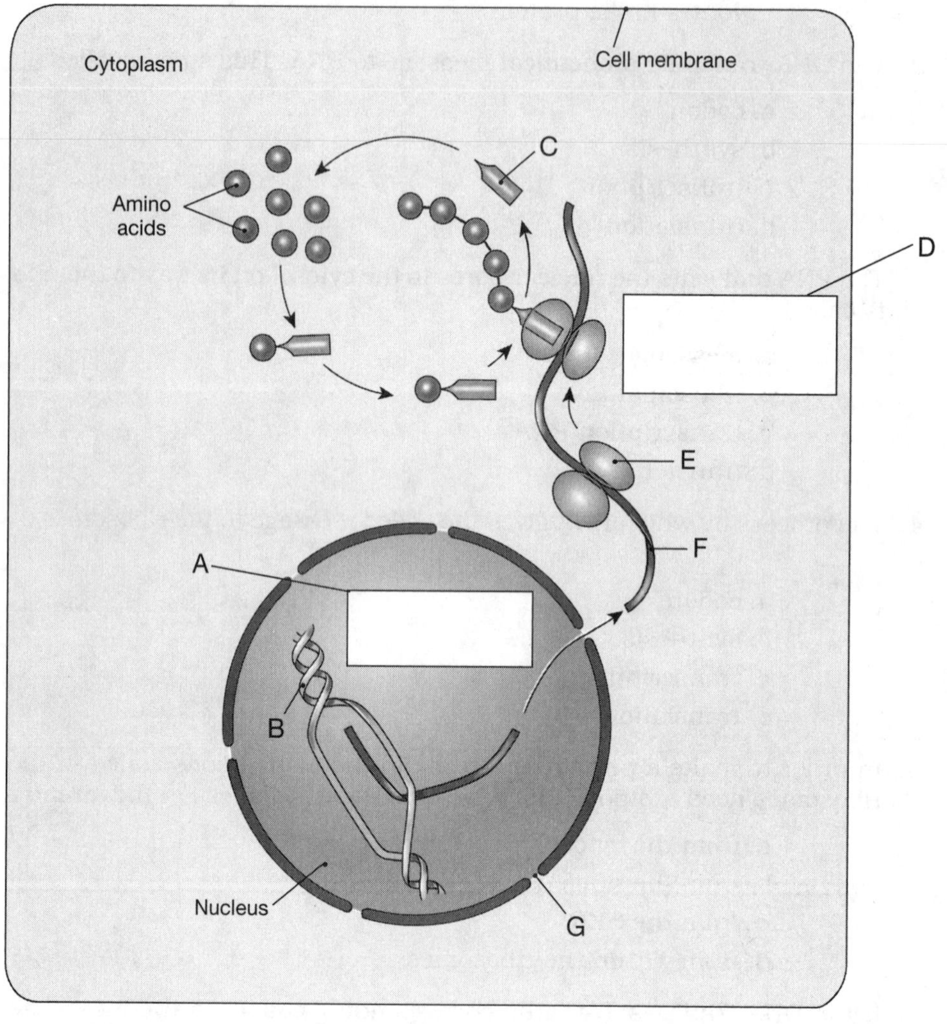

A ____________________ B ____________________

C ____________________ D ____________________

E ____________________ F ____________________

G ____________________

OBJECTIVE 6: Describe the process of mitosis and explain its significance (textbook pp. 51–52).

Cells reproduce for a variety of reasons, one of which is for tissue repair. Without tissue repair, homeostasis would be in jeopardy. After studying pp. 51–52 in the text, you should be able to describe the three stages of cell reproduction: interphase, mitosis, and cytokinesis.

_____ 1. DNA replication and cell organelle duplication occur during which phase of cell reproduction?

a. interphase
b. mitosis
c. prophase
d. meiosis

_____ 2. List the phases of mitosis in proper sequence.

a. interphase–prophase–metaphase–anaphase–telophase–cytokinesis
b. interphase–prophase–metaphase–anaphase–telophase
c. prophase–interphase–metaphase–anaphase–telophase–cytokinesis
d. prophase–metaphase–anaphase–telophase

_____ 3. In which phase of cell division do the chromatids line across the middle of the nuclear area?

a. metaphase
b. anaphase
c. prophase
d. interphase

_____ 4. The chromatids are _____ during anaphase.

a. being pulled to opposite ends of the cell by centromeres
b. coming apart at the centrioles
c. coming apart at the centromere
d. joining together at the centromere

_____ 5. The formation of two new nuclei occurs during _____, and the formation of two new cells occurs during _____.

a. anaphase, telophase
b. cytokinesis, telophase
c. telophase, cytokinesis
d. telophase, interphase

6. In Figure 3–7, identify the phases of mitosis and the various structures involved with mitosis.

Figure 3–7 Cell Reproduction: Mitosis

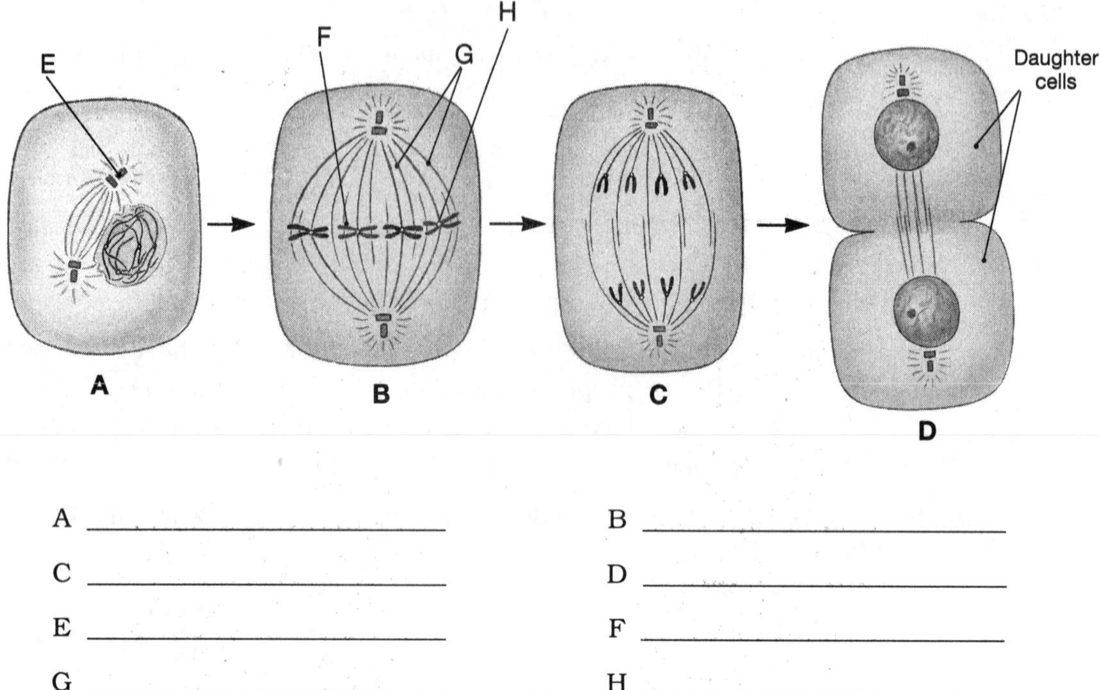

A ______________________ B ______________________

C ______________________ D ______________________

E ______________________ F ______________________

G ______________________ H ______________________

OBJECTIVE 7: Define differentiation and explain its importance (textbook p. 52).

After studying p. 52 in the text, you should be able to explain how a single cell can divide and ultimately produce several trillion cells and how these cells become specialized and form the four kinds of tissue.

_____ 1. The cells of our bodies contain the same chromosomes and the same genes. How can we have different cells if they all contain the same chromosomes and genes?

a. Some of the genes are deactivated in some cells and activated in others.

b. The cells contain the same genes but a different number of chromosomes.

c. The cells form tissues that makes them different from other cells.

d. All the above are correct.

_____ 2. Which of the following statements is true?

a. When cells differentiate they begin to develop the same protein molecules in order to develop the same tissue type.

b. When cells differentiate they begin to specialize in function, thereby forming tissues that have different functions.

c. When cells differentiate they begin to specialize in function, thereby forming tissues that have the same function.

d. None of the above is true.

____ 3. The process of differentiation, which causes cells to have different characteristics, involves ____.

a. changes in protein structure
b. gene activation and deactivation
c. the presence of genes
d. the process of fertilization

4. In order for cells to develop specific functions relating to survival, they must ________________________.

5. The process that restricts a cell to performing particular functions is known as ________________________.

OBJECTIVE 8: Describe disorders caused by abnormal mitochondria or lysosomes (textbook pp. 52–53).

The proper functioning of cellular organelles is necessary for homeostasis. Any abnormality of the cellular organelles will result in abnormal cell function, which in turn will cause the body to be out of homeostasis. After studying in the text pp. 52–53, you should be able to describe the disorders caused by abnormal mitochondria or abnormal lysosomes.

____ 1. Mitochondrial disorders are generally passed to the offspring from ____.

a. mother
b. father
c. both mother and father
d. only the X sperm cell

____ 2. A disorder of the mitochondria may result in which of the following?

a. Leber's disease
b. myoclonic epilepsy
c. Parkinson's disease
d. all the above

____ 3. A disorder of the lysosomes may result in which of the following?

a. Gaucher's disease
b. glycogen storage disease
c. Tay-Sachs disease
d. all the above

____ 4. Which of the following statements is true?

a. An abnormal lysosome may result in glycogen buildup in nerve cells.
b. An abnormal mitochondrion may result in a type of blindness.
c. An abnormal mitochondrion may result in certain types of epilepsies.
d. All the above

____ 5. Which of the following lysosome abnormalities may result in muscle weakness?

a. Gaucher's disease
b. glycogen storage disease
c. myoclonic epilepsy
d. Tay-Sachs disease

OBJECTIVE 9: Describe the development of cancer and its possible causes (textbook pp. 53–56).

After studying in the text, pp. 53–56, you should be able to identify some of the causative agents of cancer and describe how some cancers develop. You should also be able to use correctly some of the terminology associated with cancer.

_____ 1. Which of the following terms refers to the spread of abnormal cells?

a. benign
b. cancer
c. malignant
d. metastasis

_____ 2. Which of the following is true of the word *tumor*?

a. It is synonymous with the term *cancer*.
b. It is a benign cancer.
c. It is an abnormal growth of cells.
d. It is a mass of cells that will become cancerous.

_____ 3. Tumors can grow because they become vascularized, that is, new blood vessels form to actually "feed" the tumor. Which of the following terms means the formation of new blood vessels?

a. angiogenesis
b. antiangiogenesis
c. metastasis
d. secondary formation

_____ 4. Which of the following are involved in causing cancer?

a. carcinogens
b. mutagens
c. oncogenes
d. all the above

_____ 5. Radiation has some success at "killing" cancer cells because ____.

a. cancer cells undergo mitosis regularly
b. cancer cells are abnormal
c. radiation is toxic to cancer cells
d. cancer cells "attract" radiation

PART II: CHAPTER-COMPREHENSIVE EXERCISES

Match the terms in column A with the definitions in column B.

MATCHING I

	(A)	(B)
_____	1. active	A. A solution containing a greater concentration of solutes than that within the cell
_____	2. anaphase	B. A solution that will cause water to enter the cell
_____	3. centrioles	C. A type of transport that requires the use of ATP
_____	4. centromere	D. The structures that produce spindle fibers during cell reproduction
_____	5. chromosomes	E. Structure that connects chromatids to each other
_____	6. DNA	F. Strands of DNA
_____	7. hypertonic	G. The molecule that dictates what kind of protein the ribosomes should make
_____	8. hypotonic	H. The phase of mitosis during which the chromatids are pulled apart
_____	9. Tay-Sachs	I. A genetic disorder associated with abnormal lysosomes

MATCHING II

	(A)	(B)
_____	1. lysosomes	A. The process of entrapment of foreign particles
_____	2. metaphase	B. The organelles responsible for protein production
_____	3. metastasis	C. The organelles that digest material the cell has phagocytized
_____	4. mitochondria	D. The organelles responsible for producing ATP
_____	5. mutation	E. The molecule that delivers a coded message to the ribosomes
_____	6. neoplasm	F. An alteration of the DNA during cell reproduction
_____	7. phagocytosis	G. The phase of mitosis during which the chromatids line up in the middle of the cell
_____	8. ribosomes	H. The spread of cancer cells
_____	9. RNA	I. The growth of abnormal tissue

CONCEPT MAP

This concept map summarizes and organizes the structures and functions of a typical animal cell. Use the following terms to complete the map by filling in the boxes identified by the circled numbers, 1–10.

cytosol	**lysosomes**	**mitochondria**
nucleolus	**proteins**	**ribosomes**
lipid bilayer	**organelles**	**Parkinson's disease**
Tay-Sachs disease		

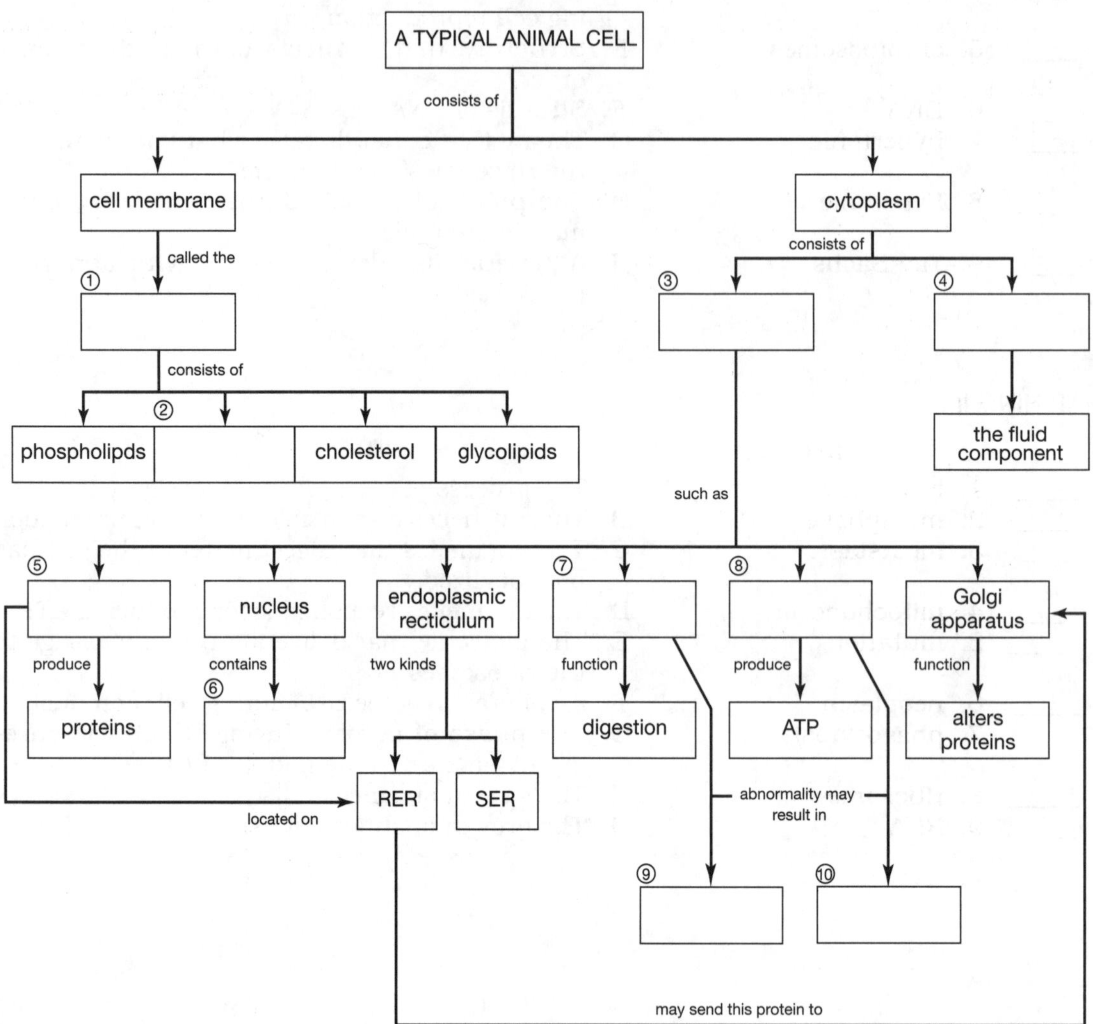

CROSSWORD PUZZLE

The following crossword puzzle reviews the material in Chapter 3. To complete the puzzle, you have to know the answers to the clues given, and you must be able to spell the terms correctly.

ACROSS

1. A solution that has a greater concentration of solutes than that within the cell
4. Molecules that bond together to produce a protein
8. The organelle responsible for making ATP
9. Tay-Sachs disease is associated with an abnormal _____.
11. The movement of water from an area of high concentration to an area of low concentration
12. A buildup of these granules in the lysosomes may cause muscle weakness.

DOWN

2. The process by which a cell engulfs particles
3. The second phase of mitosis
5. A buildup of this molecule may result in Gaucher's disease.
6. A solution that has a lower concentration of solutes than that within the cell
7. Cell organelles that produce protein
8. A type of RNA that delivers information to the ribosomes
10. A genetic change in the DNA

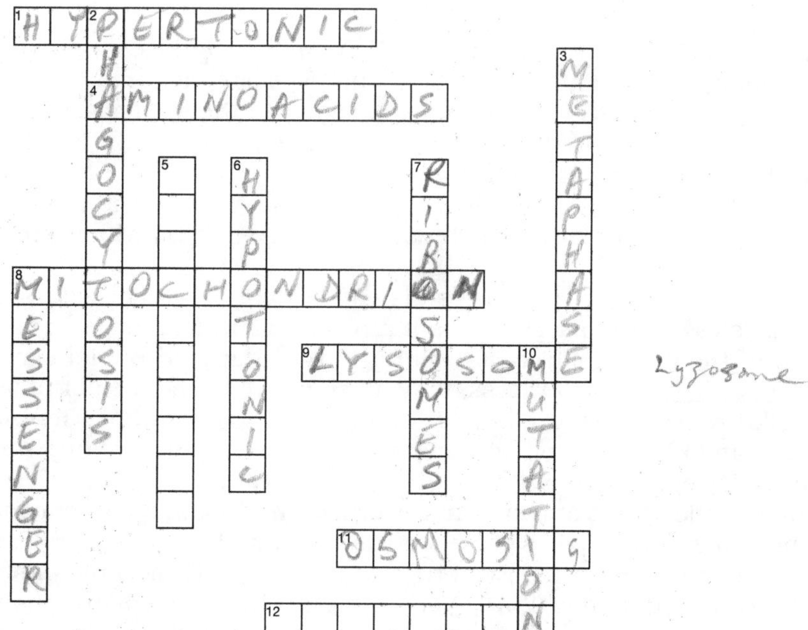

FILL-IN-THE BLANK NARRATIVE

Use the following terms to fill in the blanks of this summary of Chapter 3.

amino acids	**chromosomes**	**codons**
homeostasis	**hypertonic**	**DNA**
lose	**lysosomes**	**mRNA**
mutation	**nucleus**	**organelles**
phagocytosis	**solute**	**hypotonic**
ribosome	**mitosis**	**tumor**
genes	**carcinogenic**	

All life begins as one cell. The adult body contains roughly 75 trillion cells. In order for the body to maintain (1) ___________, each cell in the body must function properly, which means that each of the cell's (2) ___________ must do its specific job. Cells must obtain nutrients, get rid of waste products, and eliminate foreign bodies to stay alive. One way cells get rid of foreign particles is by engulfing them by the process of (3) ___________. Once the material is engulfed, organelles called (4) ___________ digest the material.

Because much of our body is composed of protein, one very important cell organelle is the protein-producing (5) ___________. Protein synthesis is controlled by (6) ___________, which is contained in the (7) ___________ of the cell. This molecule contains genes composed of sequences of nitrogenous bases called (8) ___________. In the process of transcription, (9) ___________ is formed from complementary bases that copy the information contained in a gene. This messenger molecule carries the instructions for protein synthesis from the nucleus to the ribosomes, which create proteins by translating the information and assembling (10) ___________ into specific sequences.

Water is critical to cell survival. Cells remain in homeostasis by maintaining a proper water balance between their internal and external environments. If the concentration of a (11) ___________, salt for example, is higher in the fluid outside a cell than in the cytoplasm, the cell is in a (12) ___________ solution. The cell will (13) ___________ water to balance the concentrations. If too many cells lose water, the body signals us that we are thirsty. In order for the cells to take in water, they must be exposed to a (14) ___________ solution, which occurs when we drink water and dilute the extracellular fluid.

Another important cell process is reproduction, which occurs by cell division. This process is essential to survival, as it replaces old and damaged cells. The accurate duplication of the cell's genetic material into two identical nuclei occurs in the process of (15) ___________. This process, which occurs in a sequence of phases in the nucleus of the cell, involves the movement of (16) ___________, so that each new cell has the same number and kind of chromosomes as the parent cell. When an error occurs in mitosis or in the DNA replication, a(n) (17) ___________ results.

This genetic error can result in the formation of a benign or malignant (18) ___________. The body has several types of tumor-suppressing (19) ___________ to prevent these errors from becoming cancerous. Even though the body has mechanisms to try to prevent cancer, there are numerous (20) ___________ agents in the environment that can cause cancer. Researchers are therefore in a never-ending battle with cancer.

CLINICAL CONCEPTS

The following clinical concepts apply to information presented in Chapter 3. Following the applications is a set of questions to help you understand the concepts.

A. Lysosomes digest (break down) damaged organelles or foreign bodies in the cell. If the material does not get broken down, it will begin to accumulate in the cell, thereby disrupting cellular activity by "clogging" up the cell.

In glycogen storage disease lysosomes malfunction and cannot digest glycogen material in the cell, so glycogen begins to accumulate. Glycogen fills the cytoplasm of the cell, causing the cell to malfunction.

In Tay-Sachs disease the lysosomes of nerve cells malfunction and do not break down lipids. Lipids begin to accumulate in the cell, and the nerve cell begins to malfunction, causing nerve problems in the patient.

B. Knowing how cell organelles work helps scientists cure certain illnesses. For example, the manufacture of proteins by ribosomes is essential for survival, but if protein production stops, the cell will die.

When bacteria enter the body, they produce protein, which is essential for their survival. The bacteria will ultimately make us sick. Scientists discovered that bacterial ribosomes are structurally different from human ribosomes. This knowledge led them to design the drug streptomycin. This antibiotic affects bacterial ribosomes but not human ribosomes—it causes bacterial ribosomes to stop making protein. When the bacteria stop making protein, they die. Human cells keep on making protein because their ribosomes are not affected.

1. How is glycogen storage disease similar to Tay-Sachs disease?

2. How is glycogen storage disease different from Tay-Sachs disease?

3. How does streptomycin kill bacteria?

4. Why does streptomycin affect only bacterial cells and not human cells?

5. If human ribosomes and bacterial ribosomes were structurally the same, why would streptomycin not be the drug of choice?

THIS CONCLUDES CHAPTER 3 EXERCISES

CHAPTER

4

Tissues and Body Membranes

No single cell contains the metabolic machinery and organelles needed to perform all the functions of the human body. Individual cells of similar structure and function join together to form groups called **tissues**. Tissues are identified on the basis of their origin, location, shape, and function.

In Chapter 4 the study of tissues, called **histology,** is introduced, with emphasis on the four primary types: **epithelial, connective, muscle,** and **neural** tissues. Many of the tissues are named according to the organ system in which they are found or the function that they perform. Tissue of the nervous system, for example, is referred to as *neural* tissue. Similarly, tissue of the muscular system is referred to as *muscle* tissue. Connective tissue gets its name from the function it performs—namely, that of supporting, surrounding, and interconnecting other types of tissues.

The activities and exercises in this chapter are designed to help you identify, organize, and conceptualize the interrelationships among the tissue types and the relationship between cellular organization and tissue function. You will find, after studying these exercises, that all the tissues must function together to achieve homeostasis.

Chapter Objectives:

1. Discuss the types and functions of epithelial cells.
2. Describe the relationship between form and function for each epithelial type.
3. Compare the structures and functions of the various types of connective tissues.
4. Explain how epithelial and connective tissues combine to form four different types of membranes, and specify the functions of each.
5. Describe the three types of muscle tissue and the special structural features of each.
6. Discuss the basic structure and role of nervous tissue.
7. Explain how tissues respond to maintain homeostasis after an injury.
8. Describe how aging affects the tissues of the body.
9. Describe the changes in tissue structure that can lead to a malignant tumor.
10. Describe how cancerous tumors are classified, and list the seven warning signs of cancer.
11. Describe the various forms of cancer treatment.

PART I: OBJECTIVE-BASED QUESTIONS

OBJECTIVE 1: Discuss the types and functions of epithelial cells (textbook p. 62–64).

After studying p. 62–64 in the text, you should be able to describe the types of cells that make up epithelial tissue and the function of each.

_____ 1. There are two types of epithelial cells, namely, _____.

a. glandular and layered
b. simple and stratified
c. squamous and cuboid
d. stratified and layered

_____ 2. The epithelial cells that make up several layers are called _____.

a. glandular epithelia
b. layered epithelia
c. simple epithelia
d. stratified epithelia

_____ 3. In which of the following would simple epithelial cells be found?

a. the inside of the mouth
b. the outside of the mouth
c. the outside skin layers of the hand
d. all the above

_____ 4. Which epithelial type makes up the lining of the body cavities discussed in Chapter 1?

a. serous
b. simple
c. stratified
d. none of the above

_____ 5. The type of tissue that lines exposed surfaces and also lines internal cavities and passageways is _____.

a. connective
b. epithelial
c. muscular
d. neural

6. In general, epithelial tissue provides some degree of ____________________ from either the external or the internal environment.

OBJECTIVE 2: Describe the relationship between form and function for each epithelial type (textbook pp. 62–64).

After studying pp. 62–64 in the text, you should be able to describe the seven types of cells making up epithelial tissue, the function of each, and where each type is found. You should find that all seven variations of epithelial cells have at least one thing in common, namely, they make up either the inside lining or the outside lining of external or internal body surfaces. The following questions will serve to guide you in identifying the characteristics of each epithelial type.

_____ 1. From a side view, _____ cells appear to be thin and flat.

a. columnar
b. cuboid
c. glandular
d. squamous

_____ 2. Epithelial tissue called _____ epithelium lines the trachea, epithelial tissue called _____ epithelium lines the small intestine.

a. columnar, cuboidal
b. pseudostratified columnar, simple columnar
c. simple columnar, pseudostratified columnar
d. stratified columnar, simple columnar

_____ 3. Which of the following have cilia associated with them that function to protect our lungs?

a. pseudostratified columnar
b. pseudostratified squamous
c. simple squamous
d. none of the above

_____ 4. Transitional epithelium is typically found in _____.

a. organs that require a great deal of protection
b. organs that stretch and relax a lot such as the abdomen especially during pregnancy
c. organs that stretch and relax a lot such as the heart
d. organs that stretch and relax a lot such as the urinary bladder

_____ 5. Which of the following is a function of glandular epithelia?

a. They absorb material.
b. They protect the glands of our body.
c. They secrete hormones.
d. all the above

6. The seven different types of cells making up epithelial tissue all have one common characteristic: ______________________________

7. Several types of epithelial cells are shown in Figure 4–1. Match each structure in column A with the name and function of that cell type in column B.

Figure 4–1 Epithelial Cells

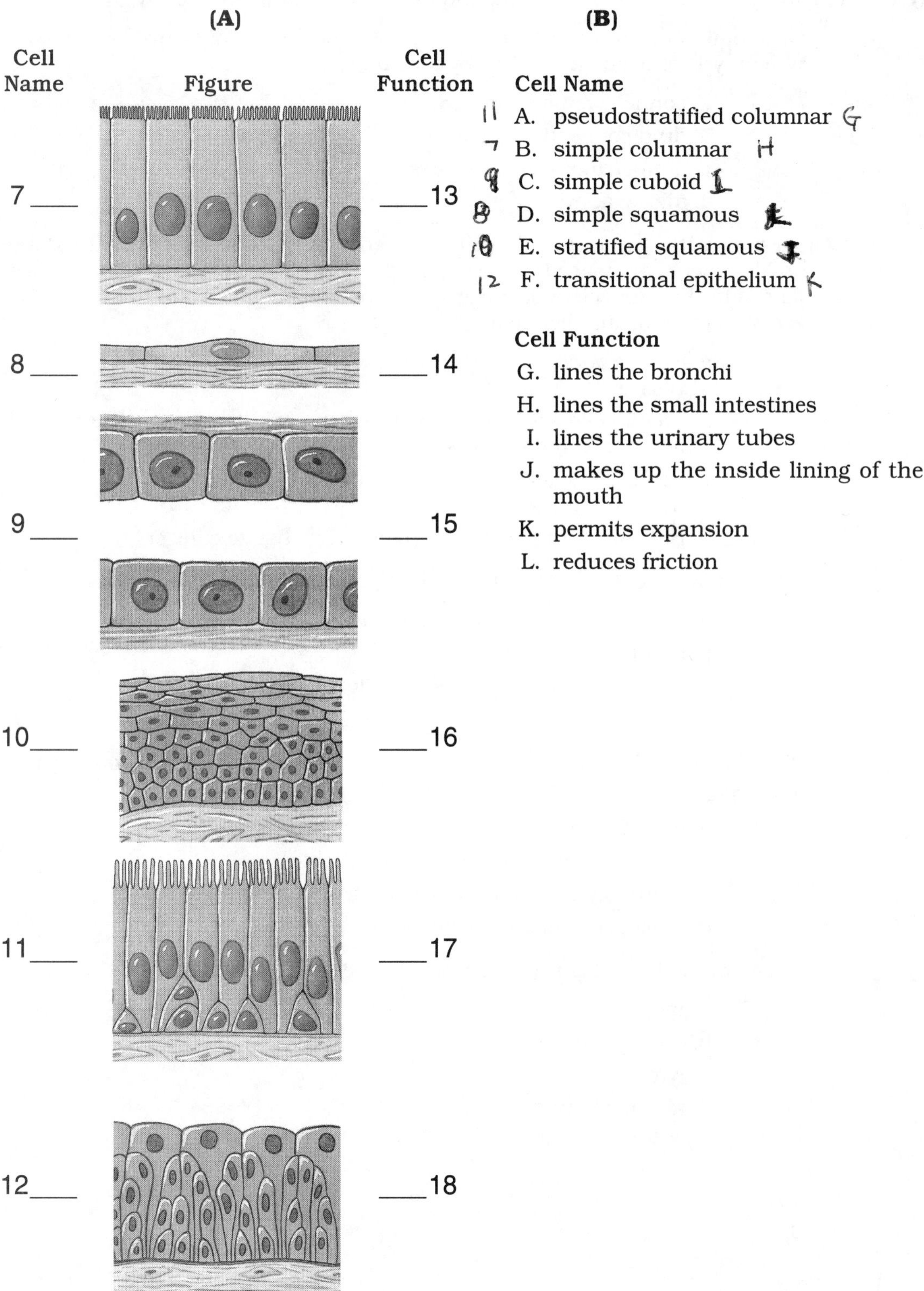

Cell Name

A. pseudostratified columnar
B. simple columnar
C. simple cuboid
D. simple squamous
E. stratified squamous
F. transitional epithelium

Cell Function

G. lines the bronchi
H. lines the small intestines
I. lines the urinary tubes
J. makes up the inside lining of the mouth
K. permits expansion
L. reduces friction

OBJECTIVE 3: Compare the structures and functions of the various types of connective tissues (textbook pp. 64–67).

After studying pp. 64–67 in the text, you should be able to describe the types of connective tissue, tell what they have in common, and identify the function of each.

_____ 1. One common characteristic of connective tissue is the presence of a matrix. Which type of tissue has a syrupy matrix?

a. connective tissue proper
b. fluid connective tissue
c. supporting connective tissue
d. all the above

_____ 2. Tendons attach muscle to bone. Ligaments attach bone to bone. Tendons and ligaments are placed in the connective tissue group. Blood cells are also placed in the connective tissue group. Why is blood placed in the same category as tendons and ligaments?

a. Blood connects or attaches one organ with another via the blood vessels.
b. Blood cells are surrounded by a matrix just as are the cells of tendons and ligaments.
c. Blood consists of lots of different cells, which is typical of connective tissue.
d. Blood is not placed in the connective tissue category.

_____ 3. Which of the following is (are) classified as supporting connective tissue?

a. areolar, ligaments, and tendons
b. cartilage and bone
c. cartilage, bone, tendons, and ligaments
d. bone only

_____ 4. Which of the following sequences of terms correlates with these words in the same sequence: red blood cells, bone cells, cartilage cells?

a. chondrocytes, osteocytes, adipocytes
b. erythrocytes, chondrocytes, osteocytes
c. erythrocytes, osteocytes, chondrocytes
d. leukocytes, chondrocytes, osteocytes

_____ 5. Which of the following is (are) in the connective tissue category?

a. blood
b. bone
c. lymph
d. all the above

_____ 6. Which of the following is *not* a characteristic of connective tissue?

a. cells close together
b. the presence of a matrix
c. the presence of ground substance
d. All the above are characteristic of connective tissue.

_____ 7. Which of the following is *not* a function of connective tissue?

a. absorption
b. provide framework
c. provide support
d. All the above are functions of connective tissue.

_____ 8. Which of the following consists of cells that set in small pockets called lacuna?

a. bone
b. blood
c. adipose
d. areolar

9. The matrix of blood is ______________________.

10. Identify the following cells and/or structures in the examples in Figure 4–2.

Figure 4–2 Examples of Connective Tissue

adipose	**bone**	**dense connective**
central canal	**collagen fibers**	**canaliculi**
hyaline cartilage	**loose connective**	**matrix**
red blood cells	**white blood cells**	

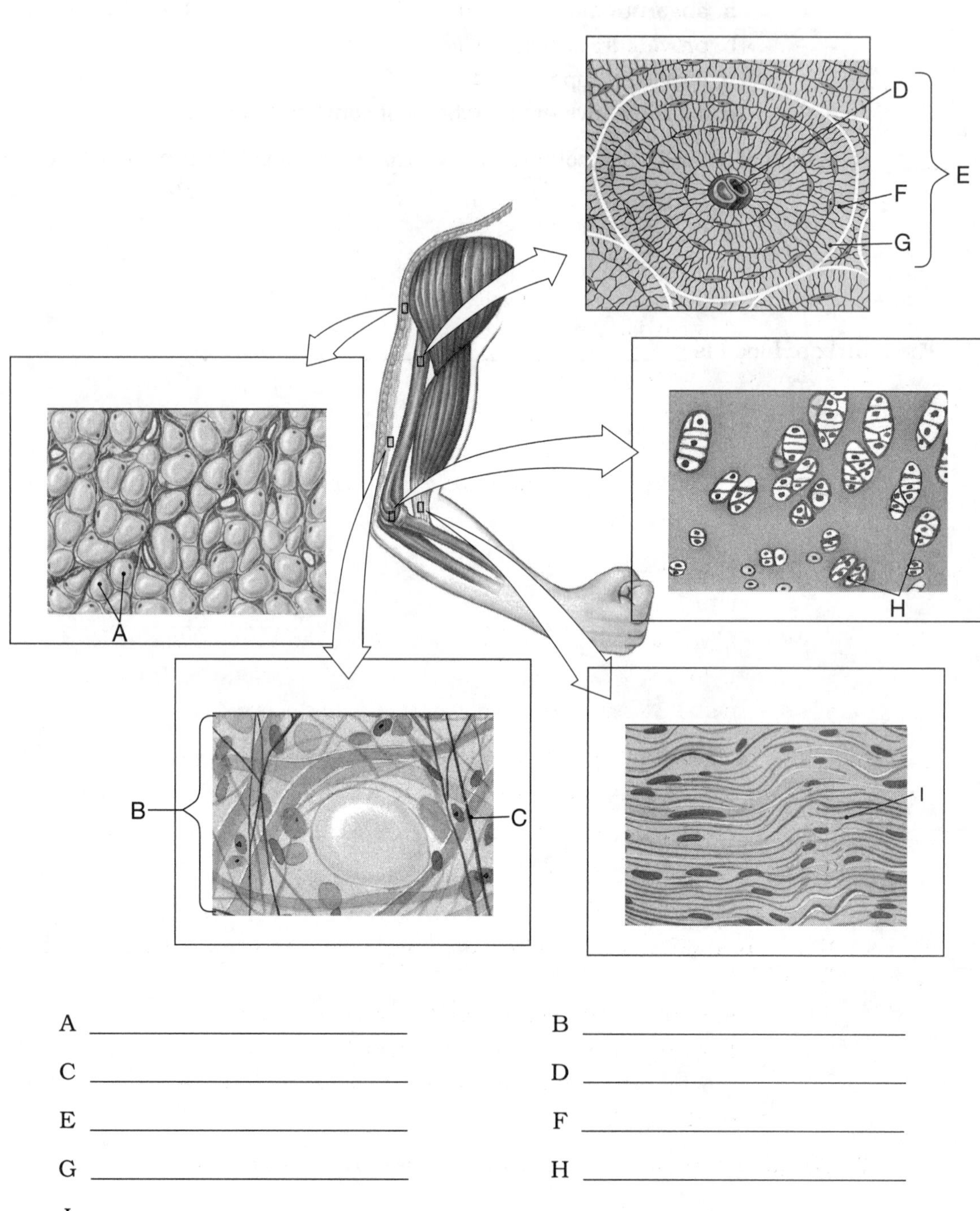

A ____________________ B ____________________

C ____________________ D ____________________

E ____________________ F ____________________

G ____________________ H ____________________

I ____________________

OBJECTIVE 4: Explain how epithelial and connective tissues combine to form four different types of membranes, and specify the functions of each (textbook pp. 68–69).

After studying pp. 68–69 in the text, you should be able to describe how the four different **tissue types work** together to accomplish the work of the four types of body membranes.

____ 1. Which of the following membranes consist of epithelial tissue and loose connective tissue?

a. mucous and cutaneous
b. cutaneous and synovial
c. mucous, serous, and synovial
d. mucous, serous, cutaneous, and synovial

____ 2. Which of the following membranes consist of epithelial tissue and adipose or areolar cells?

a. mucous, serous, and synovial
b. mucous, serous, and cutaneous
c. mucous, synovial, and cutaneous
d. synovial, serous, mucous, and cutaneous

____ 3. The visceral pleura and parietal pleura are made of ____.

a. serous membrane
b. epithelium and loose connective tissue
c. epithelium and adipose or areolar cells
d. all the above

____ 4. A serous membrane is composed of ____.

a. dense connective tissue and squamous cells
b. dense connective tissue and stratified squamous epithelium
c. loose connective tissue and simple columnar epithelium
d. loose connective tissue and squamous cells

____ 5. Which of the following can be found within the joints?

a. epithelial tissue and dense connective tissue
b. serous membrane
c. synovial membrane
d. both a and c

____ 6. Which of the following membranes produce a mucous substance and/or fluid?

a. mucous membranes and cutaneous membranes
b. serous membranes and synovial membranes
c. synovial membranes and cutaneous membranes
d. all the above

____ 7. The pleura, peritoneum, and pericardium are examples of ____.

a. mucous membranes
b. serous membranes
c. synovial membranes
d. cutaneous membranes

8. Identify the type of membrane that is located within each of the body regions identified by the arrows in Figure 4–3.

Figure 4–3 Membranes

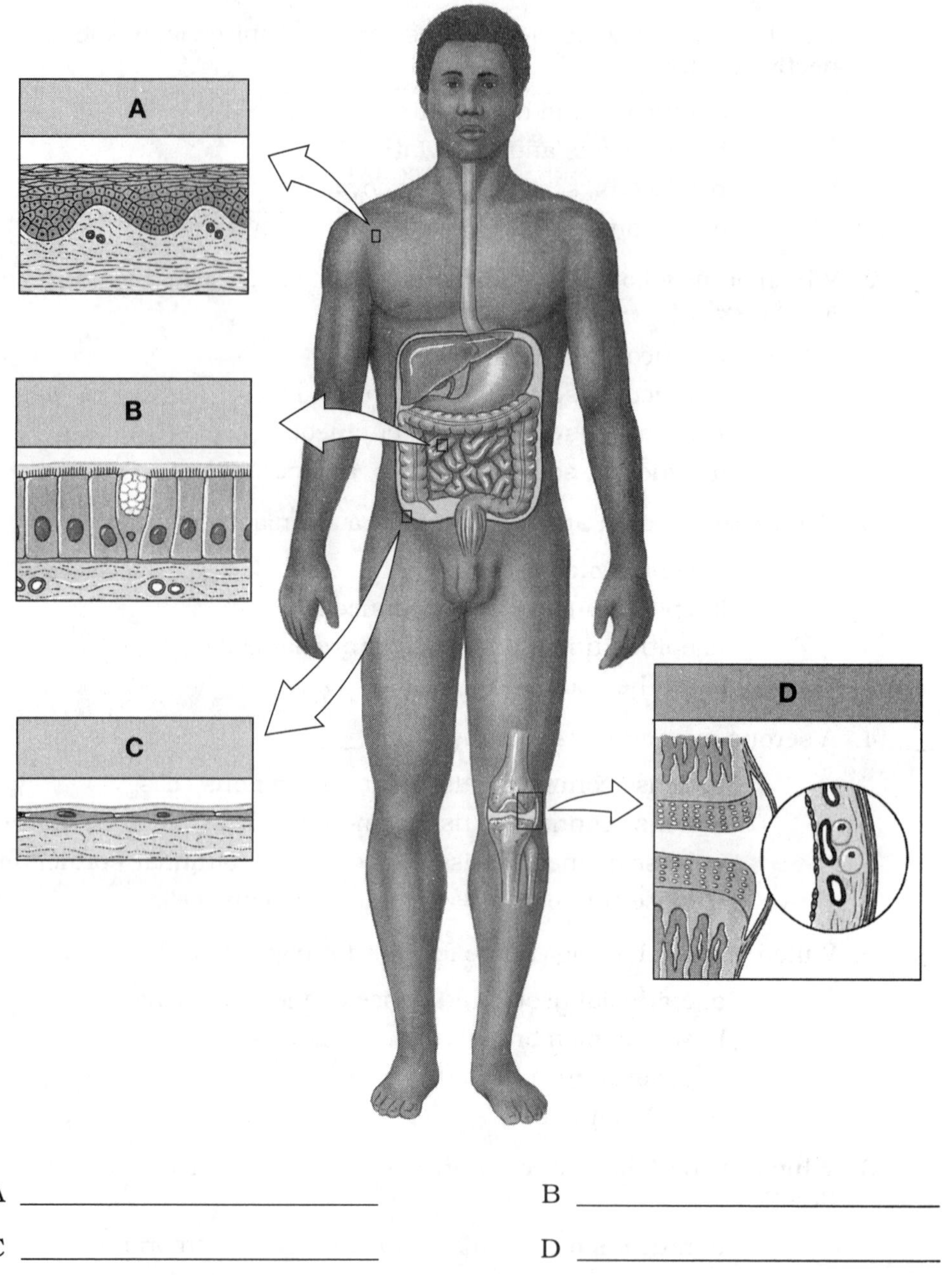

A ____________________ B ____________________

C ____________________ D ____________________

OBJECTIVE 5: Describe the three types of muscle tissue and the special structural features of each (textbook pp. 70–71).

After studying pp. 70–71 in the text, you should be able to identify the three different types of muscle cells that make up the muscle tissue group, tell what they have in common, and describe the function of each type.

_____ 1. Which of the following applies to skeletal muscle?

a. The muscle fibers are under involuntary control.
b. The muscle fibers have intercalated discs.
c. The muscle fibers have striations.
d. There is a single nucleus per cell.

_____ 2. Which of the following applies to cardiac muscle?

a. Cardiac muscle has intercalated discs.
b. Cardiac muscle has striations.
c. Cardiac muscle is under involuntary control.
d. All the above apply to cardiac muscle.

_____ 3. Which of the following applies to smooth muscle?

a. Each smooth muscle cell has multiple nuclei.
b. Smooth muscle cells are under involuntary control.
c. Smooth muscle cells have striations.
d. All the above apply to smooth muscle cells.

4. Because muscle cells are relatively long cells, they are typically referred to as ____________________.

5. Identify the following in Figure 4–4.

skeletal muscle	**smooth muscle**	**cardiac muscle**
intercalated discs	**striations**	**nucleus**

Figure 4–4 Muscle Tissue

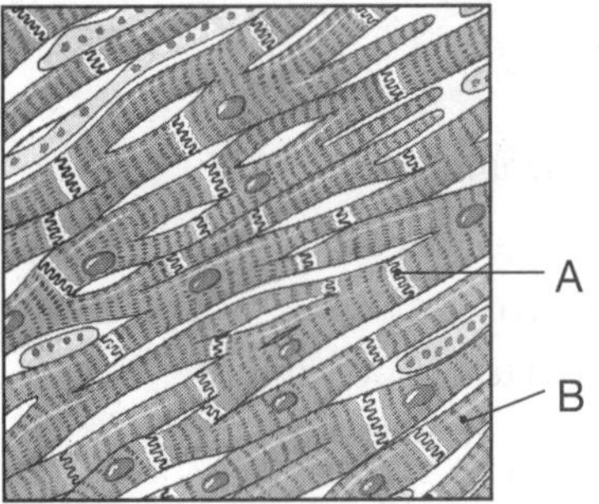

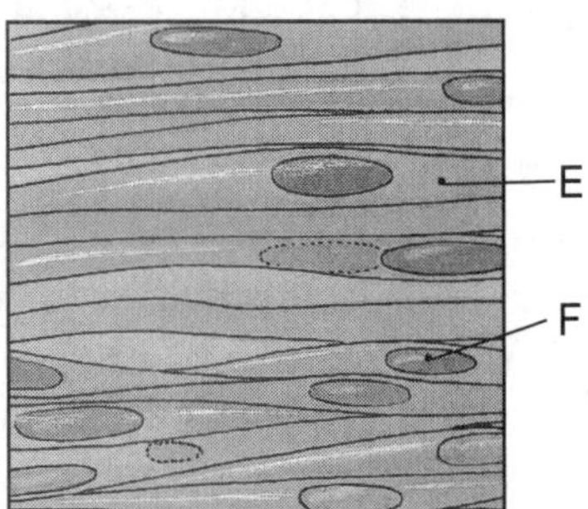

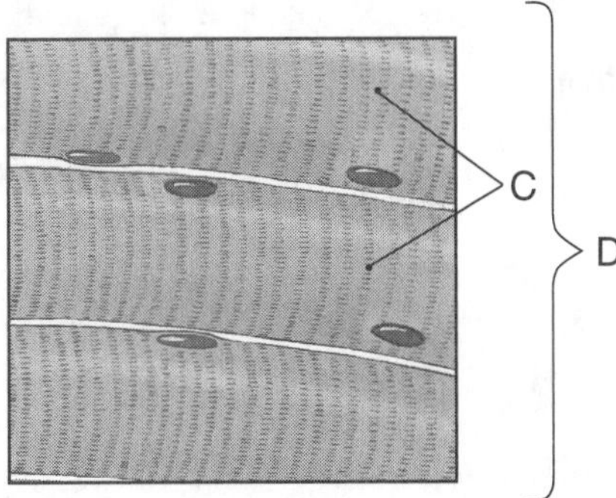

A ____________________ B ____________________

C ____________________ D ____________________

E ____________________ F ____________________

OBJECTIVE 6: Discuss the basic structure and role of nervous tissue (textbook pp. 71–72).

After studying pp. 71–72 in the text, you should be able to describe the two types of cells in nervous tissue, neuroglia and neurons, and their roles in the conduction of electrical impulses.

_____ 1. Which of the following contains the cell organelles?

a. axon
b. dendrite
c. nerve fiber
d. cell body (soma)

_____ 2. Which of the following is true about neurons?

a. They are single cells.
b. They can be microscopic in size.
c. They can be quite long in size.
d. All the above are true about neurons.

_____ 3. Which of the following represents the correct pathway for an impulse traveling through a neuron?

a. axon–soma–dendrite
b. dendrite–axon–soma
c. dendrite–soma–axon
d. soma–axon–dendrite

_____ 4. Which of the following is a function of a neuroglial cell?

a. It acts as a phagocyte to defend neural cells.
b. It provides the organelles for neurons.
c. It transmits signals to specific parts of the body.
d. All the above are functions of a neuroglial cell.

_____ 5. Cells capable of transmitting an impulse are _____.

a. neuroglia
b. neurons
c. soma
d. all the above

6. In Figure 4–5, identify the following structures: axon, dendrite, soma, and nucleus. Also, draw an arrow to indicate the direction in which an impulse will travel through the neuron.

Figure 4–5 A Typical Neuron

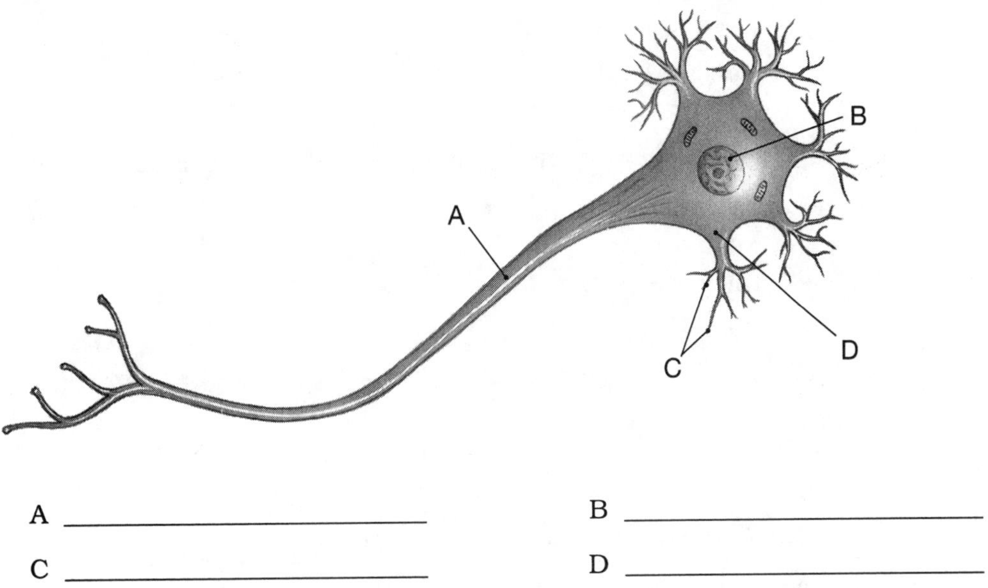

A ______________________ B ______________________

C ______________________ D ______________________

OBJECTIVE 7: Explain how tissues respond to maintain homeostasis after an injury (textbook p. 72).

After studying p. 72 in the text, you should be able to describe how injured tissue repairs itself.

_____ 1. Which of the following is (are) involved in the tissue repair process?

a. homeostatic mechanisms
b. inflammation of the tissue
c. regeneration of the tissue
d. both b and c

_____ 2. The return to homeostasis after injury involves which of the following processes?

a. inflammation
b. isolation
c. regeneration
d. all the above

_____ 3. Which of the following best describes inflammation?

a. the increased blood flow to the injured area
b. the regeneration of the tissue
c. the activation of phagocytes to protect the injured area
d. Inflammation causes injury to tissue.

4. During the repair process of tissue, which comes first, inflammation or regeneration?

____

OBJECTIVE 8: Describe how aging affects the tissues of the body (textbook p. 72).

After studying p. 72 in the text, you should be able to discuss why the repair process slows down with age and what is involved in the aging of tissues.

_____ 1. One of the major effects of aging on epithelial tissue is that it becomes _____.

a. less resilient
b. more brittle
c. thicker
d. thinner

_____ 2. One of the major effects of aging on connective tissue is that it becomes _____.

a. less resilient
b. more brittle
c. thicker
d. thinner

_____ 3. Which of the following cannot repair themselves as they age?

a. cardiac cells
b. nerve cells
c. connective tissues
d. Neither a nor b can repair themselves.

OBJECTIVE 9: Describe the changes in tissue structure that can lead to a malignant tumor (textbook p. 73).

In order to be able to identify a piece of tissue as abnormal, such as a tumor, one must know what is normal. The first part of this chapter discussed and described normal tissues. After studying p. 73 in this text, you should be able to describe changes in tissues that lead to tumor conditions.

_____ 1. A person who specializes in diagnosing disease is a(n) _____.

a. oncologist
b. pathologist
c. pediatrician
d. physician

_____ 2. Which of the following terms describes a change in cell function due to constant exposure to an irritant?

a. anaplasia
b. cancer
c. dysplasia
d. metaplasia

_____ 3. Cigarette smoke _____.

a. kills columnar cilia
b. paralyzes columnar cilia
c. disintegrates columnar cilia
d. changes the shape of columnar cilia

____ 4. In the case of a cigarette smoker, _____.

a. cilia will burn and thus be destroyed
b. goblet cells will continue to make mucus
c. goblet cells will no longer make mucus
d. cilia will continue to function until the mucus overwhelms them

____ 5. Which of the following is the most advanced form of a disease?

a. anaplasia
b. dysplasia
c. metaplasia
d. oncoplasia

OBJECTIVE 10: Describe how cancerous tumors are classified, and list the seven warning signs of cancer (textbook pp. 74–75).

After studying pp. 74–75 in the text, you should be able to list the seven warning signs of cancer and to describe how cancerous tumors are classified and the significance of those classifications.

____ 1. Anaplasia ultimately leads to the formation of _____.

a. benign tumors
b. cell death
c. malignancy
d. tumors

____ 2. Oncology is _____.

a. the study of a disease
b. the study of cancer
c. the study of malignant tumors
d. the study of oncogenes
e. the study of tumors

____ 3. If areolar tissue becomes cancerous, it is classified as a _____.

a. carcinoma
b. leukemia
c. glioma
d. sarcoma

____ 4. Myomas are cancerous cells that were derived from _____.

a. epithelial tissue
b. connective tissue
c. muscle tissue
d. neural tissue

____ 5. Why might a mole that has changed in appearance be a sign of cancer?

a. A change in a mole indicates that cell reproduction is occurring. If the cell reproduction is abnormal, it may be a sign of cancer.
b. A changing mole typically releases chemicals that may lead to cancer.
c. Because a changing mole is a warning sign of cancer.
d. The changing mole may develop a lump. A lump is generally a sign of cancer.

OBJECTIVE 11: Describe the various forms of cancer treatment (textbook pp. 75–76).

After studying pp. 75–76 in your text, you should be able to describe how some cancer treatments work and explain why some side effects occur.

_____ 1. A biopsy is an examination of ____.

a. a tumor
b. blood
c. tissue
d. urine

_____ 2. Which of the following terms mean cancerous?

a. benign
b. malignant
c. metastatic
d. tumor

_____ 3. Which of the following would not be effective against metastatic cells?

a. destroying them with chemotherapy
b. destroying them with radiation
c. surgically removing them from the body
d. All the above can be used on metastatic cells.

_____ 4. When patients undergo chemotherapy treatment, they frequently lose their hair. Why?

a. because the chemotherapy drugs slow the reproduction of epithelial cells
b. because the chemotherapy drugs kill hair follicles
c. because hair follicles have to undergo rapid and destructive cell reproduction
d. Scientists don't know for sure. This is just known as a side effect.

_____ 5. Explain why researchers are not trying to find a universal cure for cancer.

PART II: CHAPTER-COMPREHENSIVE EXERCISES

Match the terms in column A with the definitions or phrases in column B.

MATCHING I

	(A)	(B)
_____	1. adipose	A. A tissue category in which each cell type has a matrix
_____	2. blood	B. Tissue that makes up either the inside lining or the outside lining of a body surface
_____	3. bone	C. Cells that are longer than they are wide
_____	4. cardiac muscle	D. Cells that form rings around a central canal
_____	5. cartilage	E. Cells that are associated with a liquid matrix
_____	6. columnar	F. Cells that have intercalated discs
_____	7. connective	G. Cells that aid in insulation
_____	8. epithelial	H. Tissues that attach bone to bone
_____	9. ligaments	I. Cells that are also known as chondrocytes
_____	10. metaplasia	J. A change in tissue that prevents it from functioning properly

MATCHING II

	(A)	(B)
_____	1. dysplasia	A. Tissue that conducts impulses
_____	2. muscular	B. Tissue that has the ability to contract
_____	3. neural	C. Cells that appear to be thin and flat from a side view
_____	4. neuron	D. Cells that can conduct impulses
_____	5. serous	E. Cells that contract under voluntary control
_____	6. skeletal muscle	F. Cells that contract under involuntary control
_____	7. smooth muscle	G. Tissue that attaches muscle to bone
_____	8. squamous	H. Membranes that cover the lungs
_____	9. synovial	I. Membranes that line the joints
_____	10. tendons	J. A change in the shape of tissue cells

CONCEPT MAP

This concept map organizes and summarizes the concepts presented in Chapter 4. Use the following terms to complete the map filling in the boxes identified by the circled numbers, 1–13.

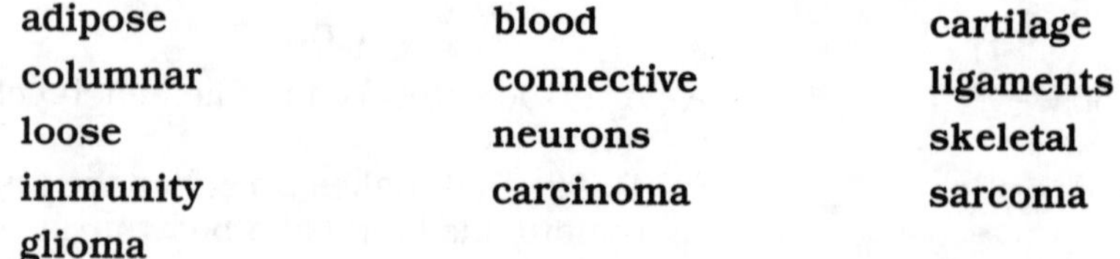

adipose	**blood**	**cartilage**
columnar	**connective**	**ligaments**
loose	**neurons**	**skeletal**
immunity	**carcinoma**	**sarcoma**
glioma		

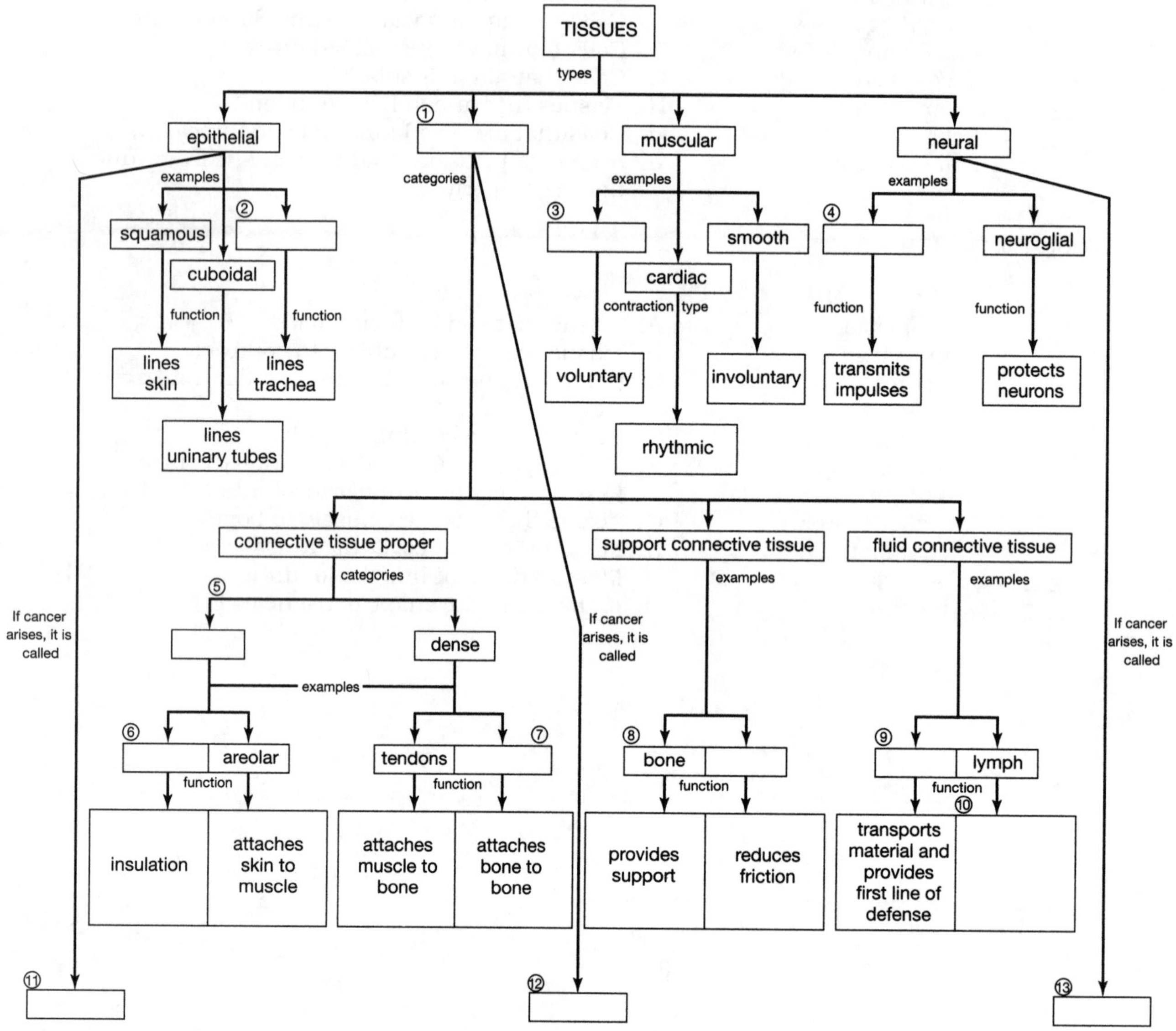

TISSUES

types

epithelial — examples: squamous, cuboidal, ② [ ]
- squamous — function — lines skin
- cuboidal — lines uninary tubes
- ② — function — lines trachea
- If cancer arises, it is called ⑪ [ ]

① [ ] — categories
- If cancer arises, it is called ⑫ [ ]

muscular — examples: ③ [ ], cardiac, smooth
- ③ — contraction — voluntary
- cardiac — type — rhythmic
- smooth — involuntary

neural — examples: ④ [ ], neuroglial
- ④ — function — transmits impulses
- neuroglial — function — protects neurons
- If cancer arises, it is called ⑬ [ ]

connective tissue proper — categories: ⑤ [ ], dense
- examples: ⑥ [ ], areolar; tendons, ⑦ [ ]
- ⑥ — function — insulation
- areolar — attaches skin to muscle
- tendons — function — attaches muscle to bone
- ⑦ — attaches bone to bone

support connective tissue — examples: bone, ⑧ [ ]
- bone — function — provides support
- ⑧ — reduces friction

fluid connective tissue — examples: ⑨ [ ], lymph
- ⑨ — function — transports material and provides first line of defense
- lymph — ⑩ [ ]

CROSSWORD PUZZLE

The following crossword puzzle reviews the material in Chapter 4. To complete the puzzle, you must know the answers to the clues given, and must be able to spell the terms correctly.

ACROSS

1. The tissue that makes up the lining of many organs
4. A person who studies the mechanisms of disease
7. Cells that have a liquid plasma
9. The tissue that has a matrix associated with the cells
11. Cells that connect skin to muscle
12. Cancer that originates from epithelial tissue
15. Cells that appear to be flat and that line some organs
16. These structures become paralyzed when exposed to cigarette smoke.

DOWN

2. Discs that are found in cardiac cells
3. Tumors that originate from connective tissue
5. Cells that form concentric rings around blood vessels
6. The portion of a neuron that contains the organelles
8. The tissues that connect bone to bone
10. Type of tissue that conducts impulses
13. An elongated cell that can conduct impulses
14. Tissue that has the ability to contract

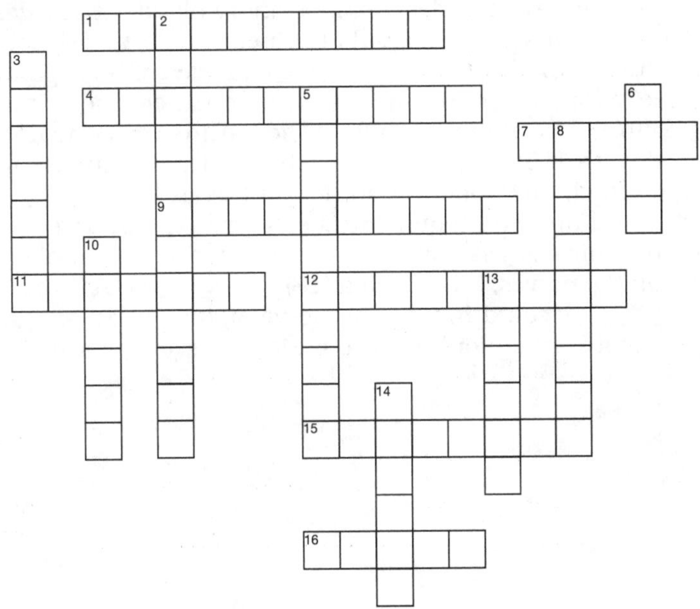

FILL-IN-THE BLANK NARRATIVE

Use the following terms to fill in the blanks of this summary of Chapter 4.

areolar	**blood**	**bone**
connective	**contract**	**epithelial**
four	**homeostasis**	**ligaments**
lining	**matrix**	**neurons**
plasma	**skeletal**	**squamous**
tendons	**voluntary**	**benign**
glioma		

The several trillion cells that make up the body can be placed into (1) ___________ major tissue categories. All tissues must work together to ensure (2) ___________.

(3) ___________ cells, such as columnar, (4) ___________, and cuboidal form the inside or outside (5) ___________ of external or internal body surfaces. Connective tissue is made of cells such as adipose, areolar, cartilage, bone, and (6) ___________. Although this group consists of a variety of cells, a common feature is that each cell type has some form of (7) ___________. This ground substance is material found in the extracellular space and can be liquid, syrupy, or solid. Adipose and (8) ___________ cells have a syrupy matrix. (9) ___________ and cartilage have a solid matrix. Blood has a liquid matrix called (10) ___________. Muscle tissue consists of skeletal muscle cells, smooth muscle cells, and cardiac muscle cells. All these cells have the ability to (11) ___________. The cells that make up neural tissue group are (12) ___________ and neuroglia. Impulse conduction is the primary function of neurons.

The following example demonstrates how the various cells and tissue work together to accomplish a task. Urine begins to fill the urinary bladder. The transitional epithelial cells permit expansion of the urinary bladder. As the urinary bladder stretches, a nerve impulse sent to the brain via nervous tissue signals the need to go to the bathroom. The brain sends signals via neurons to the (13) ___________ muscles of the legs. A person is able to make a conscience effort to go to the bathroom, since these muscles are under (14) ___________ control. The ability to walk to the bathroom is due not only to the skeletal muscles but also to (15) ___________, which attach muscle to bone, and (16) ___________, which attach one bone to another, of the (17) ___________ tissue category. All four tissue types and numerous individual cell types are involved in a task like going to the bathroom. Just think of what is involved in doing more complex tasks!

If a tumor exists on the nerves, it is called a (18) ___________. A tumor such as this, even if it were (19) ___________, could inhibit the nerve signals from arriving at the leg muscles. This would make it very difficult for the patient to even walk to the bathroom.

CLINICAL CONCEPTS

The following clinical concept applies the information presented in Chapter 4. Following the application is a set of questions to help you understand the concept.

The cells of the body will undergo cellular reproduction if they are damaged. The process of cellular reproduction is involved in the repair of damaged tissue as well. Tissue is damaged and cellular reproduction is disrupted during cigarette smoking. Smoking causes changes in the shape, size, and organization of tissues.

The trachea is lined with ciliated columnar cells and goblet cells. The goblet cells produce mucus that traps incoming foreign particles. The cilia beat back and forth to move the mucus from the trachea toward the back of the throat. Coughing expels this mucus material along with the inhaled foreign particles trapped in it. Cigarette smoke paralyzes the cilia, so they cannot move the mucus to the back of the throat to be expelled. The goblet cells continue to produce mucus, which begins to accumulate in the trachea. Because the mucus still needs to be expelled, a smoker develops a typical "hacking cough" to dislodge the excess mucus buildup.

With time, stem cell reproduction no longer differentiates into columnar cells but instead produces stratified epithelial cells, which provide a greater resistance to the drying effects and the chemical irritation of the cigarette smoke. Unfortunately, these cells do not provide protection against incoming foreign particles, nor do they provide a moist lining for the trachea.

Continuous exposure to cigarette smoke will cause the tissue cells to undergo abnormal cell reproduction. Sometimes this abnormality can result in the development of cells with abnormal chromosomes, which can lead to cancer.

1. What happens to the cilia if someone smokes cigarettes?

2. What is the purpose of the mucus produced in the lining of the trachea?

3. Extensive smoking may cause the stem cells of the trachea to develop ________________ cells instead of ________________ cells.

4. Does the process of mitosis occur normally or abnormally in the cells lining the trachea of a smoker?

5. Why do the cell types change in the trachea of a smoker?

THIS CONCLUDES CHAPTER 4 EXERCISES

CHAPTER

5

Organ Systems: An Overview

In Chapter 3 you learned that the cell is the individual structural and functional unit of the body. Chapter 4 explained how groups of cells form tissues and how the different tissue types work together to maintain homeostasis in the body.

Chapter 5 gives an overall view of how cells and tissues are integrated to form a functional unit known as an organ and how the assemblage of numerous organs ultimately forms the body's organ systems. The integration and interdependence of the body's organ systems are necessary for metabolism and homeostasis. A general review of each body system, with an emphasis on terminology and general concepts, will prepare you for the more detailed study of the systems you will encounter later.

Chapter Objectives:

1 Describe the general features of an organ system.

2 Describe the organs and primary functions of each organ system concerned with support and movement.

3 Describe the organs and primary functions of the nervous and endocrine systems.

4 Describe the organs and primary functions of the cardiovascular and lymphatic systems.

5 Describe the organs and primary functions of each organ system involved in the exchange of materials with the environment.

6 Describe the organs and primary functions of the male and female reproductive systems.

PART I: OBJECTIVE-BASED QUESTIONS

OBJECTIVE 1: Describe the general features of an organ system (textbook p. 81).

After studying p. 81 in the text, you should be able to describe the characteristics of an organ system, the largest living units of the organism.

_____ 1. Which of the following is not considered to be an organ system?

a. integumentary
b. muscular
c. urinary
d. All the above are considered to be organ systems.

_____ 2. Which of the following statements is true?

a. In order for all the organ systems to function properly, the organs have to rely only on the cardiovascular system.
b. In order for an organ system to function, all organs must work together.
c. In order for an organ system to function, the organs must work independently of each other.
d. In order for an organ system to function, all organs must be in physical contact with one another.

_____ 3. Which of the following statements is true?

a. All systems rely heavily on the function of the cardiovascular system.
b. All systems are influenced by the nervous system.
c. All systems are influenced by the endocrine system.
d. All the above are correct.

OBJECTIVE 2: Describe the organs and primary functions of each organ system concerned with support and movement (textbook pp. 81–83).

After studying pp. 81–83 in the text, you should be able to describe the organ systems involved in support and movement: integumentary, skeletal, and muscular. Most people will find it fairly obvious that the skeletal and muscular systems are involved in support and movement. However, the integumentary system is also involved. The following questions will help to clarify the role of the skeletal system, the muscular system, and the integumentary system in support and movement.

_____ 1. Body support and movement are under the control of which of the following organ systems?

a. integumentary, muscular, and neural
b. integumentary, skeletal, and muscular
c. skeletal, muscle, and respiratory
d. skeletal, muscular, and neural

_____ 2. The integumentary system is made up of _____.

a. cutaneous material
b. cutaneous material and skin
c. skin
d. skin, nails, glands, and hair

_____ 3. Which of the following organ systems is responsible for vitamin D production?

a. endocrine system
b. integumentary system
c. muscle system
d. skeletal system

_____ 4. How does the skeletal system interact with other systems?

a. It provides a place for muscle attachment (interacts with the muscle system).
b. It provides protection for the heart (interacts with the cardiovascular system).
c. The bones store minerals that can be used by other organ systems.
d. All the above are correct statements.

_____ 5. Which of the following statements is true?

a. The axial skeleton consists of the arms and legs, and the appendicular skeleton consists of the skull.
b. The axial skeleton consists of the brain and spinal cord, and the appendicular skeleton consists of the limbs.
c. The axial skeleton consists of the skull, and the appendicular skeleton consists of the arms and legs.
d. None of the above is true.

_____ 6. Which of the following is true regarding the muscle system's functioning with other organ systems?

a. Skeletal muscles function with the digestive system.
b. Skeletal muscles function with the integumentary system.
c. Skeletal muscles function with the skeletal bones.
d. All the above are true.

_____ 7. The integumentary system is involved in all the following functions. Which one relates directly to support and movement?

a. assisting in the removal of water and waste
b. protecting it from the environment
c. providing sensations to the nervous system
d. regulating body temperature

_____ 8. The ability to generate heat is one function of the _____.

a. cardiovascular system
b. integumentary system
c. muscular system
d. skeletal system

9. Explain how skeletal muscles help keep body temperature in the normal range.
____

10. The largest organ of the integumentary system is the ______________________________.

11. In Figure 5–1, identify three of the four main components of the integumentary system.

Figure 5–1 The Integumentary System

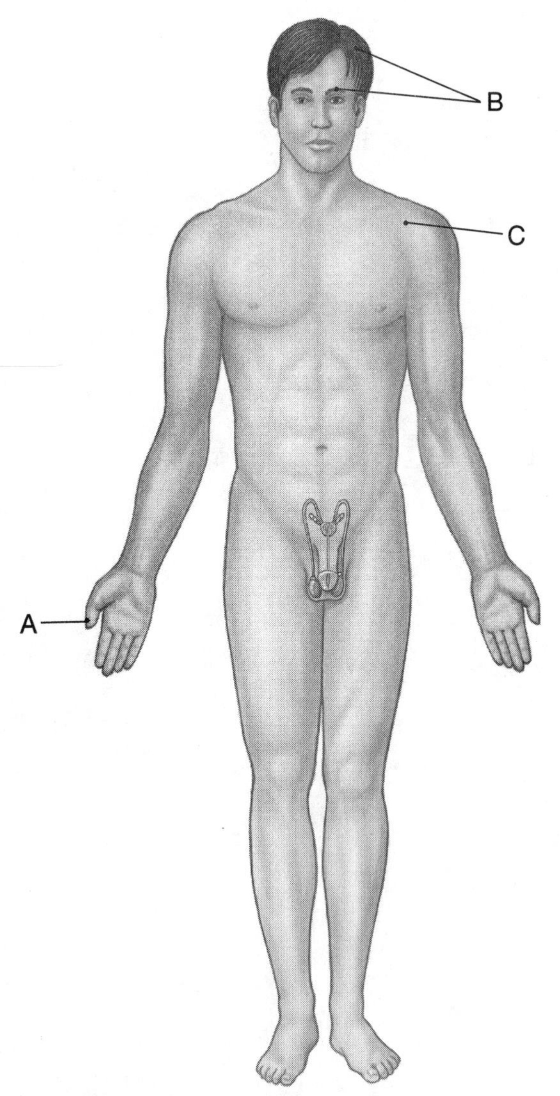

A ______________________ B ______________________

C ______________________

12. In Figure 5–2, identify the letters that represent the axial skeleton and those that represent the appendicular skeleton.

Figure 5–2 The Skeletal System

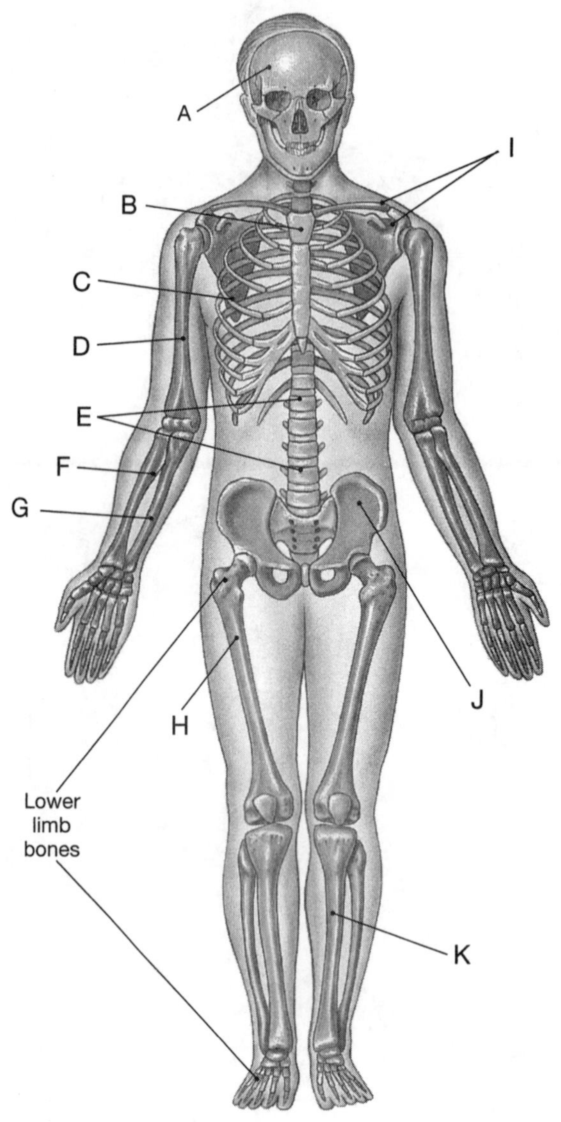

A ____________________ B ____________________

C ____________________ D ____________________

E ____________________ F ____________________

G ____________________ H ____________________

I ____________________ J ____________________

K ____________________

13. In Figure 5–3, identify the letters that represent the axial muscles and those that represent the appendicular muscles.

Figure 5–3 The Muscular System

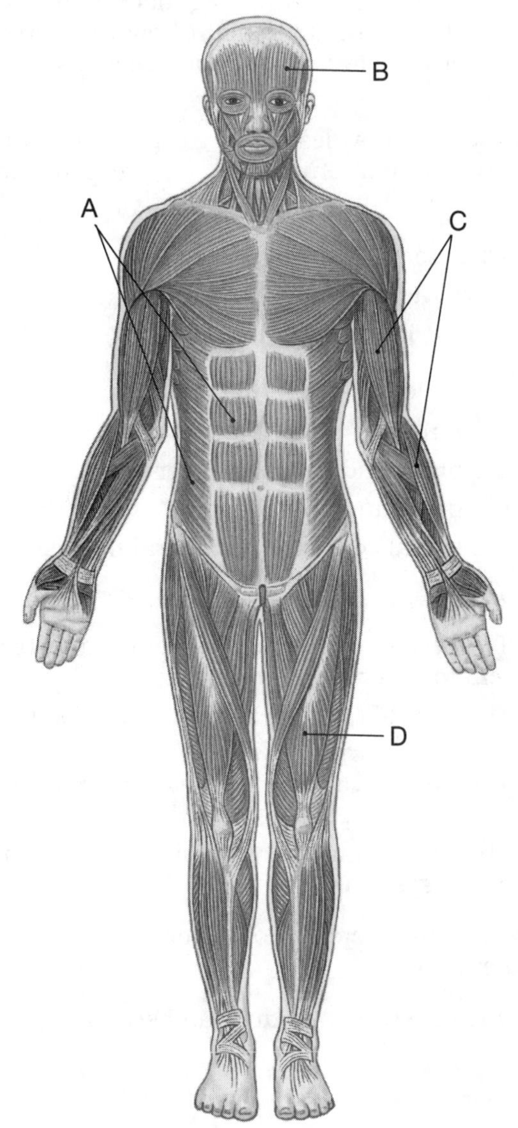

A ______________________ B ______________________

C ______________________ D ______________________

OBJECTIVE 3: Describe the organs and primary functions of the nervous and endocrine systems (textbook pp. 83–85).

In order for various organ systems to function properly, they need sensory input from the external environment and from the internal environment as well. This sensory input comes in the form of nerve impulses and in the form of endocrine hormones. After studying pp. 83–85 in the text, you should be able to explain how the nervous system and the endocrine system work together to help maintain homeostasis and control of all the other organ systems.

_____ 1. The nervous system provides a _____ response to other organ systems, whereas the endocrine system provides a _____ response to other organ systems.

a. rapid, slow
b. short term, faster
c. slow, rapid
d. slower, long term

_____ 2. The central nervous system consists of _____.

a. neurons and peripheral nerves
b. the brain and peripheral nerves
c. the brain and spinal cord
d. the spinal cord and peripheral nerves

_____ 3. Which of the following statements is true?

a. The peripheral nervous system connects the central nervous system to other organ systems.
b. The central nervous system connects the peripheral nervous system to other organ systems.
c. The peripheral nervous system connects the autonomic nerves with other organs.
d. The autonomic nerves connect the central system with other organ systems.

4. The endocrine system is connected to other organ systems because the hormones leave the glands and enter the _________________________.

5. The organs of the endocrine system produce and secrete _________________________ into the bloodstream.

6. Identify the letter that represents the central nervous system and the letter that represents the peripheral nervous system in Figure 5–4.

Figure 5–4 The Nervous System

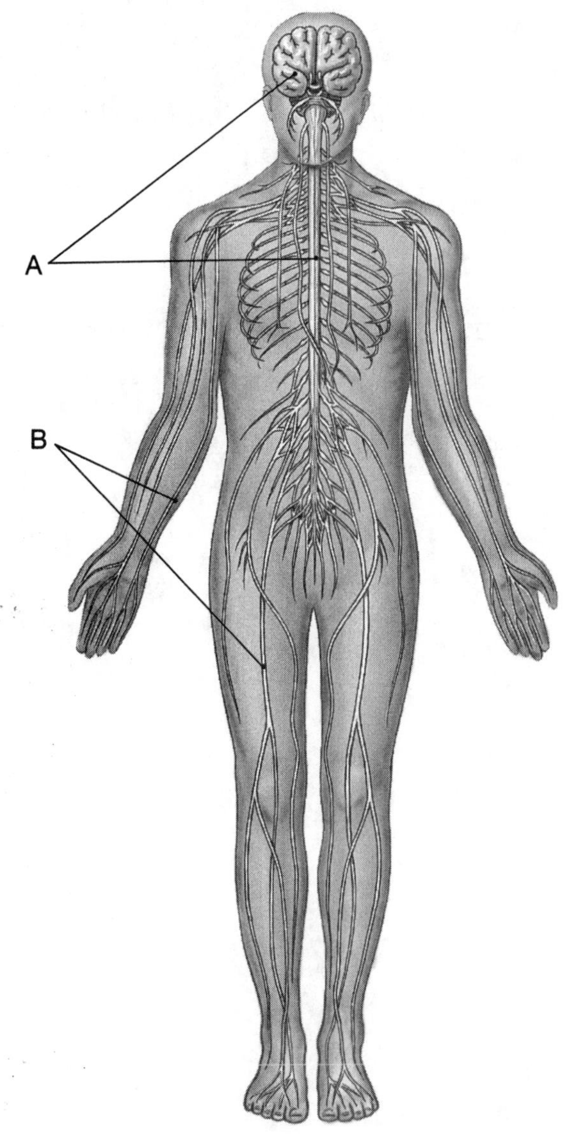

A ____________________ B ____________________

7. Identify the endocrine glands labeled in Figure 5-5.

Figure 5–5 The Endocrine System

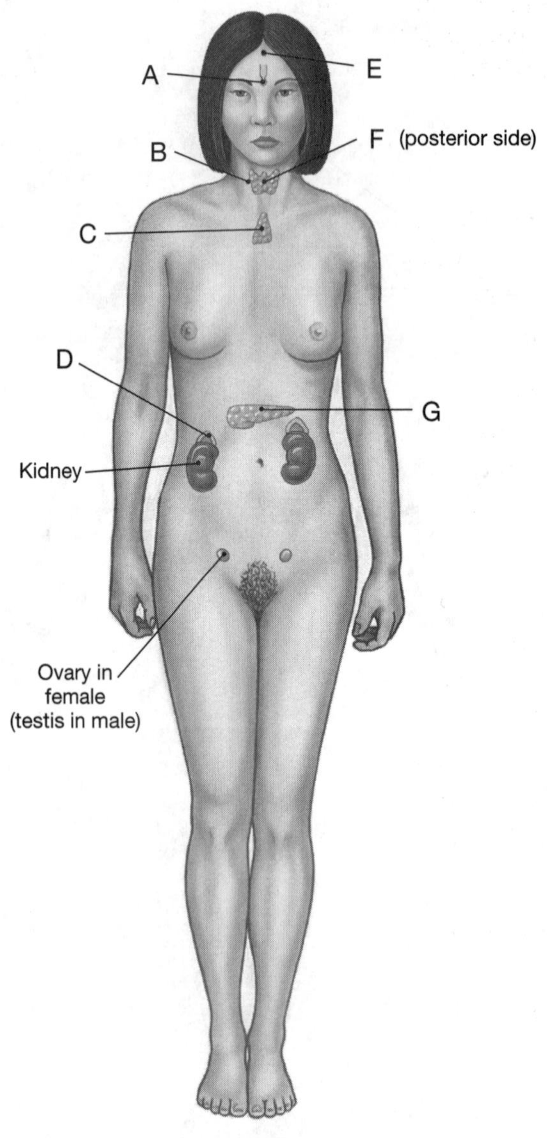

A ______________________ B ______________________

C ______________________ D ______________________

E ______________________ F ______________________

G ______________________

OBJECTIVE 4: Describe the organs and primary functions of the cardiovascular and lymphatic systems (textbook pp. 85–86).

All cells of various organs take in nutrients and produce waste products. The nutrients must travel to the cells and the waste must travel away from the cells. After studying pp. 85–86 in the text, you should be able to explain how the circulatory and lymphatic systems transport nutrients and wastes and how the lymphatic system maintains the body's immunity.

_____ 1. The cardiovascular system is linked to all other organ systems because it _____.

a. helps maintain pH for all systems
b. provides a defense mechanisms for all systems
c. transports nutrients to all systems
d. all the above

_____ 2. The lymphatic system is linked to all other organ systems because it _____.

a. helps maintain pH for all systems
b. provides a defense mechanism for all systems
c. transports nutrients to all systems
d. all the above

_____ 3. Which of the following is a characteristic of the cardiovascular system that links it to all other organ systems?

a. It transports hormones.
b. It transports nutrients.
c. It transports water.
d. All the above.

_____ 4. Which of the following is not part of the lymphatic system?

a. spleen
b. thymus
c. thyroid
d. tonsils

_____ 5. The thymus gland is part of the _____.

a. cardiovascular system
b. endocrine system
c. lymphatic system
d. respiratory system

6. Most of the body's antibodies are produced by which organ system?

____

7. In Figure 5–6, identify the heart, the capillaries associated with the lungs, the capillaries associated with the kidneys, and the capillaries associated with the liver.

Figure 5–6 The Cardiovascular System

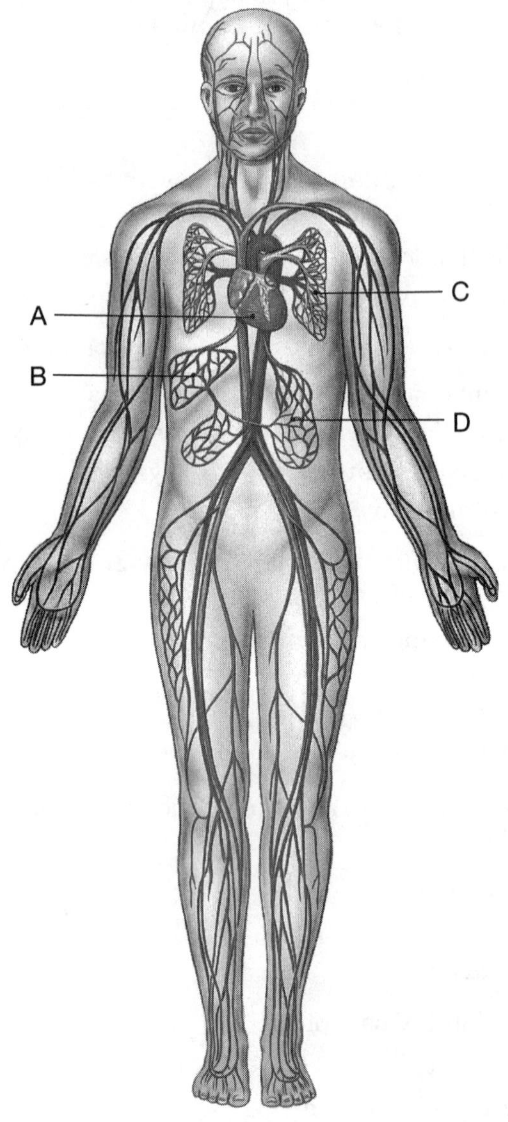

A ______________________ B ______________________

C ______________________ D ______________________

8. In Figure 5–7, identify the thymus, lymph nodes, and spleen.

Figure 5–7 The Lymphatic System

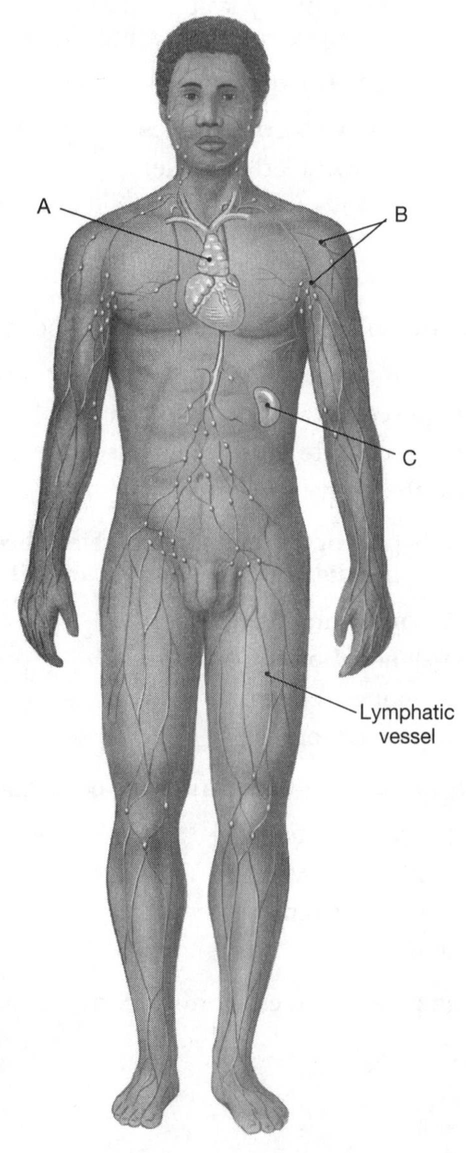

A ______________________ B ______________________

C ______________________

OBJECTIVE 5: Describe the organs and primary functions of each organ system involved in the exchange of materials with the environment (textbook pp. 87–89).

After studying pp. 87–89 in the text, you should be able to explain how the respiratory system, the digestive system, and the urinary system are linked with the other organ systems of the body in transporting nutrients and eliminating wastes.

_____ 1. The main function of the digestive system is to _____.

a. break down food particles and produce waste
b. break down food particles
c. break down food particles and absorb nutrients into the bloodstream
d. break down food particles and store it in the large intestine

_____ 2. The kidneys are linked with other systems because they _____.

a. get rid of wastes created by other systems
b. help regulate blood pressure
c. help regulate fluid levels associated with other systems
d. All the above are correct.

_____ 3. The pancreas is part of the endocrine system because it produces _____, but it is also part of the digestive system because it produces _____.

a. enzymes, metabolic waste
b. enzymes, hormones
c. hormones, enzymes
d. hormones, bile

_____ 4. Which of the following systems is involved in the elimination of waste?

a. digestive system
b. respiratory system
c. urinary system
d. all the above

5. The exchange of gas (oxygen/carbon dioxide) occurs in small sacs called ___________________________.

6. In Figure 5–8, identify the trachea, bronchi, lungs, and diaphragm.

Figure 5–8 The Respiratory System

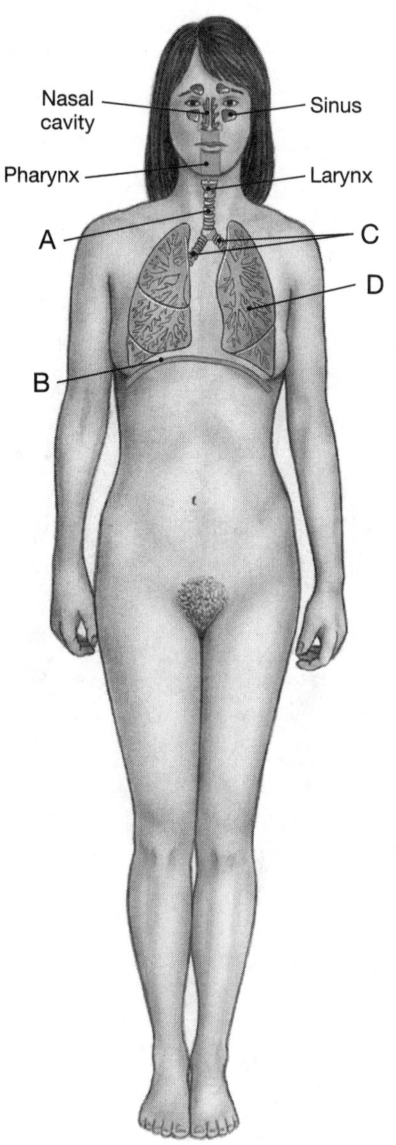

A ______________________ B ______________________

C ______________________ D ______________________

7. In Figure 5–9, identify the salivary glands, esophagus, liver, small intestine, stomach, and large intestine.

Figure 5–9 The Digestive System

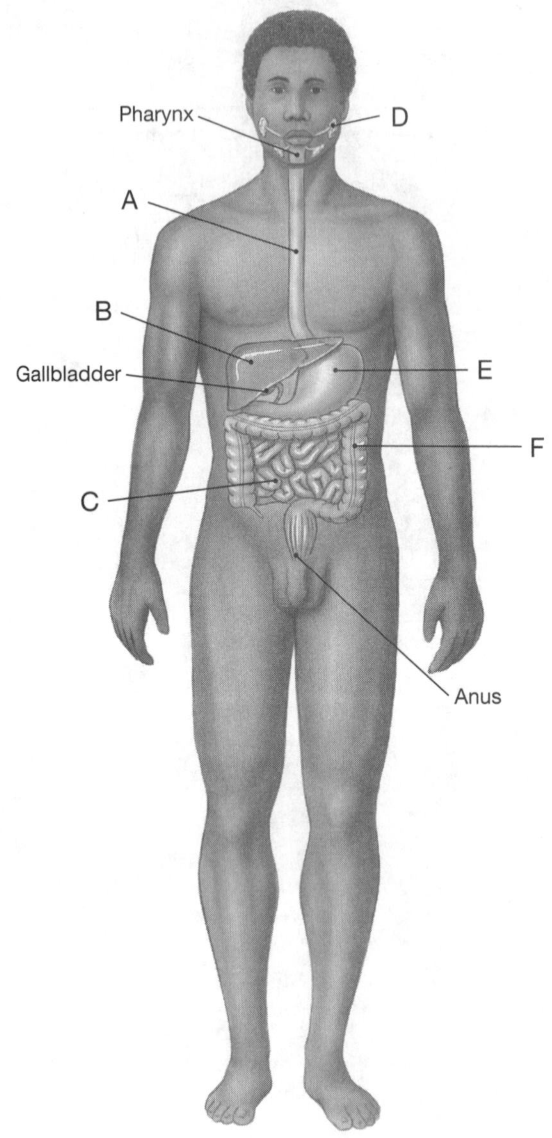

A ______________________ B ______________________

C ______________________ D ______________________

E ______________________ F ______________________

8. In Figure 5–10 identify the kidney, urinary bladder, ureter, and urethra.

Figure 5–10 The Urinary System

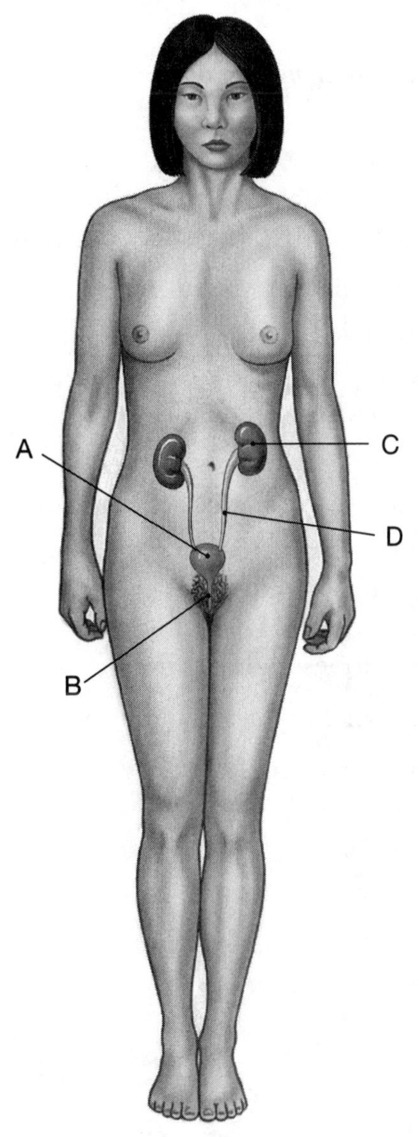

A ______________________ B ______________________

C ______________________ D ______________________

OBJECTIVE 6: Describe the organs and primary functions of the male and female reproductive systems (textbook pp. 89–90).

After studying pp. 89–90 in the text, you should be able to explain how the reproductive system is linked with other organ systems and how it is significant for the survival of the species.

_____ 1. The reproductive system is linked to other systems of the body because _____.

a. it is involved in the perpetuation of the species
b. it is made of cells, tissue, and organs
c. it produces hormones that affect other systems
d. It is not linked to any other organ system.

_____ 2. Testes and ovaries are referred to as _____.

a. gametes
b. gonads
c. reproductive systems
d. all the above

_____ 3. Which of the following are associated with the male reproductive system?

a. epididymis and urethra
b. epididymis and uterine tubes
c. uterine tubes and sperm ducts
d. uterine tubes and urethra

_____ 4. The _____ is a (are) male gamete(s), and the _____ is a (are) female gamete(s).

a. gonads, eggs
b. penis, uterus
c. sperm, eggs
d. testes, ovaries

5. In Figure 5–11, identify the following:

ductus deferens	**mammary gland**	**ovary**
penis	**prostate gland**	**scrotum**
seminal vesicle	**uterine tube**	**testis**
urethra	**uterus**	**vagina**

Figure 5–11 The Male and Female Reproductive Systems

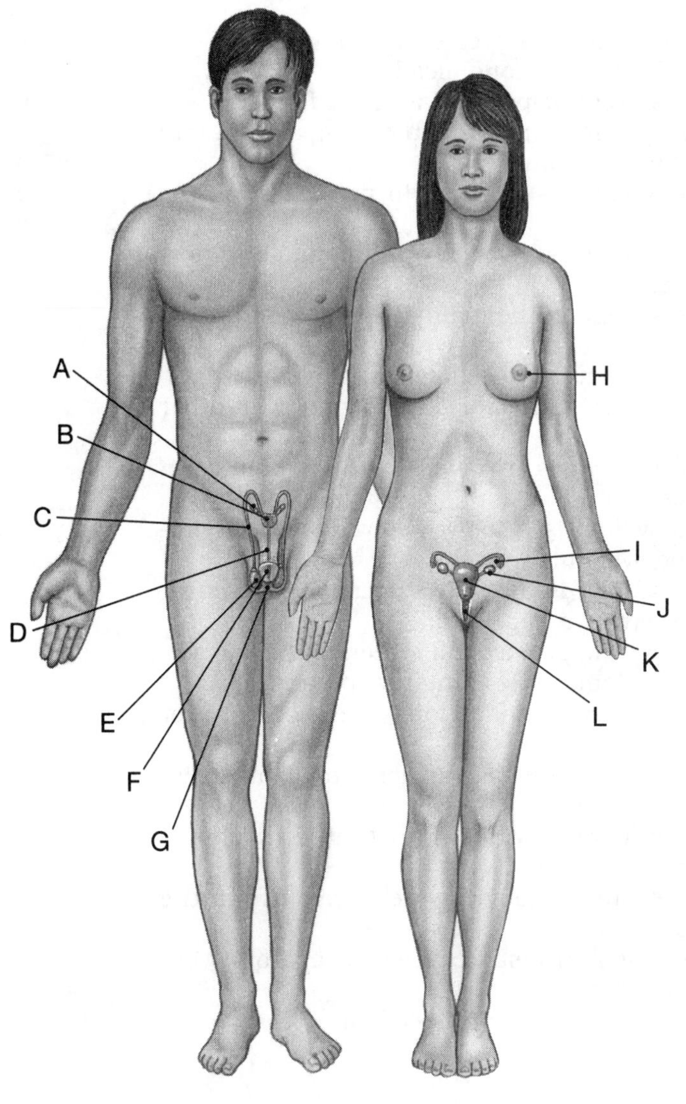

A ______________________ B ______________________

C ______________________ D ______________________

E ______________________ F ______________________

G ______________________ H ______________________

I ______________________ J ______________________

K ______________________ L ______________________

PART II: CHAPTER-COMPREHENSIVE EXERCISES

MATCHING

Match the organ systems in column B with the descriptions in column A.

(A)

_____ 1. This system absorbs nutrients into the blood.
_____ 2. This system consists of sweat glands.
_____ 3. This system controls the development of T cells.
_____ 4. We have control over this system but it can also function autonomically.
_____ 5. This system helps regulate body fluids.
_____ 6. This system is involved in getting oxygen into the bloodstream.
_____ 7. This system is involved in perpetuating the species.
_____ 8. This system is involved with glands and hormones.
_____ 9. This system is the main site for blood formation.
_____ 10. This system provides skeletal movement.
_____ 11. This system transports nutrients to all other tissues.
_____ 12. This system is involved in the exchange of gases.
_____ 13. This system is heavily involved with our immune system.
_____ 14. This system is the only system that contains cells having a flagellum.
_____ 15. This system protects all underlying tissues.
_____ 16. This system protects internal organs.
_____ 17. This system provides long-term control over other systems.
_____ 18. This system provides nutrients to all other tissues.
_____ 19. This system provides short-term control over all other systems.
_____ 20. This system removes watery waste from the blood.
_____ 21. This system transports waste away from other tissues.
_____ 22. This system helps produce heat.

(B)

A. cardiovascular
B. digestive
C. endocrine
D. integumentary
E. lymphatic
F. muscle
G. nervous
H. reproductive
I. respiratory
J. skeletal
K. urinary

CONCEPT MAP

This concept map is a review of Chapter 5. Use the following terms to complete the map by filling in the boxes identified by the circled numbers, 1–10.

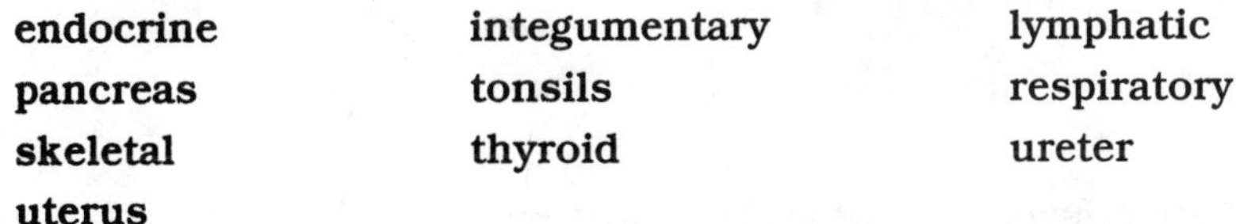

endocrine	**integumentary**	**lymphatic**
pancreas	**tonsils**	**respiratory**
skeletal	**thyroid**	**ureter**
uterus		

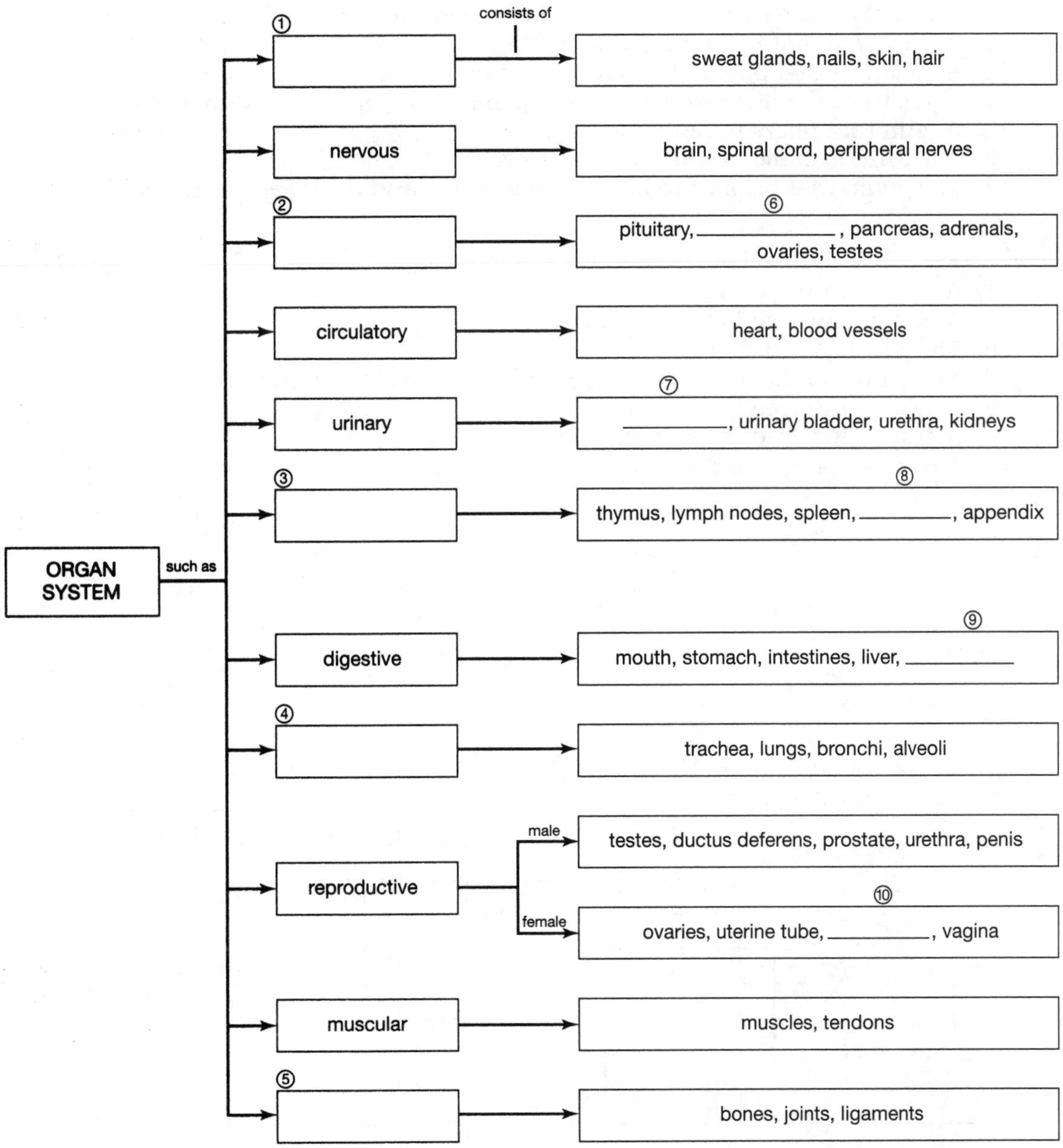

ORGAN SYSTEM — such as →

Organ system		Consists of
① ________	consists of →	sweat glands, nails, skin, hair
nervous	→	brain, spinal cord, peripheral nerves
② ________	→	pituitary, ⑥ ________, pancreas, adrenals, ovaries, testes
circulatory	→	heart, blood vessels
urinary	→	⑦ ________, urinary bladder, urethra, kidneys
③ ________	→	thymus, lymph nodes, spleen, ⑧ ________, appendix
digestive	→	mouth, stomach, intestines, liver, ⑨ ________
④ ________	→	trachea, lungs, bronchi, alveoli
reproductive	male →	testes, ductus deferens, prostate, urethra, penis
	female →	ovaries, uterine tube, ⑩ ________, vagina
muscular	→	muscles, tendons
⑤ ________	→	bones, joints, ligaments

CROSSWORD PUZZLE

The following crossword puzzle reviews the material in Chapter 5. To complete the puzzle, you have to know the answers to the clues given, and you must be able to spell the terms correctly.

ACROSS

3. The system involved in absorbing nutrients into the bloodstream
4. A system involved in defense
5. The material in bone responsible for making blood cells
7. Tubes that transport wastes from the kidneys to the urinary bladder
12. The system that includes the skin
13. A chemical that provides a slow response
14. The part of the nervous system consisting of the brain and spinal cord
15. A term that refers to reproductive cells
16. The generic name for the testes and ovaries
17. The organ that belongs to both the endocrine and the digestive systems

DOWN

1. A system involved in support
2. A system involved in transporting material
6. The site of gas exchange
8. Structures of the lymphatic system that are involved in immunity
9. A system involved in movement
10. A system involved in communication
11. A disease-causing organism

FILL-IN-THE-BLANK NARRATIVE

Use the following terms to fill in the blanks of this summary of Chapter 5.

antibodies	**calcium**	**cardiovascular**
digestive	**endocrine**	**lymphatic**
muscular	**integumentary**	**nervous**
nutrients	**oxytocin**	**plasma**
positive	**respiratory**	**skeletal**
urinary	**uterine**	**homeostasis**
species		

Chapter 5 discussed the various organ systems individually and their relationships to one another. These relationships ensure (1) ___________, survival, and perpetuation of the (2) ___________. The reproductive system is involved in the development of a child. The child receives oxygen and (3) ___________ from the mother's blood supply. The blood supply is part of the (4) ___________ system. This system receives the nutrients from the (5) ___________ system. The oxygen found in the blood supply comes from the (6) ___________ system. To ensure the survival of the baby, the mother needs to stay healthy. The (7) ___________ system helps her remain healthy because of (8) ___________ cells that produce (9) ___________ to help fight infections.

During pregnancy, the mother needs to obtain (10) ___________ ions from her diet, as the developing child needs these for bone development. If the mother's diet is deficient in these ions, her (11) ___________ system will begin to lose mass to provide the proper amount of this mineral for the developing child.

As the child develops, the child takes in food from the mother and produces waste. The waste enters the mother's cardiovascular system to be transported to the mother's (12) ___________ system. The skeletal muscles of the (13) ___________ system allow her to walk to the bathroom to eliminate these waste products. The muscles are stimulated to work by the (14) ___________ system.

During pregnancy, the mother's body undergoes many changes. One such change is indicative of the elasticity of the skin, which is a part of the (15) ___________ system. After about 9 months' gestation, the (16) ___________ system releases a hormone called (17) ___________, which creates (18) ___________ contractions. These contractions operate under a (19) ___________ feedback mechanism, and the child is born.

CLINICAL CONCEPT

The following clinical concept applies the information presented in Chapter 5. Following the application is a set of questions to help you understand the concept.

A patient goes to the doctor and complains of feeling sick, but she does not know why she feels that way. The doctor runs a few tests to determine the source of the illness, starting with a urine sample. Examining the results of tests on the urine sample, the doctor sees that the pH of the urine is fairly low. This means that the kidneys are excreting too many hydrogen ions from the blood. A loss of hydrogen ions from the blood makes the blood pH too high. When blood pH is too high, the red blood cells are not able to deliver oxygen properly. A decrease in oxygen to the tissues results in a decrease in ATP production by the cells, which reduces muscle activity. A decrease in muscle activity makes the patient feel fatigued and lethargic. The body tries to compensate for the lack of oxygen by increasing the breathing rate. As the breathing rate increases, the patient may feel a little stressed. This stressed feeling may activate the autonomic nervous system. Some of the autonomic nerves associated with the digestive system may initiate the production of excess digestive acids, which can cause an ulcer.

The doctor needs to determine which organ system to treat first. The doctor must treat the organ system that, when corrected, will ultimately cause the other organ system to return to homeostasis.

1. The pH of a normal urine sample can range from 5 to 8. A pH value of 6 is ideal. What is the ideal pH range of blood? (*Hint:* Look in Chapter 2 of the textbook).

2. How does a high blood pH affect the ability of the erythrocytes to deliver oxygen?

3. If the urine had a low pH, would its hydrogen ion concentration be too high or too low?

4. Why might this patient begin to develop ulcers just because she was feeling sick?

5. If you were the doctor, which organ system would you treat first and why?

THIS CONCLUDES CHAPTER 5 EXERCISES

CHAPTER 6

Mechanisms of Disease

One major theme of this textbook is homeostasis. This chapter introduces the concept that disease-causing agents can travel from fomite (nonliving object) to body, from body to body, and also within the body. The human body appears to be an almost perfect place for pathogens to grow. Since there are so many pathogenic organisms, it is very important for the physician to diagnose the disease-causing agent. After the diagnosis, the physician can determine what type of treatment will be the most effective.

This chapter discusses some ways that disease-causing organisms can be transmitted and how they enter the body, as well as how treatments such as antibiotics work to control the infectious disease once the disease-causing organism has been identified.

After reading Chapter 6, you should understand that there are several agents besides bacteria or viruses that can cause disease. You will find that it is rather amazing that the body is in homeostasis most of the time despite the vast array of disease-causing organisms that are present.

Chapter Objectives:

- Describe the relationship between homeostasis and disease.
- Distinguish among the various types of diseases.
- Define acute, chronic, and latent diseases.
- Distinguish among a symptom, sign, and syndrome; and a diagnosis and prognosis.
- List and describe the major types of disease-causing organisms.
- List the major pathways that allow pathogens to enter or exit the body.
- Describe three mechanisms of disease transmission and give examples of transmitted diseases.
- Distinguish between sterilization and disinfection.
- Describe the techniques involved in Universal Precautions.
- List five modes of action that antibiotics have on bacteria.
- Distinguish among endemic, epidemic, and pandemic diseases.

PART I: OBJECTIVE-BASED QUESTIONS

OBJECTIVE 1: Describe the relationship between homeostasis and disease (textbook p. 95).

Disease-causing organisms generally affect body tissues or organs. These organisms disrupt the function of the body tissues or organs. Thus, they fail to maintain homeostasis. After studying p. 95 in the text, you should be able to describe the relationship between the lack of homeostasis and disease.

_____ 1. Etiology is the study of _____.

a. disease abnormalities
b. homeostasis
c. how homeostasis fails
d. the causes and origins of diseases

2. The failure to maintain homeostasis may result in a ____________.

3. Once a disease "sets in," various organs and tissues may be affected, thus resulting in a greater failure of ______________.

OBJECTIVE 2: Distinguish among the various types of diseases (textbook pp. 95–96).

There are numerous diseases and sources of diseases that affect the human body. In order to grasp an understanding of all the diseases, the major diseases have been classified into nine different categories. After studying in the text, pp. 95–96, you should be able to describe each of the classifications of disease.

_____ 1. A disease caused by worms is classified as a/an _____.

a. iatrogenic disease
b. idiopathic disease
c. infectious disease
d. neoplastic disease

_____ 2. Abnormal cell growth is classified as a/an _____.

a. iatrogenic disease
b. idiopathic disease
c. inherited disease
d. neoplastic disease

_____ 3. Hay fever would most likely be classified as a/an _____.

a. degenerative disease
b. endocrine disease
c. immunity-related disease
d. inherited disease

_____ 4. Scurvy is classified as a/an _____.

a. iatrogeneic disease
b. idiopathic disease
c. nutritional disease
d. None of the above is correct.

_____ 5. Myasthenia gravis is a/an _____.

a. autoimmune disease
b. degenerative disease
c. inherited disease
d. neoplastic disease

OBJECTIVE 3: Define acute, chronic, and latent diseases (textbook p. 96).

The intensity and duration of a disease is defined or described as acute, chronic, or latent. After studying p. 96 in the text, you should be able to describe and compare each of these durations.

_____ 1. A disease that develops rapidly is a/an _____ disease.

a. acute
b. chronic
c. latent
d. idiopathic

_____ 2. A disease that is long lasting is a/an _____ disease.

a. acute
b. chronic
c. latent
d. iatrogenic

_____ 3. A patient had a disease that disappeared for 5 years and then reappeared. This disease is known as a/an _____ disease.

a. acute
b. chronic
c. latent
d. inherited

_____ 4. Constant high blood pressure is an example of a/an _____ disease.

a. acute
b. chronic
c. latent
d. inherited

_____ 5. A heart attack is considered a/an _____ disease, and tuberculosis is considered a/an _____ disease.

a. acute, chronic
b. acute, latent
c. chronic, acute
d. chronic, latent

OBJECTIVE 4: Distinguish among a symptom, sign, and syndrome; and a diagnosis and prognosis (textbook pp. 96–98).

To make an accurate diagnosis and prognosis of a disease, a physician needs to interpret the various signs and symptoms associated with the disease. After studying in the text, pp. 96–98, you should be able to describe how signs and symptoms are used to develop a diagnosis and prognosis of a disease.

_____ 1. A diagnosis is the _____ of a disease.

a. identification
b. treatment
c. testing
d. study

_____ 2. When a patient tells a physician that he feels nausea, he is describing to the physician a _____ of his ailment.

a. diagnosis
b. prognosis
c. sign
d. symptom

_____ 3. When a patient tells a physician that she has noticed her fingernails changing colors, she is describing to the physician a _____ of her ailment.

a. diagnosis
b. prognosis
c. sign
d. symptom

_____ 4. Cridu chat is a hereditary anomaly that is characterized by mental retardation, microcephalia, dwarfism, and laryngeal defect. When the infant cries, the cry sounds like that of a cat. Since this anomaly is characterized by a combination of several conditions, cridu chat is most accurately known as a _____.

a. disease
b. sign
c. syndrome
d. symptom

_____ 5. A prognosis is _____.

a. a prediction of the outcome resulting from a disease
b. a prediction about the origin of the disease
c. an estimate of how long the patient will have the disease
d. an estimate of how long the patient has had the disease

OBJECTIVE 5: List and describe the major types of disease-causing organisms (textbook pp. 98–106).

After studying pp. 98–106 in the text, you should be able to name numerous disease-causing organisms.

_____ 1. An organism that causes an infection is a _____.

a. disease
b. germ
c. parasite
d. pathogen

_____ 2. Which of the following choices consists of only plural terms?

a. coccus, bacillus, spirillus, bacterium
b. diplococcus, streptococcus, staphylococcus, bacteria
c. diplococcus, bacilli, spirilla, bacteria
d. spirilla, cocci, bacilli, bacteria

_____ 3. All the following pathogens are cells except for _____.

a. amoeboids
b. bacteria
c. protozoa
d. viruses

_____ 4. Researchers are now injecting _____ into the body with the hopes of correcting genetic defects.

a. bacteria
b. protozoa
c. vectors
d. viruses

_____ 5. Athlete's foot is a disease caused by _____.

a. aflatoxins
b. bacteria
c. fungi
d. viruses

_____ 6. The filarial worm (a species of roundworm) causes a condition known as elephantiasis. This worm enters the body via the bite of a mosquito or fly. In this scenario, the mosquito is a _____.

a. helminth
b. host
c. vector
d. victim

OBJECTIVE 6: List the major pathways that allow pathogens to enter or exit the body (textbook pp. 107–108).

Pathogens can enter the body via any one of its many orifices. After studying in the text, pp. 107–108, you should be able to describe the openings and exits of the body.

_____ 1. Which of the following is classified as a portal of exit for pathogens?

a. tears
b. semen
c. feces
d. all the above

_____ 2. Which of the following is classified as a portal of entry for pathogens?

a. mouth
b. eyes
c. ears
d. all the above

OBJECTIVE 7: Describe three mechanisms of disease transmission and give examples of transmitted diseases (textbook pp. 108–109).

After studying in the text, pp. 108–109, you should be able to describe how pathogens utilize the three major modes of transmission to spread.

_____ 1. Imagine that someone gave you $100. A few days after receiving the money, you became ill. The physician attributed the illness to pathogens that passed to you via the money. This type of transmission of diseases is known as _____.

a. direct contact
b. environmental transmission
c. indirect contact
d. vector transmission

_____ 2. You and your friend share a can of pop. You later come down with a disease. You then find out that your friend also had the same disease a few days earlier. Your friend therefore transmitted the disease to you. This type of transmission is known as _____.

a. direct contact
b. environmental transmission
c. indirect contact
d. vector transmission

_____ 3. Some pathogens must first undergo development in a host, an insect, for example, before they can be passed on to humans. This mode of transmission is known as a/an _____.

a. biological vector
b. environmental transmission
c. indirect contact
d. mechanical vector

_____ 4. You travel to another country and develop diarrhea after drinking the water. The water in that country contains a protozoan called *Entamoeba histolytica.* This mode of transmission is called _____.

a. biological vector
b. environmental transmission
c. indirect contact
d. mechanical vector

OBJECTIVE 8: Distinguish between sterilization and disinfection (textbook pp. 109–110).

In order to maintain homeostasis, we cannot afford to let pathogens "take over." After studying in the text, pp. 109–110, you should be able to describe how aseptic techniques are used to control pathogens.

_____ 1. Which of the following appears to be the most effective in killing pathogens?

a. antiseptics
b. disinfectants
c. hydrogen peroxide
d. isopropyl alcohol

_____ 2. Which of the following techniques utilizes high pressure and high temperature to kill microbes ?

a. asepsis
b. disinfection
c. sterilization
d. all the above

_____ 3. Which of the following statements is true?

a. Antiseptics are a type of antibiotic.
b. Disinfectants are a type of antiseptic.
c. Sterilization requires the application of antiseptics.
d. Using disinfectants is an aseptic technique.

OBJECTIVE 9: Describe the techniques involved in Universal Precautions (textbook p. 110).

After studying in the text, p. 110, you should be able to describe what Universal Precautions are and how they are implemented to maintain homeostasis.

_____ 1. Universal Precautions apply to which of the following?

a. working with blood
b. working with semen
c. working with AIDS patients
d. all the above

_____ 2. Which of the following would fit under the category of Universal Precautions?

a. wearing splash shields
b. thoroughly washing hands
c. wearing gloves
d. all the above

OBJECTIVE 10: List five modes of action that antibiotics have on bacteria (textbook p. 110).

Antibiotics play a major role in controlling bacteria. After studying in the text p. 110, you should be able to explain how antibiotics work at controlling bacteria and maintaining homeostasis.

_____ 1. Which of the following are true in reference to antibiotics?

a. Microorganisms make antibiotics.
b. The human's immune system can make antibiotics.
c. Both a and b are true.
d. None of the above are true.

_____ 2. Streptomycin acts on bacterial ribosomes to prevent bacteria from producing _____.

a. cell membrane material
b. cell wall material
c. DNA
d. Protein

_____ 3. Antibiotics are effective against _____.

a. bacteria
b. germs
c. viruses
d. worms

OBJECTIVE 11: Distinguish among endemic, epidemic, and pandemic diseases (textbook p. 112).

One of the major concerns regarding the control of disease is that in our highly mobile society, diseases can spread easily. The spread of diseases can be endemic, epidemic, or pandemic. After studying in the text, p. 112, you should be able to distinguish among these three disease incidences.

_____ 1. Chickenpox that is spreading through the local day-care center would be classified as _____.

a. endemic
b. epidemic
c. pandemic

_____ 2. Which of the following terms refers to the widest spread of a disease?

a. endemic
b. epidemic
c. pandemic

_____ 3. Which of the following terms best describes a disease that is spreading from Europe to the United States?

a. endemic
b. epidemic
c. pandemic

_____ 4. Which of the following terms best describes a disease that seems to affect people in only one locale?

a. endemic
b. epidemic
c. isolation
d. pandemic

_____ 5. Which of the following terms best describes a disease that is spreading throughout the midwest United States?

a. endemic
b. epidemic
c. isolation
d. pandemic

PART II: CHAPTER-COMPREHENSIVE EXERCISES

Match the terms in column A with the descriptions or statements in column B.

MATCHING I

	(A)	(B)
_____	1. anaerobic	A. The study of the causes of diseases
_____	2. chickenpox	B. An endocrine disease
_____	3. diabetes mellitus	C. A latent disease
_____	4. etiology	D. A communicable disease
_____	5. eukaryotic	E. A noncommunicable disease
_____	6. mycosis	F. An organism that has a membrane-bound nucleus
_____	7. prokaryotic	G. An organism that does not have a membrane-bound nucleus
_____	8. shingles	H. An organism that does not require the use of oxygen to survive
_____	9. tetanus	I. A disease that is produced by a fungus
_____	10. vector	J. An organism that carries and transmits a disease-causing organism

CONCEPT MAP

This concept map summarizes and organizes the information in Chapter 6. Use the following terms to complete the map by filling in the boxes identified by the circled numbers, 1–10.

biological	**foodborne**	**fungi**
protozoans	**semen**	**sneezing**
utensils	**urethra**	**vectors**
viruses		

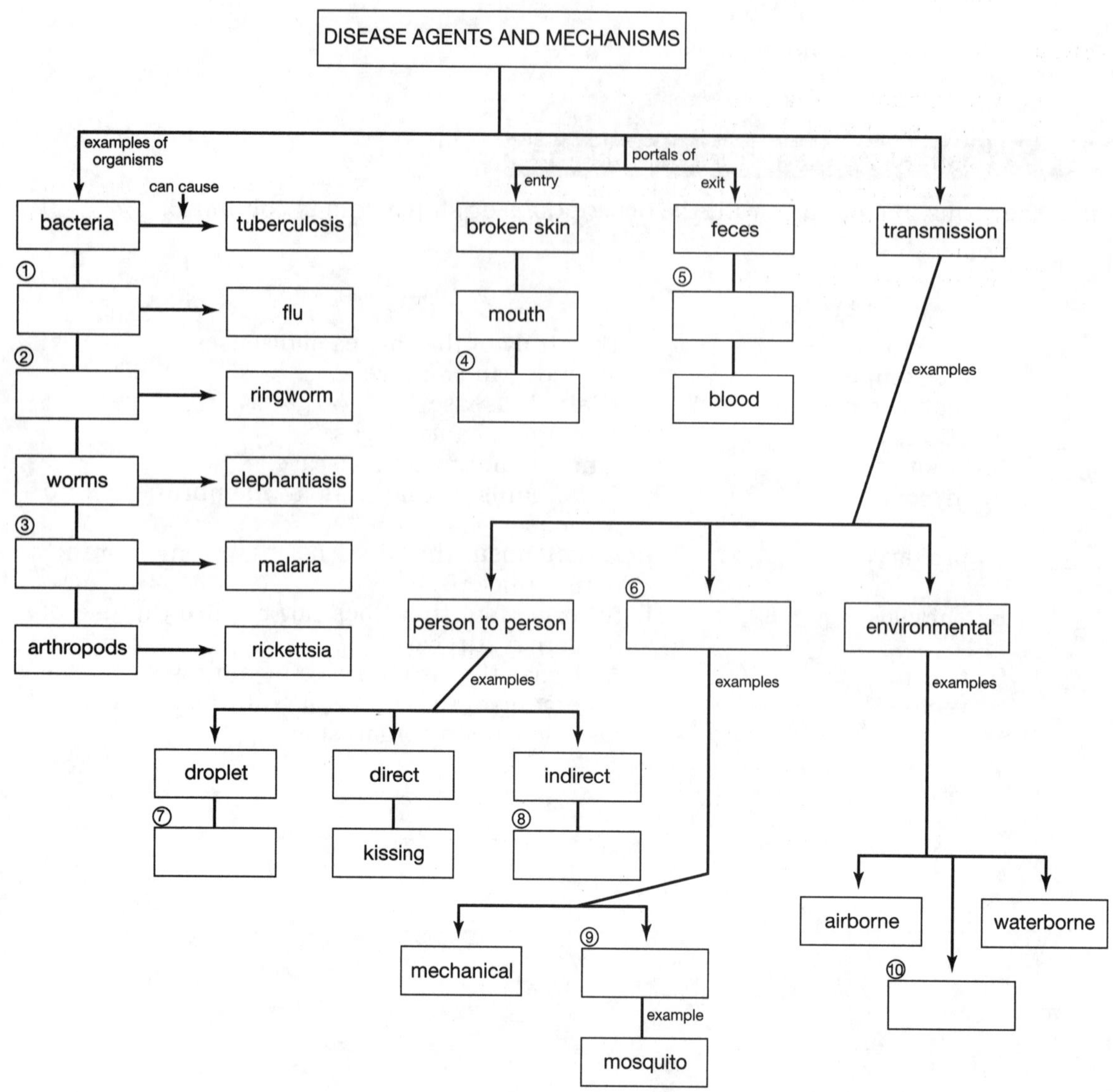

CROSSWORD PUZZLE

The following crossword puzzle reviews the material in Chapter 6. To complete the puzzle, you have to know the answers to the clues given, and you must be able to spell the terms correctly.

ACROSS

1. A disease that is occurring at the same time in different parts of the world
6. A disease-causing organism
7. A disease due to unknown reasons or causes
8. A disease that occurs continuously in a particular population
10. A disease with rapid onset
11. Smaller than bacteria

DOWN

1. A prediction of the course of a disease
2. A disease of long duration
3. Singular form of the word bacteria
4. Goodpasture's ____ is a disease that affects the kidneys and lungs at the same time
5. An organism that lives within or upon the body, causing it to be out of homeostasis
9. Spherical-shaped bacteria

FILL-IN-THE BLANK NARRATIVE

Use the following terms to fill in the blanks of this summary of Chapter 6.

arthropod	**antibiotic**	**bacterial**
diagnosis	**skin**	**homeostasis**
neoplasm	**ringworm**	**virus**

There are many organisms that simply "love" our body. Often, the conditions are just right in terms of temperature and pH. Since the (1) ___________ is the most observable organ of the body, this narrative is limited to skin infections.

(2) ___________ is a fungus infection. A round rash is found at the site of the infection. Warts are a type of benign (3) ___________. Warts are caused by a (4) ___________ and can be spread by direct contact. Boils typically are due to an inflammation that has resulted from a (5) ___________ infection. Scabies is a skin condition caused by a mite. The mite is an example of an (6) ___________.

Because of the wide range of organisms that can cause abnormal conditions, there is no single treatment. For example, an (7) ___________ is not effective against a fungus infection. Because of the variety of infectious organisms, it is critical for the physician to diagnose the type of infectious organism. After making the (8) ___________, the doctor can take the appropriate steps to rid the body of the disease-causing organism so that the body can get back to (9) ___________.

CLINICAL CONCEPTS

The following clinical concept applies the information presented in Chapter 6. Following the applications is a set of questions to help you understand the concepts.

Nadine has had a sore throat for two days. She becomes rather concerned when she notices that her uvula is red and slightly swollen, so she makes an appointment to see her physician.

The physician examines her oral cavity and notices that the palate of her mouth is quite red and that her uvula is indeed swollen and is also red. The physician makes a quick diagnosis that Nadine has strep throat. However, to be positive, the physician swabs Nadine's pharynx area to obtain a sample to culture the bacteria causing Nadine's strep throat.

The physician smears the swab on a culture dish containing blood and in just a few hours the physician notices that there is a clear area surrounding the colony of organisms. This indicates that the organisms are lysing the blood cells. Upon further examination of the dish with a microscope, the physician notices that the organisms are spherical in shape and group together in chains. The physician thus confirms her diagnosis that Nadine indeed has strep throat. Bacteria called *streptococci* cause strep throat. The physician prescribes an antibiotic, either penicillin or a penicillin product such as amoxicillin. These drugs affect the bacteria's ability to make cell walls. The cell walls become weak, and the bacteria die.

One week later, Nadine observes that one of her children also has a red throat and is experiencing difficulty in swallowing. She takes her child to the physician and the physician makes the same diagnosis – strep throat. The physician explains to Nadine that streptococci can be spread by droplets from sneezing or by using contaminated utensils. The child needs to stay home from school, as the streptococci can easily be spread in the school.

1. What are the identifiable characteristics of streptococci?

2. What would happen if the physician were to prescribe an antiviral medication instead of an antibiotic?

3. Sometimes the uvula swells with a strep throat infection. What is the uvula?

4. Why do you suppose the pharynx region appears red with a strep infection?

5. What might be some other forms of transmission of streptococci?

THIS CONCLUDES CHAPTER 6 EXERCISES

CHAPTER 7

The Integumentary System

Many may wonder why anatomists use such big words as integumentary when they refer to skin. After studying this chapter, you will find that the integumentary system is more than "just skin." The integumentary system consists of the skin and associated structures, including hair, nails, and a variety of glands. Of all the body systems, the integument is the only one seen every day. Because the skin and its associated structures are readily seen by others, a lot of time is spent caring for the skin, to enhance its appearance and to prevent skin disorders that may alter desirable structural features on and below the skin's surface. Washing the face, brushing and/or trimming the hair, taking showers, and applying deodorant are activities that modify the appearance or properties of the integument.

The four primary tissue types making up the skin constitute what is considered to be a large, highly complex **organ,** or a structurally integrated **organ system.** The integument manifests many of the functions of living matter, including protection, excretion, secretion, absorption, synthesis, storage, sensitivity, and temperature regulation.

The exercises in Chapter 7 emphasize important facts and concepts about the skin and its associated structures, and they will enhance your understanding of the important roles the skin, hair, and nails play in our lives.

Chapter Objectives:

1. Describe the general functions of the skin.
2. Compare the structures and functions of the different layers of the skin.
3. Explain what accounts for individual and racial differences in skin, such as skin color.
4. Discuss the effects of ultraviolet radiation on the skin.
5. Discuss the functions of the skin's accessory structures.
6. Describe the mechanisms that produce hair and determine hair texture and color.
7. Summarize the effects of the aging process on the skin.
8. Describe the major causes, signs, and symptoms of several skin disorders.
9. Explain how the skin repairs itself after an injury.
10. Describe what distinguishes first-, second-, and third-degree burns.

PART I: OBJECTIVE-BASED QUESTIONS

OBJECTIVE 1: Describe the general functions of the skin (textbook pp. 117–122).

After studying pp. 117–122 in the text, you should be able to describe how the integumentary system is more than "just skin."

_____ 1. Which of the following structures of the integumentary system is (are) involved in protection?

a. hair
b. nails
c. skin
d. all the above

_____ 2. Which of the following structures of the integumentary system is (are) involved in the synthesis of vitamin D?

a. glands
b. nails
c. skin
d. all the above

_____ 3. Which of the following structures of the integumentary system is (are) involved in the formation of sweat?

a. glands
b. nails
c. skin
d. all the above

_____ 4. Which of the following is (are) considered an accessory structure of the integumentary system?

a. integument
b. nails
c. skin
d. all the above

5. Identify the various parts of the integumentary system in Figure 7–1.

Figure 7–1 Components of the Integumentary System

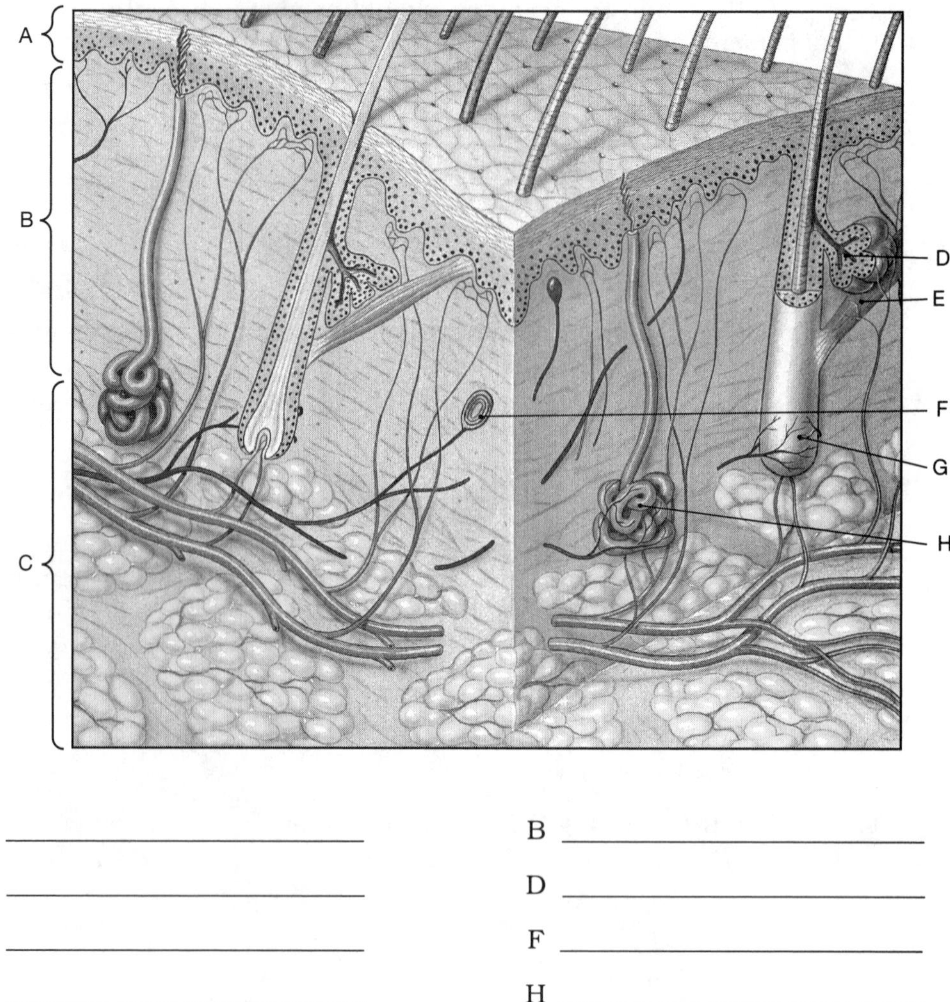

A ______________________ B ______________________

C ______________________ D ______________________

E ______________________ F ______________________

G ______________________ H ______________________

OBJECTIVE 2: Compare the structures and functions of the different layers of the skin (textbook pp. 117–120).

After studying pp. 117–120 in the text, you should be able to describe the two major divisions of skin layers—the epidermis and dermis—and the divisions of each layer.

_____ 1. Which of the following is a synonym for skin?

a. cutaneous membrane
b. dermis
c. epidermis
d. integumentary system

_____ 2. The outermost layer of the skin is called _____, which is made of stratum _____ and stratum _____ layers.

a. epidermis, corneum, germinativum
b. epidermis, corneum, melanocyte
c. dermis, corneum, germinativum
d. epithelium, corneum, germinativum

_____ 3. Which of the following is (are) *not* a part of the integumentary system?

a. arrector pili muscles
b. hypodermis
c. subcutaneous layer
d. Both b and c are correct.

_____ 4. Which layer of skin is (are) responsible for vitamin D production?

a. dermis
b. keratinized layer
c. lower epidermal layers
d. stratum corneum

_____ 5. Which layer of skin creates our fingerprints?

a. dermis
b. epidermis
c. stratum corneum
d. stratum germinativum

_____ 6. The outermost layer of skin is constantly sloughed; therefore these cells need to be replaced. The replacement cells come from _____.

a. the dermal layer
b. the hypodermis
c. the stratum corneum layer
d. the stratum germinativum layer

_____ 7. Most of the accessory structures are located in the _____.

a. dermis
b. epidermis
c. hypodermis
d. subcutaneous

_____ 8. Most of a person's body fat is located in the _____.

a. dermis
b. epidermis
c. hypodermis
d. none of the above

_____ 9. A person's body fat does which of the following?

a. protects to a certain degree
b. reduces heat loss
c. stores energy
d. all the above

OBJECTIVE 3: Explain what accounts for individual and racial differences in skin, such as skin color (textbook pp. 118–119).

After studying pp. 118–119 in the text, you should be able to describe the factors that affect skin color, including the activity of the melanocytes.

_____ 1. Melanocytes are cells found in the _____.

a. dermis
b. hypodermis
c. stratum corneum
d. stratum germinativum

_____ 2. Which of the following statements regarding albinos and non-albinos is true?

a. Albinos have fewer melanocytes than do non-albinos.
b. Albinos have no melanocytes; non-albinos have many melanocytes.
c. Albinos have the same number of melanocytes as non-albinos.
d. The melanocytes of albinos are overactive.

_____ 3. Which of the following create darker skin color?

a. active melanocytes
b. excessive blood supply to the skin
c. exposure to sunlight
d. nonactive melanocytes

_____ 4. People of a race with dark-colored skin get their dark color from _____.

a. DNA that created very active melanocytes
b. excessive numbers of melanocytes
c. overexposure to sunlight
d. all the above

_____ 5. Some people who try to get a tan in the summer never tan but "burn." Which of the following reasons might explain why they can't tan?

a. Their melanocytes are simply not very active.
b. They have a partial albino gene.
c. They lack enough melanocytes to create the tan.
d. All the above are correct.

6. The pigment produced by melanocytes is ______________________.

7. In Figure 7–2, identify the stratum germinativum layer, the nucleus of the cells, the melanocytes, and melanin pigment.

Figure 7–2 Melanocytes

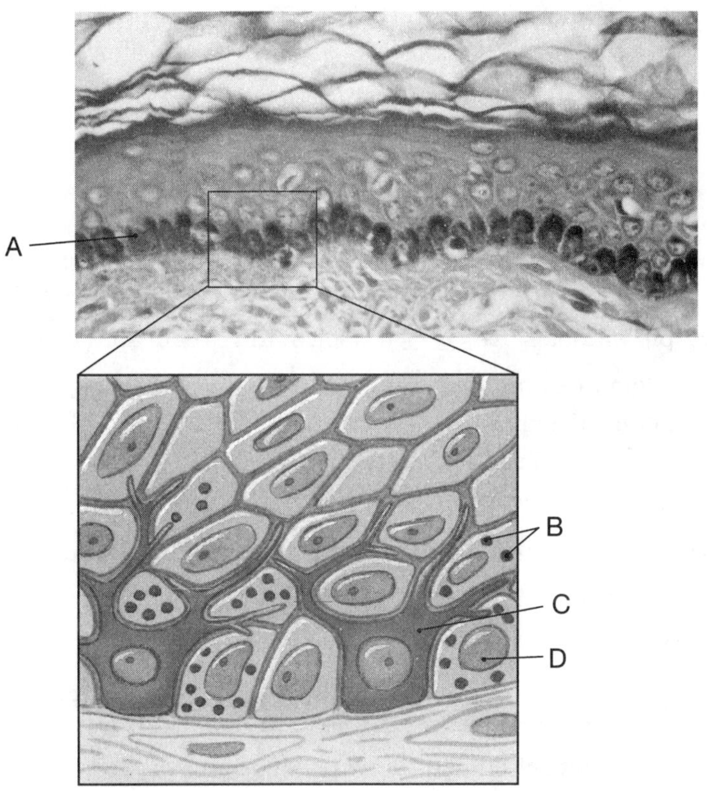

A ______________________ B ______________________

C ______________________ D ______________________

OBJECTIVE 4: Discuss the effects of ultraviolet radiation on the skin (textbook p. 119).

After studying p. 119 in the text, you should be able to describe how melanocytes will protect the body against limited exposure to sunlight and discuss the dangers of overexposure to sunlight.

_____ 1. Melanin prevents skin damage due to UV light by _____.

a. absorbing UV light
b. covering the epidermal layers
c. protecting the nucleus of epidermal cells
d. both a and c

_____ 2. Excess exposure to UV light may _____.

a. damage connective tissue, resulting in wrinkling
b. damage cellular DNA, resulting in mutation
c. damage chromosomes, resulting in cancer
d. all the above

_____ 3. Excess exposure to UV light may _____.

a. cause a decrease in the number of melanocytes
b. cause a decrease in vitamin D production
c. cause an increase in the number of melanocytes
d. cause damage to the DNA of cells in the stratum germinativum area

4. Melanocytes begin producing melanin when they are exposed to ______________________.

5. Melanocytes prevent skin damage due to UV light by protecting the ______________________ within the epidermal cell's nucleus.

OBJECTIVE 5: Discuss the functions of the skin's accessory structures (textbook pp. 120–122).

After studying pp. 120–122 in the text, you should be able to discuss the roles of the glands, hair, and nails, the accessory structures of the integumentary system.

_____ 1. The arrector pili muscles are _____ muscles and are located in the _____.

a. skeletal, dermis
b. smooth, dermis
c. smooth, epidermis
d. smooth, hypodermis

_____ 2. Which glands are responsible for producing natural body odor?

a. apocrine
b. eccrine
c. merocrine
d. sebaceous

_____ 3. Which glands are associated with acne production?

a. apocrine
b. eccrine
c. sebaceous
d. sweat

_____ 4. Which glands are responsible for assisting in cooling the body?

a. apocrine
b. eccrine
c. sebaceous
d. sweat

5. What is the function of nails?

6. How does the hair in the nose provide protection?

7. What is the purpose of eyelashes?

8. What structures, found in the integumentary system, are involved in creating goosebumps?

9. In Figure 7–3 identify the following accessory structures: sebaceous gland, sweat gland, arrector pili muscle, and the hair follicle.

Figure 7–3 Accessory Structures of the Integumentary System

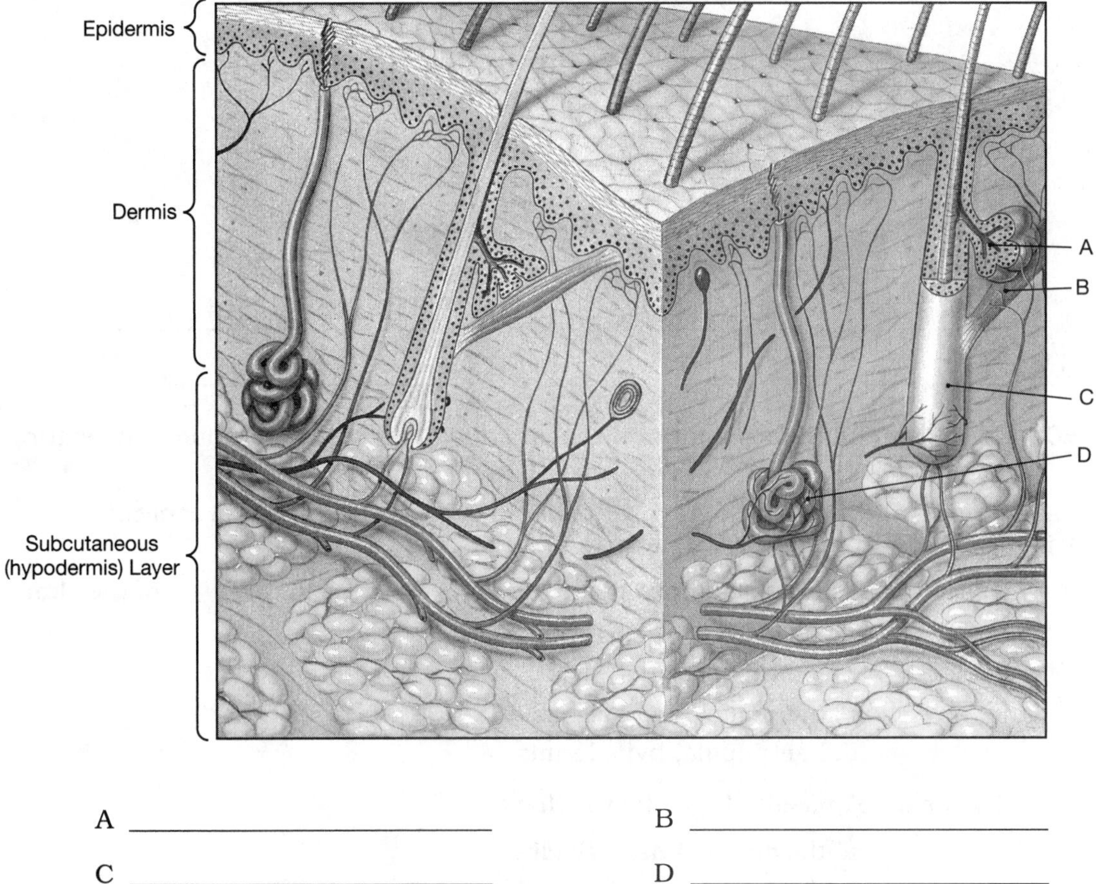

A ______________________ B ______________________

C ______________________ D ______________________

10. In Figure 7–4 identify the lunula, nail bed, cuticle, and free edge.

Figure 7–4 Accessory Structures of the Integumentary System: The Nail

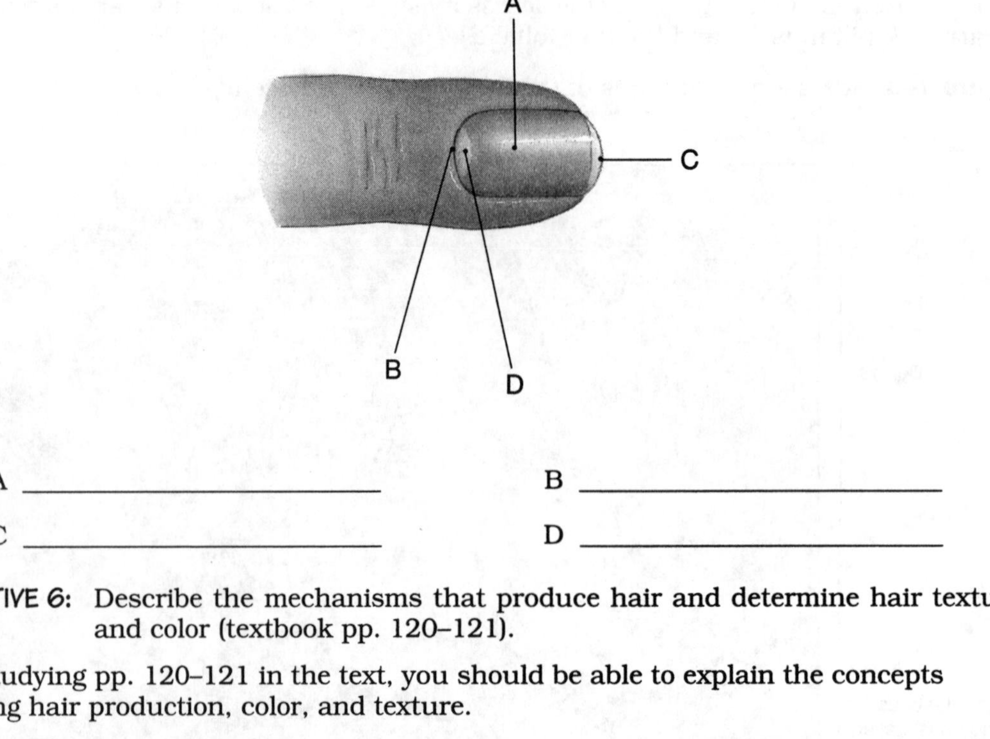

A ____________________ B ____________________

C ____________________ D ____________________

OBJECTIVE 6: Describe the mechanisms that produce hair and determine hair texture and color (textbook pp. 120–121).

After studying pp. 120–121 in the text, you should be able to explain the concepts involving hair production, color, and texture.

_____ 1. Hair arises from structures called the _____ located in the _____ of the skin.

a. arrector pili, epidermis
b. hair follicles, dermis
c. hair follicles, epidermis
d. hair papilla, hypodermis

_____ 2. The development of gray hair is due to _____.

a. the death of hair follicles
b. the production of air bubbles in the hair
c. the production of gray pigments
d. the reduction of melanocyte activity

_____ 3. The various textures of hair are due to _____.

a. the arrector pili
b. the follicles
c. the hair papilla
d. the melanocytes

4. What material makes the shaft of the hair stiff? ____________________

OBJECTIVE 7: Summarize the effects of the aging process on the skin (textbook p. 123).

Aging has an effect on all the organ systems of the body. The aging effects on the integumentary system are quite noticeable because the skin is so visible to the naked eye. After studying p. 123 in the text, you should be able to describe the aging effects on the integumentary system.

_____ 1. The elderly appear to be paler than young people because _____.

a. their epidermis is thinner
b. their hair follicles have stopped functioning
c. their melanin production has decreased
d. their melanocyte numbers have decreased

_____ 2. Which of the following statements regarding the effects of aging on the integumentary system is true?

a. The epidermis gets thinner.
b. The dermis gets thinner.
c. Sebaceous gland activity decreases.
d. all the above

_____ 3. As the _____ becomes thinner, the skin begins to wrinkle.

a. dermis
b. epidermis
c. hypodermis
d. stratum germinativum

_____ 4. Which of the following statements regarding aging skin is true?

a. Blood supply to the skin increases, thereby making it easy to lose heat.
b. Blood supply to the skin increases, thereby making it hard to lose heat.
c. Blood supply to the skin decreases, thereby making it easy to lose heat.
d. Blood supply to the skin is reduced, thereby making it hard to lose heat.

OBJECTIVE 8: Describe the major causes, signs and symptoms of several skin disorders (textbook pp. 124–132).

There are numerous skin disorders and each is characterized by specific signs and symptoms. After studying pp. 124–132 in the text, you should be able to describe some of the disorders associated with the skin.

_____ 1. All the following are due to excess keratin production except _____.

a. calluses
b. corns
c. psoriasis
d. xerosis

_____ 2. Which of the following skin conditions is commonly known as dandruff?

a. eczema
b. psoriasis
c. seborrheic dermatitis
d. urticaria

_____ 3. Which of the following is caused by a virus and stays dormant in the body only to possibly reappear as shingles?

a. chicken pox
b. human papilloma
c. impetigo
d. urticaria

_____ 4. Ringworm is caused by a _____.

a. flatworm
b. fungus
c. nematode
d. virus

_____ 5. Athlete's foot is caused by a _____.

a. bacterium
b. fungus
c. ringworm
d. virus

OBJECTIVE 9: Explain how the skin repairs itself after an injury (textbook pp. 132–133).

The skin provides physical protection for the body. It covers the entire body and is therefore vulnerable to injury. It is essential that the skin repairs itself in order to continue providing protection. After studying pp. 132–133 in the text, you should be able to explain the process of skin repair.

_____ 1. After a cut into the dermis, the stratum germinativum cells begin to migrate _____.

a. away from one another to make room for fibroblasts
b. deeper into the dermis around the wound
c. toward one another to close the wound
d. toward the stratum corneum

_____ 2. During the repair process of a wound, the number of capillaries at the wound site _____.

a. decreases in number
b. increases in number
c. remains the same as before

_____ 3. Which cells are mostly responsible for scar formation?

a. fibroblast
b. epidermal
c. stratum germinativum
d. macrophages

_____ 4. Which of the following statements in reference to scab formation is not true?

a. The wound is healed when a scab finally forms.
b. A scab will dissolve.
c. A scab forms from the bottom of the wound to the top.

_____ 5. Which of the following statements is true in reference to tissue healing in a fetus?

a. The tissue will heal without leaving a scar.
b. The tissue is vulnerable to scar formation as it heals.
c. The tissue may form abnormal scars such as keloids.
d. The healed area will become a little darker than the rest of the skin.

_____ 6. Which of the following statements about wound repair is true?

a. Stratum germinativum cells will replace the epidermal cells.
b. Stratum germinativum cells will replace the dermal cells.
c. Stratum germinativum cells will replace both epidermal and dermal cells.
d. Stratum germinativum cells will replace only the stratum germinativum layer.

OBJECTIVE 10: Describe the distinguishing features of first-, second-, and third-degree burns (textbook p. 134)?

After studying p. 134 in the text, you should be able to identify the characteristics of first-, second-, and third-degree burns.

_____ 1. Which of the following surface area burns would be the worst based on percentage values alone?

a. anterior torso and genitals
b. hand and one arm
c. one leg and genitals
d. one leg and one arm

_____ 2. Sun-damaged tissue would be rated as a _____ burn.

a. first-degree
b. second-degree
c. third-degree

_____ 3. Burn damage from the stratum corneum to the stratum germinativum is considered to be a _____ burn.

a. first-degree
b. second-degree
c. third-degree

_____ 4. Which of the following burns is most likely to be the least painful?

a. first-degree
b. second-degree
c. third-degree

_____ 5. A sunburn usually causes the skin to appear red due to _____.

a. adverse chemical reactions in the skin
b. damaged blood vessels
c. the inflammatory response of blood vessels
d. the intense heat

PART II: CHAPTER-COMPREHENSIVE EXERCISES

MATCHING

Match the terms in column A with the definitions or phrases in column B.

	(A)	(B)
_____	1. adipose	A. The outermost layer of skin consisting of several layers
_____	2. apocrine	B. The layer of skin consisting of the accessory structures
_____	3. arrector pili	C. Special cells that produce a pigment designed to protect the body from harmful UV light
_____	4. cutaneous	D. The pigment produced by some cells in the stratum germinativum layer
_____	5. dermis	E. The skin consisting of the epidermis and dermis
_____	6. eccrine	F. Another name for the hypodermis layer
_____	7. epidermis	G. The name of the tissue that is below the dermis
_____	8. hemangioma	H. There are many of these cells located in the hypodermis.
_____	9. hirsutism	I. The smooth muscles responsible for creating goosebumps
_____	10. hypodermis	J. There are many of these cells located in the hypodermis
_____	11. melanin	K. The smooth muscles responsible for creating goosebumps
_____	12. melanocytes	L. The gland that produces material to lubricate the skin
_____	13. sebaceous	M. A tumor of a blood vessel resulting in a birthmark
_____	14. subcutaneous	N. Growth of hair on women that is characteristic of men

CONCEPT MAP

This concept map summarizes and organizes some of the ideas in Chapter 7. Use the following terms to complete the map by filling in the boxes identified by the circled numbers, 1–9.

dermis	**glands**	**lubrication**
production of secretions	**sensory reception**	**vitamin D synthesis**
first-degree	**second-degree**	**third-degree**

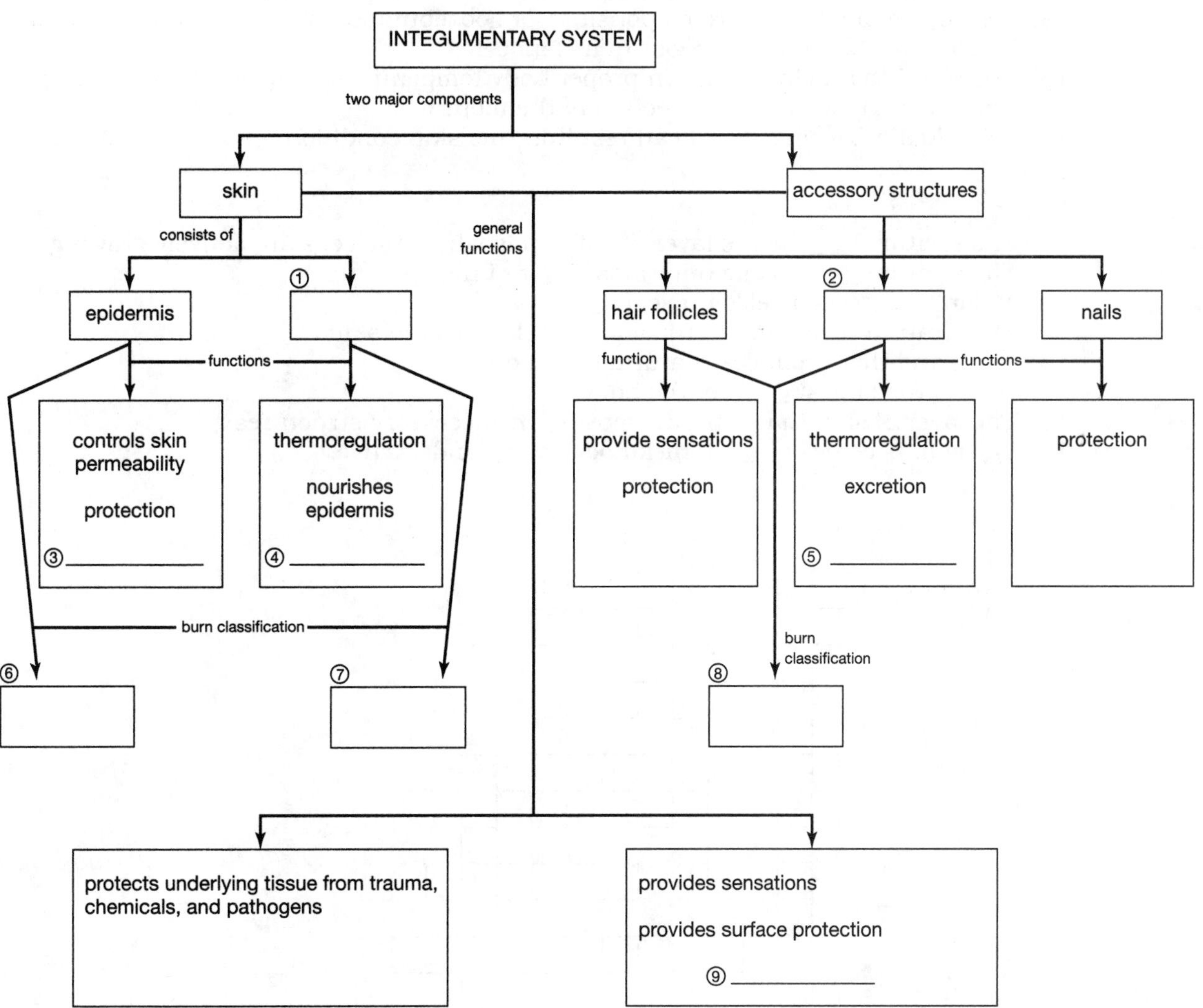

CROSSWORD PUZZLE

This crossword puzzle reviews the material in Chapter 7. To complete the puzzle, you have to know the answers to the clues given, and you must be able to spell the terms correctly.

ACROSS

3. Albinos have the same number of _____ as non-albinos.
4. Freckles are spot concentrations of _____.
6. A term that refers to skin and its accessory structures
9. The _____ pili muscles are responsible for goosebumps.
10. The loss of skin due to the loss of melanocytes
13. The gland that helps maintain proper body temperature
14. Thrush is a type of _____ infection of the mouth.
16. A blocked sebaceous gland can result in this skin condition.

DOWN

1. The stratum _____ is the layer of epidermis where the cells are actively growing.
2. The stratum _____ is the outermost layer of the epidermis.
5. Melanin protects a cell's nuclear _____.
7. The stratum corneum layer is a part of this layer of skin.
8. The gland that produces "natural body odor"
11. Skin is the largest _____ of the body.
12. The layer of skin that contains most of the accessory structures
15. The skin becomes _____ if melanocytes are underactive.

FILL-IN-THE-BLANK NARRATIVE

Use the following terms to fill in the blanks of this summary of Chapter 7.

cancer	**cooling**	**DNA**
mutate	**stratum germinativum**	**lubricates**
melanin	**melanocytes**	**organ**
organ systems	**homeostasis**	**pathogens**
sebaceous	**eccrine**	**fibroblasts**
scar	**epidermis**	

The skin is the largest (1) ___________ of the body and is probably the organ that is taken most for granted. The skin is actively involved in helping us maintain (2) ___________. We are frequently exposed to sunlight, which can cause burns, dry skin, and damage to our skin.

The integumentary system has (3) ___________ glands that produce a watery substance. This watery substance evaporates from the body, thus (4) ___________ it. The (5) ___________ glands produce an oily substance called sebum that (6) ___________ the skin to reduce dryness. (7) ___________ produce (8) ___________, which is a pigment that protects the cell's (9) ___________ from overexposure to sunlight. Harmful UV rays may cause the skin cell's DNA to (10) ___________, which may lead to skin (11) ___________.

The skin also serves as protection against infectious agents. If the skin is damaged, it becomes vulnerable to the entrance of (12) ___________. But skin can repair itself. Cells from the (13) ___________ layer will undergo cell reproduction to replace the lost cells. During the repair process, (14) ___________ begin to repair the dermal region while cells from the stratum germinativum layer repair the (15) ___________. As the fibroblasts work they lay down a layer of fibers that may result in (16) ___________ formation.

All these events are essential not only for the homeostasis of the integumentary system but for the homeostasis of all the (17) ___________ of the body.

CLINICAL CONCEPT

The following clinical concept applies the information in Chapter 7. Following the application is a set of questions to help you understand the concept.

The role of the sebaceous glands is to produce sebum to lubricate the skin. Sometimes a sebaceous duct leading to the surface of the skin gets blocked with dirt and bacteria. Even though the duct is blocked, the sebaceous gland continues to produce sebum.

The autonomic nerves, one portion of the nervous system, activate the sebaceous glands. A number of stimuli will activate the autonomic nervous system, including environmental conditions (dry weather) and stress. The autonomic nerves continue to activate the sebaceous gland even when the duct is blocked. Sebum continues to build within the duct. Pressure begins to build, and soon the surface of the skin begins to bulge. After a while, the bulge starts to turn red because the body recognizes that bacteria (foreign invaders) are trying to enter the body. To get white blood cells to the infected site to fight off these infectious agents, the blood capillaries in the dermal region dilate. This allows a greater flow of blood to the surface. Because the majority of the cells making up blood are red blood cells, the area appears red.

The white blood cells begin the battle. Many bacteria die, and so do many white blood cells. The combination of dead bacteria and dead white blood cells creates pus. As pus accumulates, a whitehead develops. If you were to squeeze the pimple, white material (pus) and blood would squirt out, but this is not advisable. Squeezing the tissue would force some of the bacteria out, but at the same time would force some bacteria into deeper tissues, which could cause more serious infection.

In time, the white blood cells accomplish their task, the fluid pus is reabsorbed into the tissues, and the pimple goes away. However, this process takes about two weeks.

1. What causes the bulging of the skin when pimples are forming?

2. Why are pimples typically red?

3. Why do some pimples develop a whitehead?

4. Why should pimples not be squeezed?

5. Why are pimples extremely common during early teenage years?

THIS CONCLUDES CHAPTER 7 EXERCISES

CHAPTER 8

The Skeletal System

The skeletal system consists of bones and related connective tissues, including cartilage, tendons, and ligaments. Although bones from different parts of the body share microscopic characteristics and are of the same basic, dynamic nature, each one has its own characteristic pattern of ossification and growth, characteristic shape, and identifiable surface features that reflect its functional relationship to other bones as well as other systems throughout the body. It is important to understand that even though bone tissue is structurally stable, it is made of cells and therefore is **living** tissue that is functionally **dynamic**.

The skeletal system is divided into two divisions: the **axial** (which consists of 80 bones) and the **appendicular** (which consists of 126 bones). The skeletal system performs many functions, including (1) structural support, (2) storage of minerals and lipids, (3) blood cell production, (4) protection, (5) leverage, (6) assistance in respiratory movements, (7) providing sites for the attachment of muscles, and (8) involvement with movement and mobility.

Almost all daily activities require that the skeletal system perform in a coordinated, dynamic way to allow an individual to be active and mobile. Various disorders associated with bones such as osteoperosis or arthritis can hinder daily activities.

The activities in Chapter 8 review not only the identification and location of bones and bone markings but also the functional anatomy of the bones and articulations that constitute the skeletal system.

Chapter Objectives:

- Describe the functions of bones and the skeletal system.
- Compare the structure and function of compact and spongy bones.
- Discuss the processes by which bones develop and grow.
- Describe how the activities of the bone cells constantly remodel bone.
- Discuss the effects of aging on the bones of the skeletal system.
- Contrast the structure and function of the axial and appendicular skeletons.
- Identify the bones of the skull.
- Discuss the differences in structure and function of the various vertebrae.
- Identify the bones of the limbs.
- Describe the three basic types of joints.
- Relate the body movements to the action of specific joints.
- Describe the different causes of bone disorders and give examples of each.
- Distinguish between degenerative and inflammatory arthritis diseases.

PART I: OBJECTIVE-BASED QUESTIONS

OBJECTIVE 1: Describe the functions of bones and the skeletal system (textbook p. 143).

After studying p. 143 in the text, you should be able to explain the dynamic nature of bones and the numerous functions they serve in the human body.

_____ 1. Bones work with muscles to create movement. This statement best describes which function of bones?

a. leverage
b. protection
c. support
d. none of the above

_____ 2. Bones contain calcium ions, phosphate ions, and energy reserves. This statement best reflects which function of bones?

a. blood cell production
b. storage
c. support
d. none of the above

_____ 3. Which of the following are considered supporting tissue?

a. bone and muscle
b. osseous and bone
c. osseous and cartilage
d. osteocytes and osseous

_____ 4. The mass of bone is primarily due to _____.

a. bone marrow
b. collagen fibers
c. the matrix
d. the osteocytes

OBJECTIVE 2: Compare the structure and function of compact and spongy bones (textbook pp. 144–145).

After studying pp. 144–145 in the text, you should be able to describe the differences between compact bone and spongy bone, namely, that compact bone is composed of osteons and provides a great deal of strength, and spongy bone forms within the marrow cavity of the bone and is found where bones are not heavily stressed.

_____ 1. Compact bone is found _____, whereas spongy bone is found _____.

a. along the edge of the diaphysis, in the marrow area
b. close to the periosteum, close to the endosteum
c. in the epiphysis, in the diaphysis
d. in the marrow area, along the edge of the diaphysis

_____ 2. Which of the following statements is true?

a. Osteoblasts form osteocytes.
b. Osteocytes form osteoblasts.
c. Osteocytes form osteoclasts.
d. Osteocyte and osteoblast are interchangeable terms.

_____ 3. The basic functional unit of compact bone is the _____.

a. osteon
b. osteocyte
c. osteoblast
d. osteoclast

_____ 4. The central canal is _____.

a. the area of the osteon where blood vessels are located
b. the area of the osteon where the nucleus is located
c. the area of the osteon where the osteoclasts are located
d. the area of the osteon where the bone marrow is located

_____ 5. Blood vessels are found in the _____.

a. central canal
b. diaphysis
c. epiphysis
d. all the above

_____ 6. Osteocytes are found within the _____.

a. osteons of compact bone
b. osteons of spongy bone
c. osteons of the periosteum
d. all the above

_____ 7. Bones store minerals to be used at later times by the body. These stored minerals are released into the bloodstream by the action of _____.

a. osteoblasts
b. osteoclasts
c. osteocytes
d. osteons

8. In Figures 8–1 and 8–2, identify the proximal epiphysis, marrow cavity, periosteum, compact bone, spongy bone, and osteon.

Figure 8–1 Structure of a Long Bone

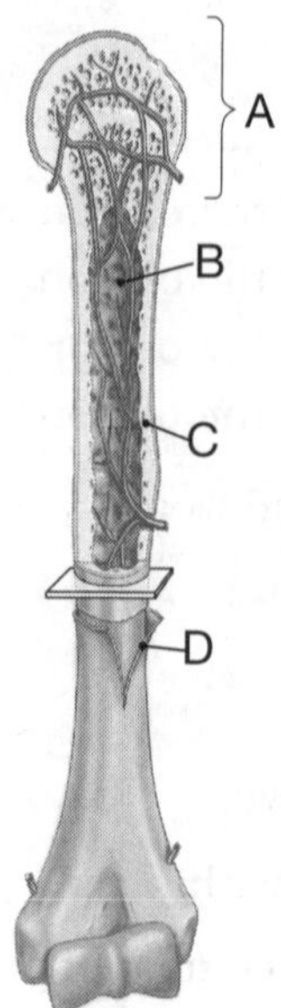

A ______________________ B ______________________

C ______________________ D ______________________

Figure 8–2 Internal Structure of a Long Bone

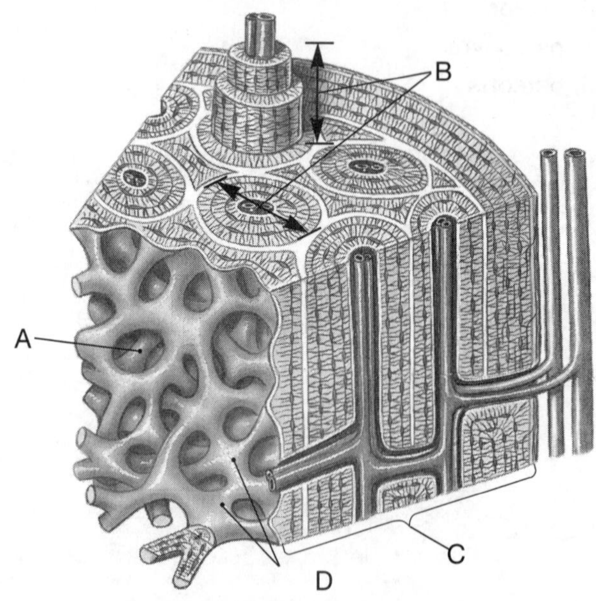

A ______________________ B ______________________

C ______________________ D ______________________

OBJECTIVE 3: Discuss the processes by which bones develop and grow (textbook pp. 146–147).

All our bones begin as cartilage and are later replaced by bone by specialized cells. Our bones are alive and actively grow until about the age of 25. At that age, the bone growth rate ceases but maintains its size. After studying pp.146–147 in the text, you should be able to explain how bones grow and develop from cartilage by the actions of specialized cells.

_____ 1. Which of the following cause the replacement of cartilage by bone?

a. osteoblasts
b. osteoclasts
c. osteocytes
d. osteons

_____ 2. Which bone cells are responsible for "carving" out the marrow areas?

a. osteoblasts
b. osteoclasts
c. osteocytes
d. osteons

_____ 3. Ossification first occurs in the _____.

a. diaphysis
b. distal ends of the bone
c. epiphysis
d. both a and c

_____ 4. In the epiphyseal plate region (known as the growth plate) osteoblasts replace cartilage with bone on the _____ while new cartilage is produced on the _____.

a. side closest to the diaphysis, side farthest from the diaphysis
b. side closest to the shaft, side farthest from the shaft
c. side farthest from the diaphysis, side closest to the diaphysis
d. both a and b

5. In Figure 8–3, identify the epiphysis, diaphysis, marrow cavity, blood vessel, and epiphyseal plate.

Figure 8–3 Endochondral Ossification

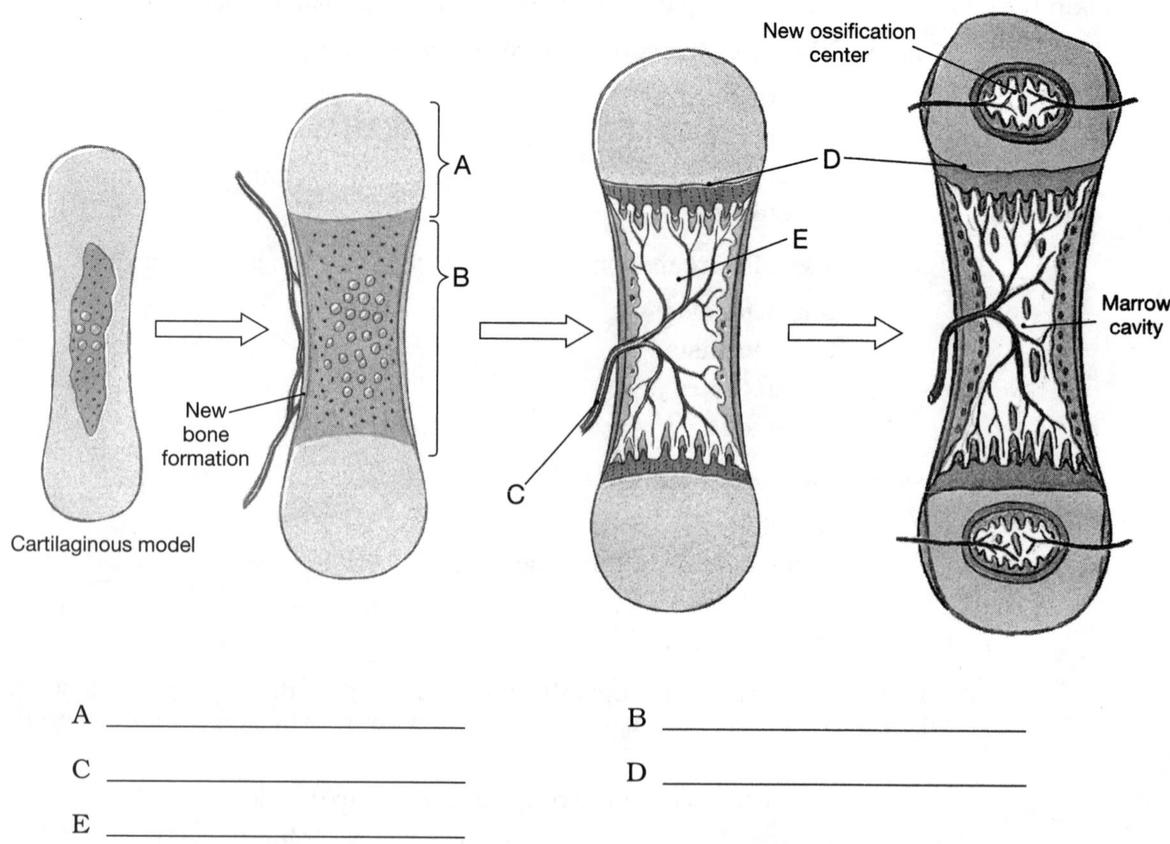

A ______________________ B ______________________

C ______________________ D ______________________

E ______________________

OBJECTIVE 4: Describe how the activities of the bone cells constantly remodel bone (textbook p. 147).

Bones are subject to damage like any other tissue and because they are living tissue, they are able to repair themselves. After studying p. 147 in the text, you should be able to explain how bones can repair themselves when damaged.

_____ 1. In order for bone repair to occur, osteoblast cells are needed. These cells come from _____.

a. spongy bone
b. the blood
c. the osteon
d. the periosteum

_____ 2. Bones cannot be repaired if _____.

a. cells of the endosteum die
b. cells of the periosteum die
c. circulatory supply has been blocked
d. all the above

____ 3. Heavily stressed bones will ____.

a. become brittle
b. become stronger
c. lose calcium ions at a faster rate than normal
d. develop osteoporosis at a faster rate than normal

____ 4. Exercise seems to stimulate the activity of ____.

a. estrogen
b. osteoblasts
c. osteoclasts
d. osteoporosis

____ 5. In order for bone to repair itself, ____.

a. osteoblast activity must be greater than osteoclast activity
b. osteoclast activity must be greater than osteoblast activity
c. osteoblast activity must equal osteoclast activity
d. osteocyte activity must increase

OBJECTIVE 5: Discuss the effects of aging on the bones of the skeletal system (textbook p. 148).

After studying p. 148 in the text, you should be able to describe the changes that take place in bones with age.

____ 1. The rate of bone breakdown increases when the rate of ____ activity increases.

a. osteocyte
b. osteoblast
c. osteoclast
d. all the above

____ 2. If bone mass begins to decrease in women at the age of 40, how much will be lost by the age of 80?

a. 8%
b. 16%
c. 28%
d. none of the above

____ 3. If a male at the age of 40 has 23 kg of bone mass, approximately how much will he have left at the age of 80?

a. 3 kg
b. 7 kg
c. 18 kg
d. 20 kg

OBJECTIVE 6: Contrast the structure and function of the axial and appendicular skeletons (textbook pp. 148 and 157).

After studying pp. 148 and 157 in the text, you should be able to describe the differences between the axial and the appendicular divisions of the skeletal system.

_____ 1. Which of the following is part of the axial division?

a. femur
b. hips
c. humerus
d. skull

_____ 2. Which of the following is (are) part of the axial division?

a. mandible
b. pectoral girdle
c. vertebrae
d. both a and c

_____ 3. Which of the following is (are) part of the appendicular division?

a. coxae
b. pectoral girdle
c. pelvic girdle
d. all the above

_____ 4. Which of the following is (are) part of the appendicular skeleton?

a. coxae
b. ethmoid
c. ribs
d. sternum

_____ 5. The axial skeleton protects _____.

a. abdominal cavities
b. dorsal cavities
c. ventral cavities
d. both b and c

_____ 6. The axial skeleton protects all the following except the _____.

a. brain
b. heart
c. spinal cord
d. liver

OBJECTIVE 7: Identify the bones of the skull (textbook pp. 149–153).

After studying pp. 149–153 in the text, you should be able to identify 22 bones of the skull, which can be divided into those of the face and those of the cranium.

_____ 1. Which of the following face bones occur in pairs?

a. mandible
b. nasal
c. vomer
d. both a and c

____ 2. Which of the following cranial bones occur in pairs?

a. occipital
b. parietal
c. sphenoid
d. all the above

____ 3. The anatomical name for the cheekbone is the ____.

a. ethmoid
b. mastoid process
c. maxilla
d. zygomatic

____ 4. The nasal septum is made of ____.

a. the nasal bones
b. the nasal concha and the perpendicular plate of the ethmoid
c. the vomer and the nasal concha
d. the vomer and the perpendicular plate of the ethmoid

____ 5. The roof of the mouth is the ____.

a. mandible
b. maxilla
c. occipital
d. vomer

____ 6. Nasal conchae are ____.

a. lateral to the vomer
b. medial to the perpendicular plate of the ethmoid
c. medial to the vomer
d. both b and c

____ 7. The suture that joins the occipital bone with the parietal bones is called the ____ suture.

a. coronal
b. lambdoidal
c. sagittal
d. squamosal

____ 8. The depression for the pituitary gland is actually a part of the ____ bone.

a. ethmoid
b. frontal
c. occipital
d. sphenoid

____ 9. The sphenoid bone can be seen ____.

a. inside the eye socket
b. inside the nasal cavity
c. on the posterior skull
d. posterior to the temporal bone

_____ 10. The upper jaw is anatomically called the _____ bone.

a. frontal
b. mandible
c. maxilla
d. vomer

_____ 11. The lacrimal bone is located _____.

a. anterior to the ethmoid
b. lateral to the ethmoid
c. medial to the ethmoid
d. posterior to the ethmoid

12. Identify the skull parts in Figure 8–4.

Figure 8–4 Cranial and Facial Skull Bones

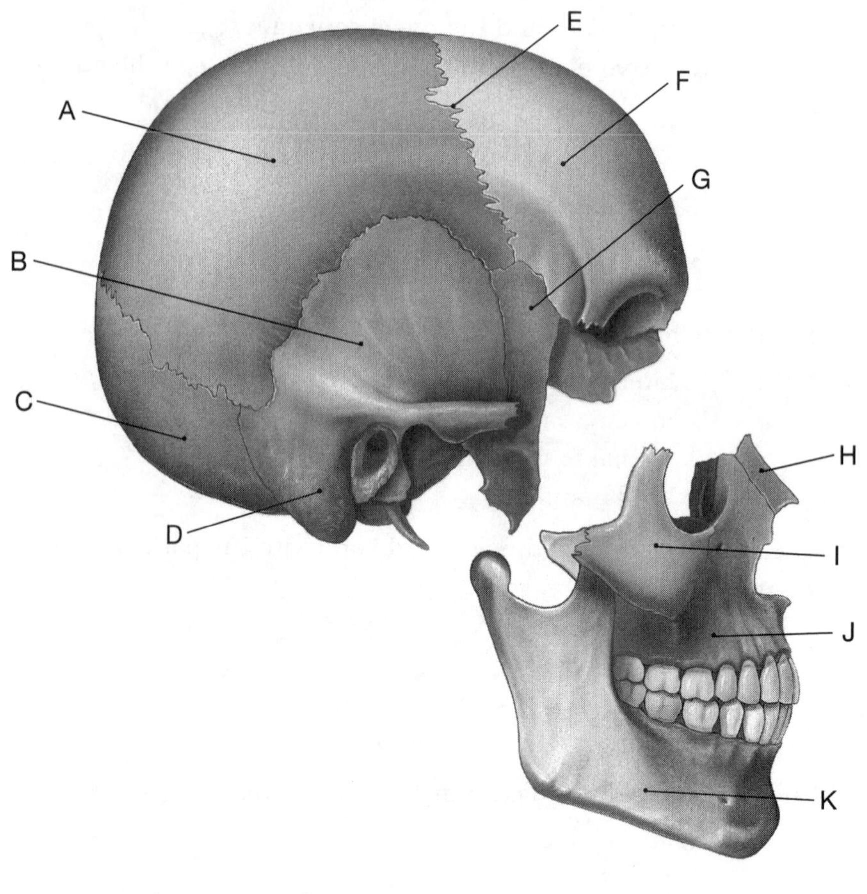

A ____________________ B ____________________

C ____________________ D ____________________

E ____________________ F ____________________

G ____________________ H ____________________

I ____________________ J ____________________

K ____________________

13. In Figure 8–5, identify the various skull parts.

Figure 8–5 Anterior and Inferior Views of the Adult Skull

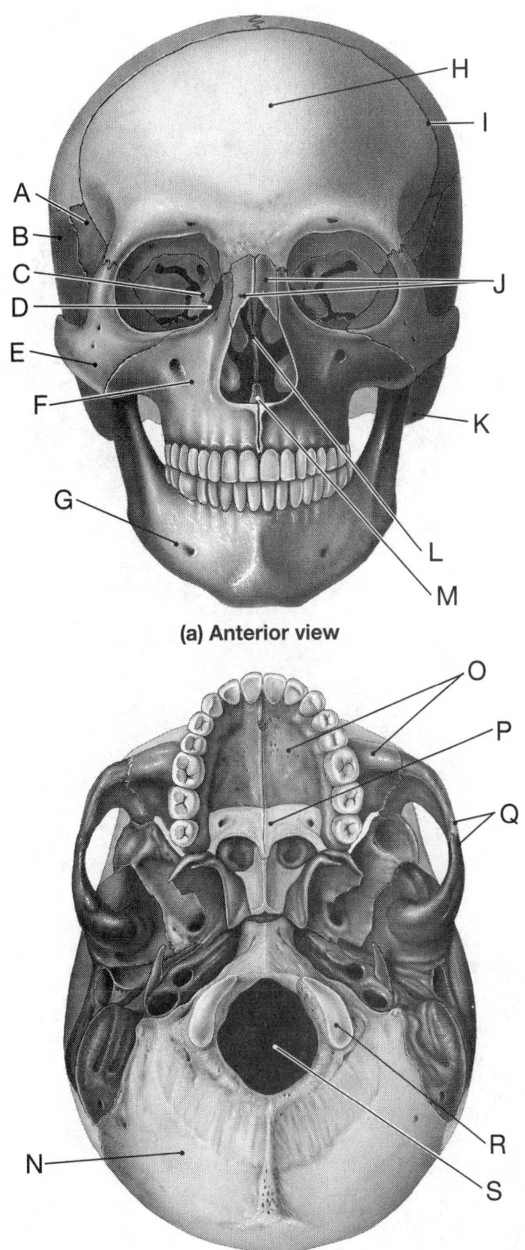

(a) Anterior view

(b) Inferior view

A ____________________ B ____________________ C ____________________

D ____________________ E ____________________ F ____________________

G ____________________ H ____________________ I ____________________

J ____________________ K ____________________ L ____________________

M ____________________ N ____________________ O ____________________

P ____________________ Q ____________________ R ____________________

S ____________________

14. In Figure 8–6, identify the various skull parts.

Figure 8–6 Internal and Sagittal Views of the Adult Skull

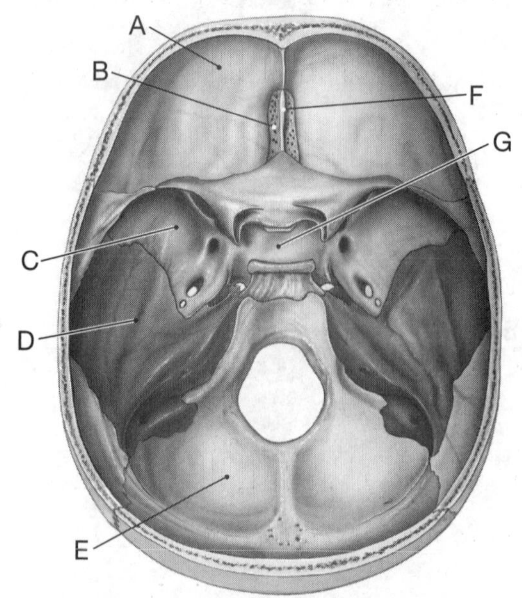

(a) Horizontal section

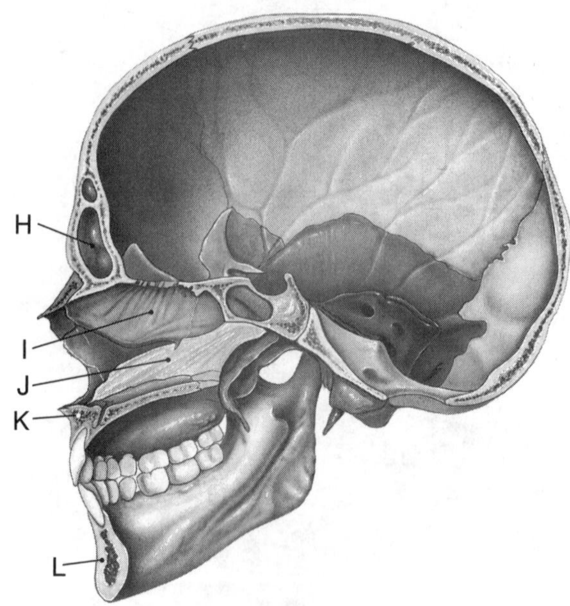

(b) Sagittal section

A ______________________ B ______________________

C ______________________ D ______________________

E ______________________ F ______________________

G ______________________ H ______________________

I ______________________ J ______________________

K ______________________ L ______________________

OBJECTIVE 8: Discuss the differences in structure and function of the various vertebrae (textbook pp. 154–156).

After studying pp. 154–156 in the text, you should be able to identify the five regions of the adult vertebral column, which consists of 26 vertebrae, and the specific function of each region.

_____ 1. There are _____ cervical vertebrae, _____ thoracic vertebrae, and _____ lumbar vertebrae.

a. 5, 7, 12
b. 7, 5, 12
c. 7, 12, 5
d. 12, 7, 5

_____ 2. The ribs are attached to _____.

a. the cervical vertebrae
b. the lumbar vertebrae
c. the thoracic vertebrae
d. some cervical, most thoracic, and some lumbar vertebrae

_____ 3. Look at Figure 8–8 in the textbook. If you were to draw a straight line from C_1 to L_5, the line (representing the center of gravity) would miss the thoracic vertebrae but would pass through the lumbar vertebrae. It would appear that the _____ vertebrae support the bulk of the body's weight.

a. cervical
b. lumbar
c. thoracic
d. all the above

_____ 4. Based on question 3, which vertebrae support the head?

a. cervical
b. lumbar
c. thoracic
d. all of the above

_____ 5. What part of the vertebrae do you feel when you run your thumbnail down the "middle of your back?"

a. body
b. spinous process
c. transverse process
d. superior articulating surfaces

_____ 6. The intervertebral discs are located between the _____ of adjacent vertebrae.

a. bodies
b. spinous processes
c. transverse processes
d. vertebral foramina

_____ 7. How many pairs of ribs are there in the body?

a. 12
b. 14
c. 24
d. The number depends on gender.

_____ 8. The ribs can be placed into two categories, namely, _____.

a. true ribs and costal ribs
b. true ribs and false ribs
c. true ribs and floating ribs
d. Both b and c are correct.

_____ 9. The floating ribs have _____.

a. a direct attachment to the sternum
b. an attachment to the cartilage of the rib superior to it
c. no anterior attachment
d. no posterior attachment

10. List the parts of the sternum from superior to inferior. ______________________

11. In Figure 8–7, identify the cervical vertebrae, thoracic vertebrae, lumbar vertebrae, sacrum, and coccygeal.

Figure 8–7 The Vertebral Column

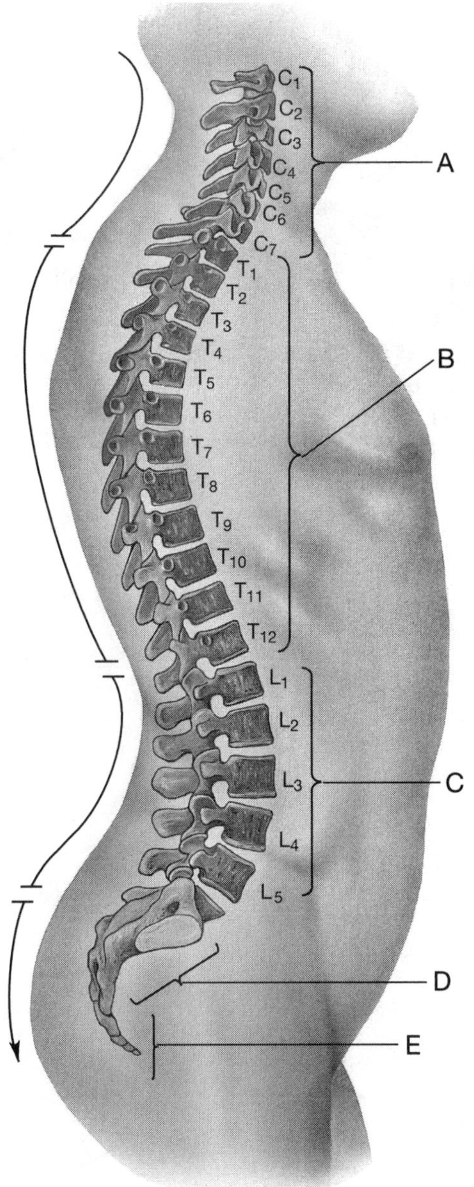

A ______________________ B ______________________

C ______________________ D ______________________

E ______________________

12. In Figure 8–8, identify the vertebral body, vertebral foramen, transverse process, lamina, spinous process, and pedicle.

Figure 8–8 Typical Vertebrae

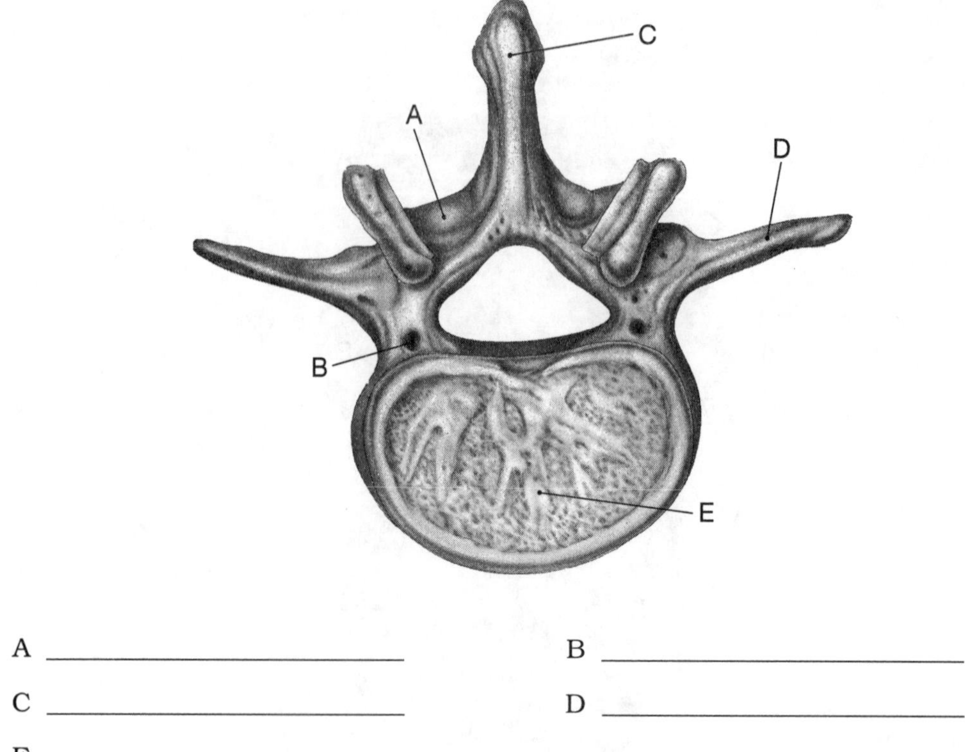

A ______________________ B ______________________

C ______________________ D ______________________

E ______________________

13. In Figure 8–9, identify the true ribs, false ribs, and floating ribs. Indicate how many pairs of true ribs, false ribs, and floating ribs there are. Also, identify the three sternal parts.

Figure 8–9 The Thoracic Cage

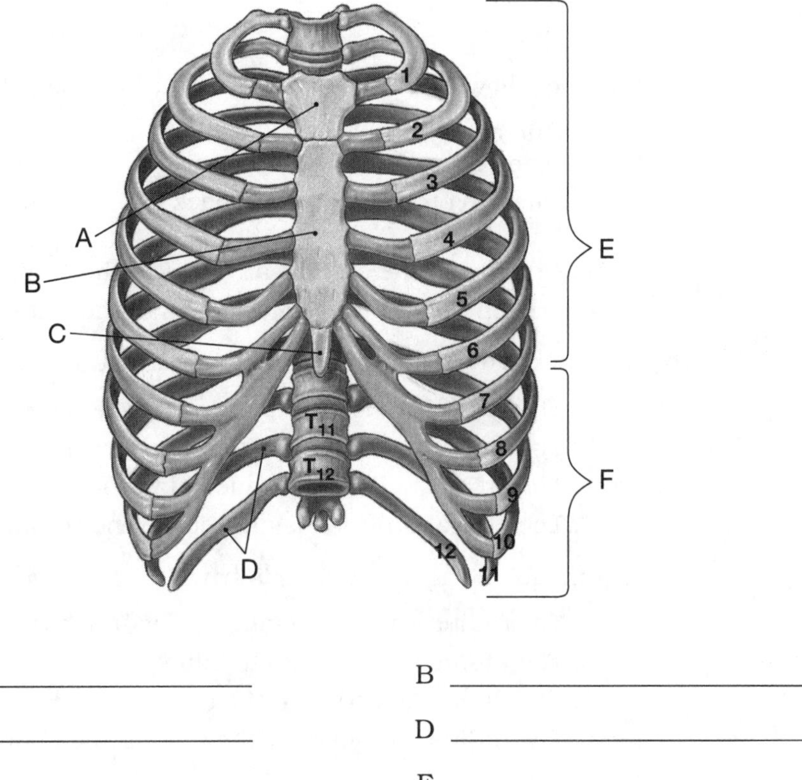

A ______________________ B ______________________

C ______________________ D ______________________

E ______________________ F ______________________

OBJECTIVE 9: Identify the bones of the limbs (textbook pp. 157–162).

After studying pp. 157–162 in the text, you should be able to name the bones of the upper and lower limbs and those of the pectoral and pelvic girdles, which together make up the appendicular skeleton.

_____ 1. The glenoid cavity is a _____ structure of the scapula and is where the _____ attaches.

a. lateral, clavicle
b. lateral, humerus
c. medial, humerus
d. medial, radius

_____ 2. The acromion process is _____ to the coracoid process.

a. anterior
b. lateral
c. medial
d. posterior

_____ 3. Which of the following structures of the scapula is nearest the vertebral column?

a. acrimion process
b. lateral border
c. medial border
d. spine

_____ 4. Which of the following statements about the clavicle is true?

a. One end of the clavicle attaches to the coracoid process of the scapula.
b. One end of the clavicle attaches to the glenoid cavity of the scapula.
c. One end of the clavicle attaches to the acromion.
d. One end of the clavicle attaches to the humerus.

_____ 5. Which of the following statements about the humerus is true?

a. The capitulum is lateral to the trochlea.
b. The capitulum is medial to the trochlea.
c. The greater tubercule is medial to the lesser tubercule.
d. The olecranon fossa is anterior to the coronoid fossa.

_____ 6. Which of the following statements is true?

a. The radius and ulna make up the brachium.
b. The radius is lateral to the ulna.
c. The radius is medial to the ulna.
d. The radius crosses over the ulna in the anatomical position.

_____ 7. What is the anatomical name for the elbow?

a. the ulnar head
b. medial epicondyle of the humerus
c. olecranon process of the ulna
d. the head of the radius

_____ 8. The head of the radius hinges with the _____ side of the ulna at the _____ end.

a. lateral, distal
b. lateral, proximal
c. medial, distal
d. medial, proximal

_____ 9. How many carpal bones are there in each wrist?

a. 5
b. 7
c. 8
d. none of the above

_____ 10. How many phalanges does each hand have?

a. 3
b. 12
c. 14
d. 19

_____ 11. Which carpal hinges with the radius?

a. capitate
b. hamate
c. scaphoid
d. trapezium

_____ 12. Which structure of the pelvic bone provides attachment for the head of the femur?

a. acetabulum
b. ischium
c. obturator foramen
d. pubis

_____ 13. Which of the following statements about the femur is true?

a. The greater trochanter hinges to the acetabulum of the hip.
b. The greater trochanter is lateral to the lesser trochanter.
c. The greater trochanter is located at the distal end of the femur.
d. All the above are true.

_____ 14. The fibula is _____ to the tibia.

a. inferior
b. lateral
c. medial
d. superior

_____ 15. The tibial tuberosity is a(n) _____ structure.

a. anterior
b. lateral
c. medial
d. posterior

_____ 16. How many tarsal bones does each foot have?

a. 3
b. 7
c. 8
d. none of the above

_____ 17. Which tarsal is attached to the tibia?

a. calcaneus
b. cuboid
c. navicular
d. talus

_____ 18. What is the anatomical name for the heel bone?

a. calcaneus
b. cuboid
c. navicular
d. talus

19. Identify the bone structures in Figure 8–10.

Figure 8–10 Anterior View of the Pectoral Girdle; The Bones of the Upper Limb and the Wrist

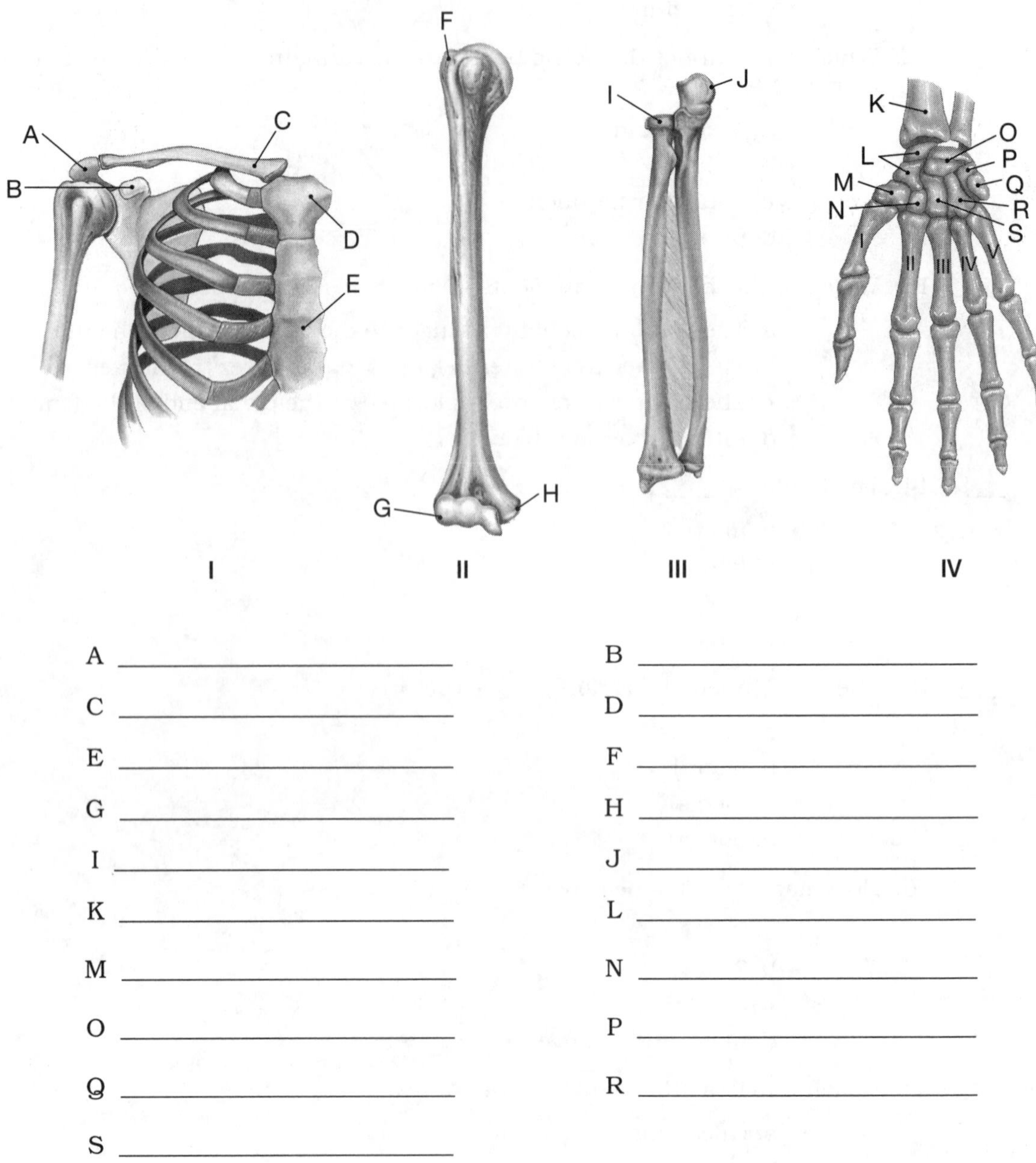

A ______________________ B ______________________

C ______________________ D ______________________

E ______________________ F ______________________

G ______________________ H ______________________

I ______________________ J ______________________

K ______________________ L ______________________

M ______________________ N ______________________

O ______________________ P ______________________

Q ______________________ R ______________________

S ______________________

20. Identify the bone structures in Figure 8–11.

Figure 8–11 Anterior View of the Pelvic Girdle; the Bones of the Lower Limb

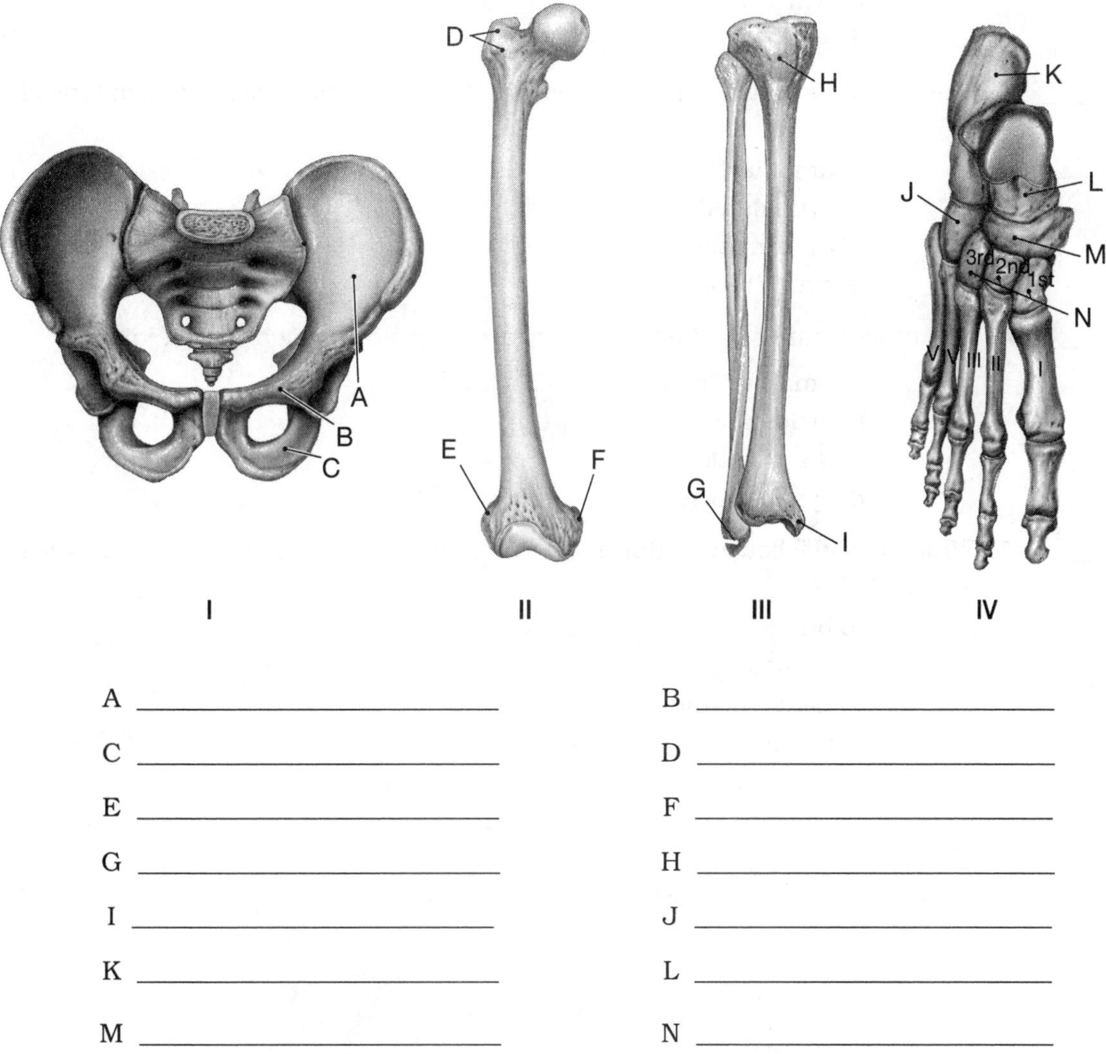

A ______________________ B ______________________

C ______________________ D ______________________

E ______________________ F ______________________

G ______________________ H ______________________

I ______________________ J ______________________

K ______________________ L ______________________

M ______________________ N ______________________

OBJECTIVE 10: Describe the three basic types of joints (textbook pp. 163–164).

After studying pp. 163–164 in the text, you should be able to name the three types of joints and describe how they aid in support and movement of the bones. Some joints provide extreme flexibility but limited support while other joints provide limited movement but extreme support.

_____ 1. The ball-and-socket joint of the humerus and scapula is which of the following types of joints?

a. amphiarthrosis

b. articulated

c. diarthrosis

d. synarthrosis

____ 2. The sagittal suture is an example of which type of joint?

a. amphiarthrosis
b. articulated
c. diarthrosis
d. synarthrosis

____ 3. The joints that contain the intervertebral discs are examples of which type of joint?

a. amphiarthrosis
b. articulated
c. diarthrosis
d. synarthrosis

____ 4. The hinge joint of the knee region is an example of which type of joint?

a. amphiarthrosis
b. articulated
c. diarthrosis
d. synarthrosis

____ 5. Which of the following reduces the rubbing of tendons against other tissues as a joint moves?

a. bursa
b. fat pad
c. joint capsule
d. meniscus

6. In Figure 8–12, identify the bursa, meniscus, patella, joint cavity, and articular cartilage.

Figure 8–12 The Structure of a Synovial Joint

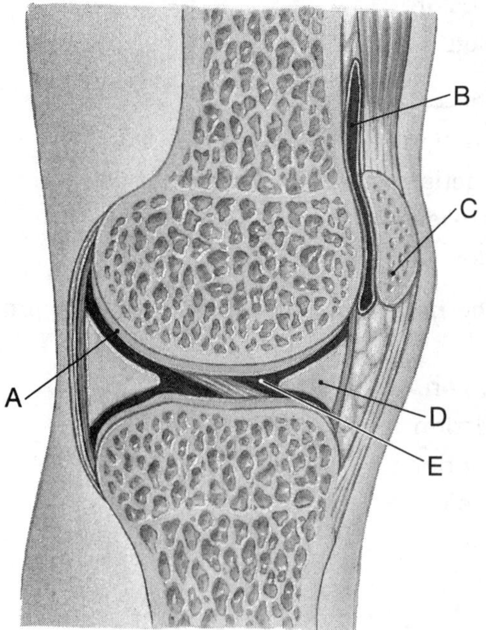

A ____________________ B ____________________

C ____________________ D ____________________

E ____________________

OBJECTIVE 11: Relate the body movements to the action of specific joints (textbook pp. 165–168).

After studying pp. 165–168 in the text, you should be able to describe accurately the movements of different parts of the skeleton using proper terms and explain the various ranges of motion permitted by each type of joint.

_____ 1. Movement of the arm laterally from the body is called _____.

a. abduction
b. adduction
c. extension
d. flexion

_____ 2. Lifting of the lower arm so that it moves only at the elbow is called _____.

a. abduction
b. adduction
c. extension
d. flexion

____ 3. To do the splits, the initial movement of the legs would be ____.

a. abduction
b. adduction
c. extension
d. flexion

____ 4. Stooping over is ____ of the torso.

a. abduction
b. adduction
c. extension
d. flexion

____ 5. Movement of the hand from palm facing up to palm facing down is called ____.

a. eversion
b. inversion
c. pronation
d. supination

6. In Figure 8–13, identify the following types of movements: flexion, extension, abduction, adduction, supination, and pronation.

Figure 8–13 Body Movements

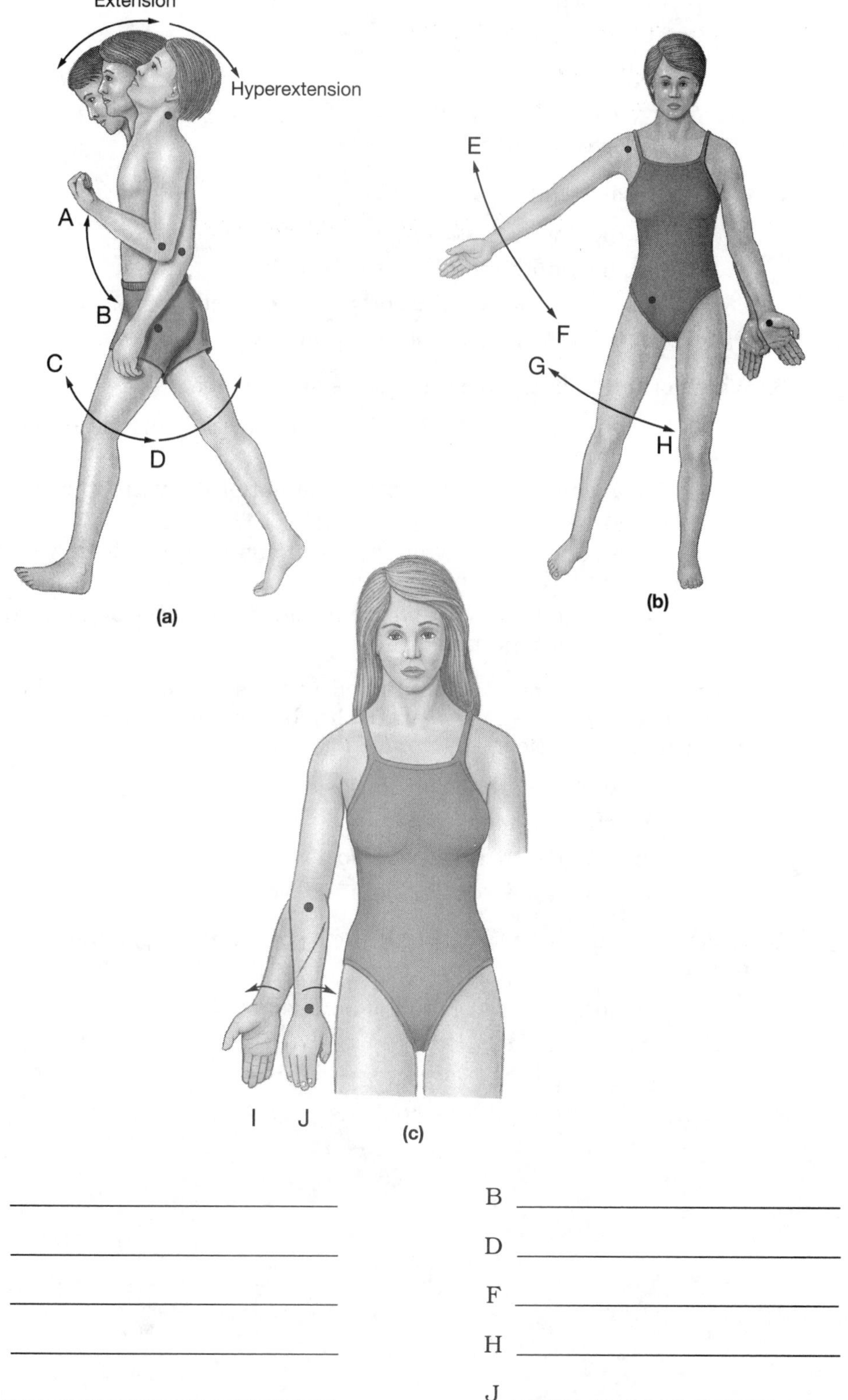

A ____________________ B ____________________

C ____________________ D ____________________

E ____________________ F ____________________

G ____________________ H ____________________

I ____________________ J ____________________

OBJECTIVE 12: Describe the different causes of bone disorders and give examples of each (textbook pp. 168–176).

After studying pp. 168–176 in the text, you should be able to name and describe several bone disorders.

____ 1. Which of the following is the name of the condition in which there is excessive cartilage formation at the epiphyseal plate?

a. achondroplasia
b. Marfan's syndrome
c. osteogenesis imperfecta
d. osteomyelitis

____ 2. Cleft palate is a condition in which ____.

a. the lingual frenulum does not form properly
b. the mandibular bones do not fuse together
c. the maxillary bones do not fuse together
d. the vomer does not fuse properly

____ 3. Which of the following best describes spina bifida?

a. a condition in which the body of the vertebrae does not form properly
b. a condition in which the lamina of the vertebrae does not form properly
c. a condition in which the spinous process of the vertebrae does not form properly
d. a condition in which the transverse process of the vertebrae does not form properly

____ 4. There are three general abnormal curvatures of the spine. A lateral curvature is called ____, an exaggerated lumbar curve is called ____, and an exaggerated thoracic curve is called ____.

a. kyphosis, scoliosis, lordosis
b. lordosis, scoliosis, kyphosis
c. scoliosis, kyphosis, lordosis
d. scoliosis, lordosis, kyphosis

____ 5. Rickets is a condition in which the long bones, under pressure, begin to bend due to ____.

a. abnormal formation of bone matrix
b. osteomalacia
c. osteoporosis
d. overactivity of osteoblasts

OBJECTIVE 13: Distinguish between degenerative and inflammatory arthritic diseases (textbook pp. 177–179).

There are numerous arthritic conditions of the joints which are discerned by their origin. After studying pp. 177–179 in the text, you should be able to describe the differences between degenerative and inflammatory arthritis.

____ 1. Some osteosarcomas can cause severe osteoporosis because ____.

a. the cancer cells may increase the activity of osteoclasts
b. the cancer cells may increase the activity of osteoblasts
c. the cancer cells may inhibit the uptake of calcium for bone usage
d. the cancer cells affect hormones that regulate bone development

____ 2. Arthritis that originates with a pathogen is called ____.

a. degenerative arthritis
b. inflammatory arthritis
c. rheumatoid arthritis
d. synovial arthritis

____ 3. Arthritis that is due to the breakdown of cartilage within the joints is called ____.

a. degenerative arthritis
b. inflammatory arthritis
c. osteoarthritis
d. gout

____ 4. Which of the following arthritic conditions is an autoimmune disease?

a. ankylosis
b. gout
c. rheumatic fever
d. rheumatoid

____ 5. Arthritis can be very painful so patients may restrict their movements. However, this restriction may result in abnormal fusion of the bones called ____.

a. ankylosis
b. gout
c. osteoarthritis
d. synovitis

____ 6. As the cartilage within the joints begin to wear down due to age, friction at the joints may occur. Which of the following best describes this condition?

a. Degenerative arthritis has occurred, which may lead to inflammatory arthritis at the joint.
b. Inflammatory arthritis has occurred, which will lead to degenerative arthritis.
c. Osteoporosis has occurred, which will lead to degenerative arthritis.
d. Osteoporosis has occurred, which will lead to inflammatory arthritis.

PART II: CHAPTER-COMPREHENSIVE EXERCISES

Match the terms in column A with the descriptions or statements in column B.

MATCHING I

	(A)	(B)
_____	1. appendicular	A. The shaft of the bone
_____	2. adduction	B. The ends of the bone
_____	3. axial	C. The division of the skeleton that includes the skull
_____	4. coronal	D. The division of the skeleton that comprises the limbs
_____	5. carpals	E. The suture that exists between the frontal bone and the two parietal bones
_____	6. clavicle	F. The vertebrae that supports the weight of the body
_____	7. diaphysis	G. The anatomical name for the collarbone
_____	8. epiphysis	H. The bone making up the brachium
_____	9. humerus	I. The collective name for the wrist bones
_____	10. lumbar	J. Movement of the limbs toward the long axis of the body
_____	11. rheumatoid arthritis	K. Inflammation of the joints
_____	12. rheumatic fever	L. Inflammation of the joints due to an automatic disorder

MATCHING II

	(A)	(B)
_____	1. Marfan's syndrome	A. Bone-remodeling cells
_____	2. menisci	B. Bone-forming cells
_____	3. osteoblast	C. Cartilage pads associated with the knee joint
_____	4. osteoclast	D. Concentric rings of osteocytes
_____	5. osteon	E. The inferior bone of the nasal septum
_____	6. Paget's disease	F. The vertebrae that have rib attachments
_____	7. tarsals	G. The spinal cord passes through this area of the vertebrae
_____	8. thoracic	H. The inferior portion of the sternum
_____	9. tibia	I. The collective name for the ankle bones
_____	10. vertebral foramen	J. The medial bone of the lower leg
_____	11. vomer	K. Accelerated osteoclast activity
_____	12. xiphoid	L. Accelerated cartilage formation

CONCEPT MAP

This concept map is a review of Chapter 8. Use the following terms to complete the map by filling in the boxes identified by circled numbers, 1–12.

axial	**lacunae**	**lumbar**
osteons	**pectoral girdle**	**sternum**
support	**tibia**	**true**
osteocytes	**osteosarcoma**	**myeloma**

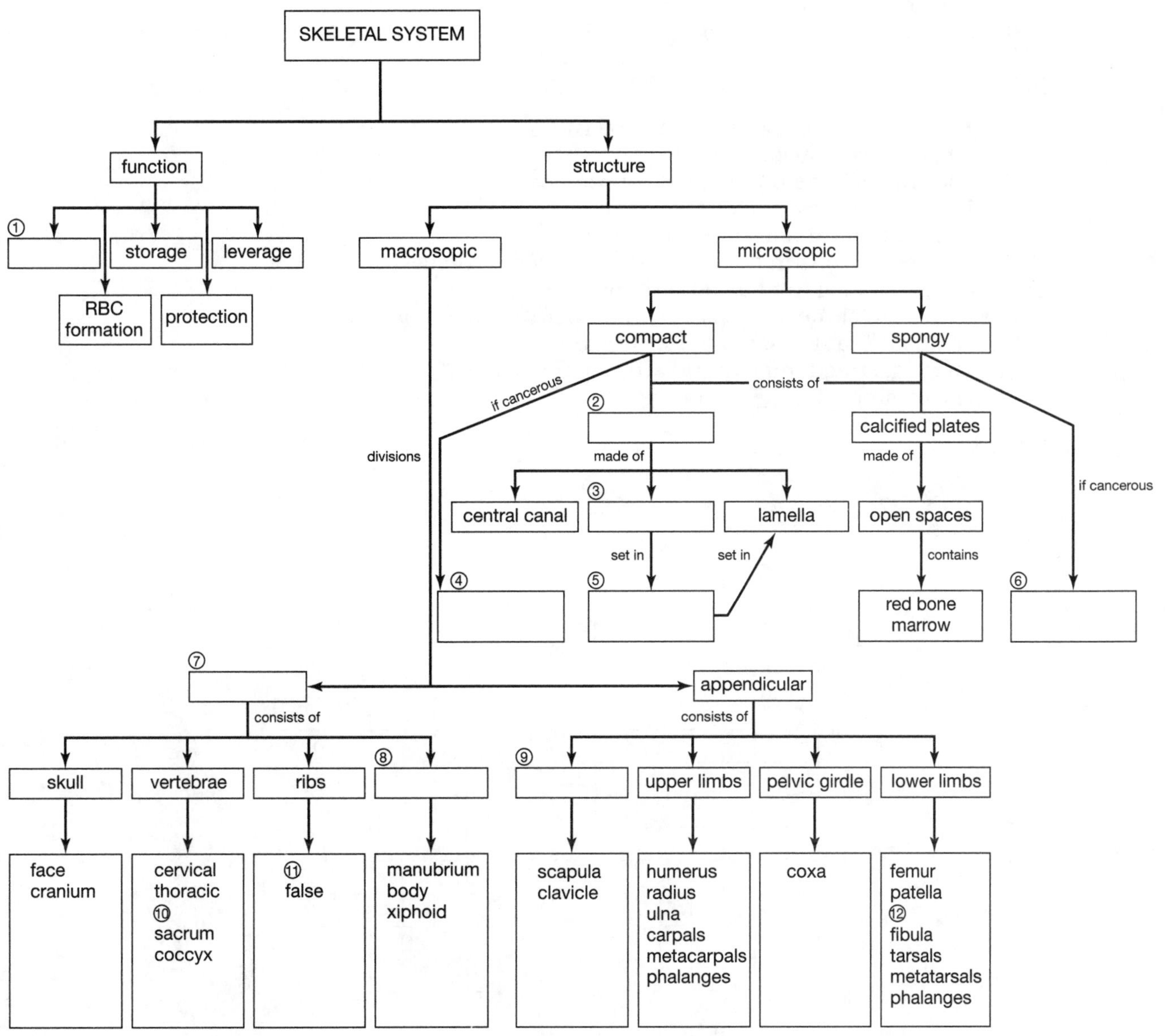

CROSSWORD PUZZLE

The following puzzle reviews the material in Chapter 8. To complete the puzzle, you have to know the answers to the clues given, and you must be able to spell the terms correctly.

ACROSS

6. This term refers to the shaft of the bone.
7. A malformation of the _____ can result in spina bifida.
8. Tumor within the epiphyseal plate
10. The cavity (fossa) that serves as the "socket" into which the femur fits
13. The cavity (fossa) that serves as the "socket" of the scapula
16. The fusion of articulating surfaces resulting in a lack of movement
17. This term refers to the end of a bone.
18. Because osteoclasts remodel bone, they are probably responsible for creating _____.

DOWN

1. The most inferior portion of the sternum
2. A baby's "soft spots"
3. The area of bone that is made of osteons
4. The bone-forming cells
5. The anatomical name for the heel bone
7. The depressions in the osteons in which the osteocytes set
9. Cancer arising from blood-forming cells in bone
11. A fracture of the _____ plate can hinder the growth of bones.
12. The fluid that is associated with many joints
14. An exaggerated lumbar curvature of the spine
15. The capitate is a _____ bone.

FILL-IN-THE-BLANK NARRATIVE

Use the following terms to fill in the blanks of this summary of Chapter 8.

acromion	**homeostasis**	**synarthrosis**
sagittal	**attachments**	**osteoblast**
supinate	**storage**	**extend**
osteoclast	**tendons**	**calcium**
lambdoidal	**osteoporosis**	**diarthrosis**
movement	**pronated**	**greater trochanter**
greater tubercle	**olecranon process**	**red marrow**
red blood	**tibial tuberosity**	**transverse processes**
arthritis	**rheumatoid**	**cartilage**

Before bony tissue forms in the embryo, all supporting tissues are made of cartilage. (1) ___________ cells gradually replace the cartilage with bone. (2) ___________ cells erode the inner surface of the bone and create the marrow cavity. A balance between the actions of these two types of cells results in (3) ___________. A decline of osteoblast activity with age coupled with normal osteoclast activity may lead to (4) ___________.

Healthy bones provide protection, leverage, (5) ___________, and support. Bones are dynamic tissues that maintain homeostasis by providing the body with (6) ___________ ions and (7) ___________ for (8) ___________ cell production. Leverage permits (9) ___________. Bones move as a result of muscle contractions. Muscles attach to bone via (10) ___________. These structures attach to bones at various bony projections. Some major sites of muscle (11) ___________ are the (12) ___________ at the humerus, the (13) ___________ at the ulna, the coracoid process and the (14) ___________ at the scapula, the (15) ___________ at the femur, the (16) ___________ at the tibia, and the (17) ___________ at the vertebrae.

In order to move, the bones must have moveable joints. The bones are held together at these joints by ligaments. The joints at the elbow, knee, hip, and shoulder are freely moveable. This type of joint is a(n) (18) ___________. Some joints are designed for strength and do not move, such as the (19) ___________ and the (20) ___________ sutures of the skull. This type of joint is a(n) (21) ___________. (22) ___________ is a name given to a variety of joint inflammations. These inflammations can hinder movement at the joints. (23) ___________ arthritis is an autoimmune disorder of the joints that affects the (24) ___________, thus causing crippling.

For every movement there is an antagonistic movement. For example, to return a flexed forearm to its original position, we must (25) ___________ it. If a hand is in the (26) ___________ position, we cannot hold a cup of soup. However, if we (27) ___________ the hand, we can then hold a cup of soup.

CLINICAL CONCEPT

The following clinical concept applies the information in Chapter 8. Following the application is a set of questions to help you understand the concept.

Christine is a very weight conscious person who exercises every day and watches her diet very closely, and she is pregnant. Christine knows the importance of eating a well-balanced diet for herself and for her unborn child, but she is concerned about putting on excess weight during the pregnancy. The doctor encourages Christine to continue eating properly to ensure the survival of her unborn child, but she restricts her diet in hopes of not gaining too much weight.

At a routine prenatal appointment, the doctor runs a few tests and determines that Christine's bones are beginning to deteriorate. The doctor explains that the unborn child is developing bone material, for which it needs calcium. Because Christine has limited her diet, she is not supplying enough calcium for her unborn child to use for making bone, so it is obtaining calcium from Christine's bones, which are consequently beginning to deteriorate. The human body does whatever it can to ensure the survival of the unborn child. Christine agrees to follow closely the doctor's recommendations for her diet.

1. Identify one place where calcium is stored in the body.

2. What mineral do osteoblasts need in order to replace cartilage with bone in the developing fetus?

3. Why do doctors often prescribe calcium supplements to pregnant women?

4. Why were Christine's bones beginning to deteriorate during her pregnancy?

5. Due to the excessive loss of calcium, Christine's bones were becoming porous. This condition is called ____________.

THIS CONCLUDES CHAPTER 8 EXERCISES

CHAPTER 9

The Muscular System

Human life as we know it would cease to exist without muscle tissue. Sitting, standing, walking, speaking, and grasping objects would be impossible. Blood would not circulate, the lungs would cease to function, and food could not move through the digestive tract. Many of our physiological processes, and virtually all our dynamic interactions with the environment, involve muscle tissue.

The body consists of three types of muscle tissue: (1) **cardiac,** which is found in the heart; (2) **smooth,** which forms a substantial part of the walls of hollow organs; and (3) **skeletal,** which is attached to bones. Smooth and cardiac muscles are involuntary and they are primarily responsible for transport of materials within the body. Skeletal muscles are *voluntary* and are the main focus of this chapter. These muscles are responsible for more than just body movement. They are indirectly related to heat regulation and protection. A select number of skeletal muscles are included in this chapter for study. A great deal of memorization is necessary to master this material. Arranging the information in an organized way and learning to recognize characteristics and clues will make the task easier to manage and will facilitate the memorization process.

The muscle disorders discussed in this chapter affect numerous muscles. Some disorders are due to infections, and some are associated with immune malfunctions. Muscle disorders can make some simple tasks such as sitting and walking difficult and limited at best.

apter Objectives:

- Describe the properties and functions of muscle tissue.
- Describe the organization of skeletal muscle.
- Describe the structure of a sarcomere.
- Describe the sliding-filament model of muscle contraction.
- Describe the relationship between muscle contractions and motor units .
- Contrast isotonic and isometric muscle contractions.
- Distinguish between aerobic and anaerobic exercise.
- Identify the major axial muscles of the body.
- Identify the major appendicular muscles of the body.
- Describe the effects of exercise and aging on muscle tissue.
- Describe the two common symptoms of muscle disorders.
- Describe the major disorders of the muscular system.

PART I: OBJECTIVE-BASED QUESTIONS

OBJECTIVE 1: Describe the properties and functions of muscle tissue (textbook p. 187).

After studying p. 187 in the text, you should be able to describe the three types of muscle tissue in the human body and explain the ability to contract, which is a common feature of all of them.

_____ 1. Which of the following is not a *function* of muscle tissue but is a *property* of muscle tissue?

a. excitability
b. maintains body temperature
c. maintains posture
d. produces movement

_____ 2. The ability to respond to stimuli is which property of muscle?

a. contractibility
b. elasticity
c. excitability
d. production of movement

_____ 3. When muscles contract, they _____.

a. extend
b. shorten
c. stretch
d. both a and c

_____ 4. Which of the following muscle types pulls on various body parts when it contracts?

a. cardiac
b. skeletal
c. smooth
d. both b and c

OBJECTIVE 2: Describe the organization of skeletal muscle (textbook pp. 187–189).

After studying pp. 187–189 in the text, you should be able to describe the molecular and cellular level of organization of individual muscles.

_____ 1. Several fasciculi make a _____.

a. muscle body
b. muscle fiber
c. myofibril
d. sarcomere

_____ 2. Several muscle fibers make a _____.

a. muscle body
b. muscle fascicle
c. myofibril
d. sarcomere

_____ 3. The structure that is made of actin and myosin is the _____.

a. muscle body
b. muscle fiber
c. myofibril
d. sarcomere

_____ 4. Which one of the following structures is multinucleated?

a. muscle body
b. muscle fiber
c. myofibril
d. sarcomere

_____ 5. Which one of the following structures is responsible for the contraction of a muscle?

a. actin and myosin
b. muscle fascicle
c. muscle fiber
d. none of the above

6. In Figure 9–1, identify the muscle fascicle, muscle fiber, and myofibril.

Figure 9–1 Organization of a Skeletal Muscle

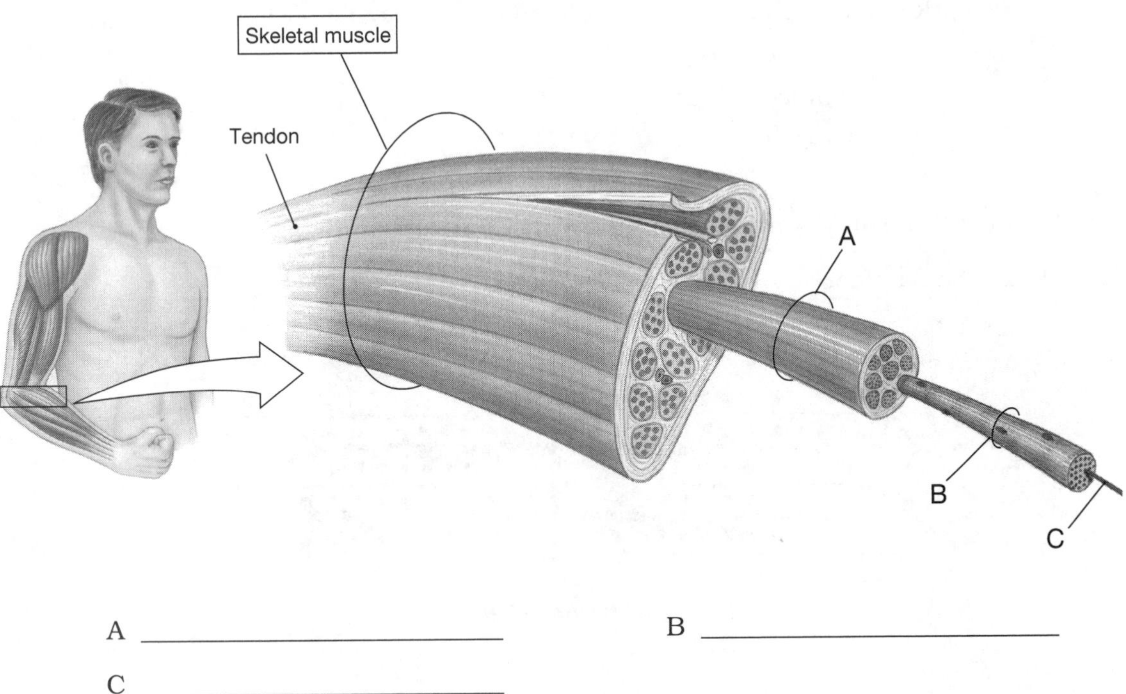

A ______________________ B ______________________

C ______________________

OBJECTIVE 3: Describe the structure of a sarcomere (textbook p. 189).

After studying p. 189 in the text, you should be able to describe the structure of a sarcomere, the basic functional unit of muscle, and explain how muscles contract.

_____ 1. The thick filaments of a sarcomere are _____.

a. actin
b. myofibrils
c. myosin
d. both b and c

_____ 2. The thin filaments of a sarcomere are _____.

a. actin
b. myofibrils
c. myosin
d. both a and c

_____ 3. Cross-bridges are a part of _____.

a. actin
b. myofibrils
c. myosin
d. fascicles

_____ 4. Which of the following structures is responsible for the alternating light and dark bands seen in skeletal muscles under the microscope?

a. muscle cells
b. muscle fiber
c. myofibril
d. sarcomere

5. In Figure 9–2, identify actin, myosin, the z line, and cross-bridges.

Figure 9–2 The Sarcomere

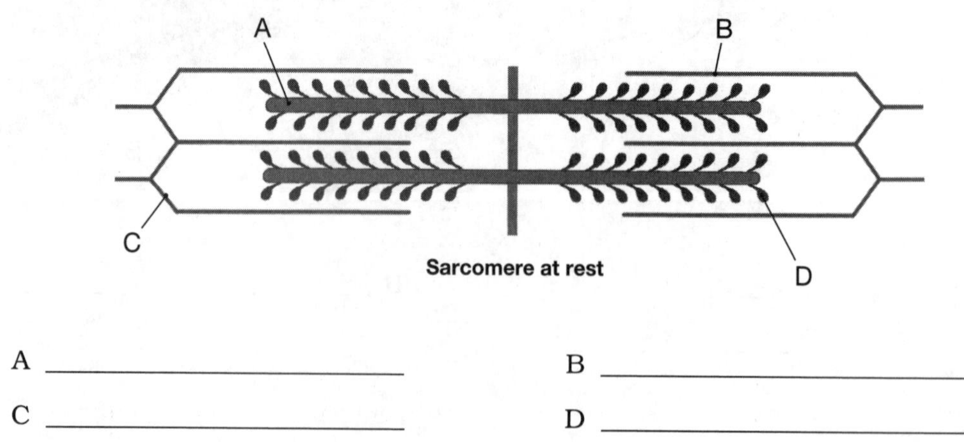

A ______________________ B ______________________

C ______________________ D ______________________

OBJECTIVE 4: Describe the sliding-filament model of muscle contraction (textbook pp. 189–190).

After studying pp. 189–190 in the text, you should be able to explain muscle contraction based on the sliding filament model.

____ 1. A muscle contracts when the ____.

a. actin filaments slide toward each other
b. free ends of actin come closer to each other
c. free ends of actin get farther away from each other
d. both a and b

____ 2. When a muscle contracts, which of the following remains stationary?

a. cross-bridges
b. myosin
c. sarcomere
d. free ends of actin

____ 3. Which of the following is the correct description of muscle contraction?

a. A nerve impulse causes the cross-bridges to attach to the actin filaments. ATP then causes the cross-bridges to pivot, thus sliding the actin filaments toward each other. The muscle then shortens.
b. A nerve impulse causes the muscle cell's endoplasmic reticulum to release calcium ions. The calcium ions indirectly cause the cross-bridges to attach to the actin filaments. Powered by the breakdown of ATP, the cross-bridges pivot, thus sliding the actin filaments toward each other. The muscle then shortens.
c. A nerve impulse causes the muscle cell's endoplasmic reticulum to release calcium ions. The calcium ions indirectly cause the cross-bridges to attach to the myosin filaments. Powered by the breakdown of ATP, the cross-bridges pivot, thus sliding the myosin filaments toward each other. The muscle then shortens.
d. A nerve impulse causes the muscle cell's endoplasmic reticulum to release calcium ions. The calcium ions indirectly cause the cross-bridges to attach to the actin filaments. Powered by the breakdown of ATP, the cross-bridges pivot, thus sliding the actin filaments toward each other. The muscle then extends.

____ 4. A (an) ____ will cause the release of calcium ions from the ____ of the muscle fiber. ____

a. nerve impulse, endoplasmic reticulum
b. nerve impulse, cytoplasm
c. muscle contraction, endoplasmic reticulum
d. ATP molecule, cytoplasm

____ 5. In order for a muscle to relax, ____.

a. the calcium ions must go back to normal levels
b. the cross-bridges must detach from the actin filaments
c. the nerve impulse must stop
d. all the above

6. In Figure 9–3, identify actin, myosin, cross-bridges, a contracted sarcomere, and a relaxed sarcomere.

Figure 9–3 Sarcomere Contractions and Relaxation

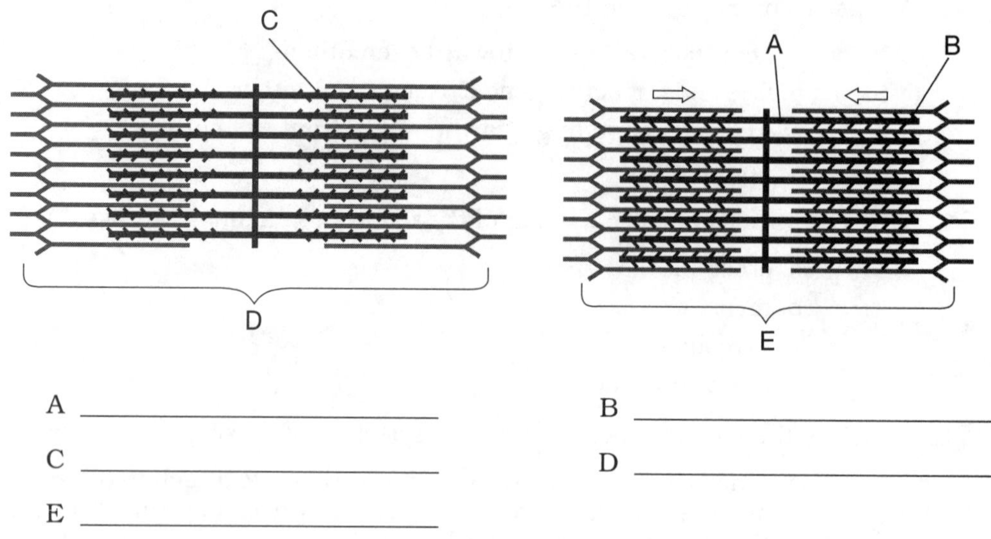

A ____________________ B ____________________

C ____________________ D ____________________

E ____________________

OBJECTIVE 5: Describe the relationship between muscle contractions and motor units (textbook pages 190–191).

After studying pp. 190–191 in the text, you should be able to explain how the strength and duration of muscle contraction depends on nerve impulses from the motor neurons.

_____ 1. A motor unit is _____.

a. a muscle fiber
b. a muscle fiber or fibers that are controlled by a single motor neuron
c. a neuromuscular junction
d. the entire muscle that is controlled by a single motor neuron

_____ 2. Which of the following provides the muscle with the best control?

a. a motor neuron that causes an entire muscle to contract
b. a motor neuron that controls several muscle fibers
c. a motor neuron that controls very few muscle fibers
d. a muscle fiber that is controlled by several neurons

_____ 3. A muscle is able to be contracted to maximum contraction because _____.

a. several motor units are activated at one time
b. several neurons are activated at one time
c. a stronger impulse is created going to a muscle
d. a single neuron is fully activated

_____ 4. Which of the following statements is true?

a. Nerve impulses to the motor units can cause a muscle to relax.
b. When a muscle is creating movement, the muscle is actually creating muscle tone.
c. When a muscle is relaxed, all the motor units for that muscle are "turned off."
d. When a muscle is relaxed, not all the motor units are "turned off."

5. In Figure 9–4, identify the nerve, the muscle, and the motor unit.

Figure 9–4 A Neuromuscular Junction

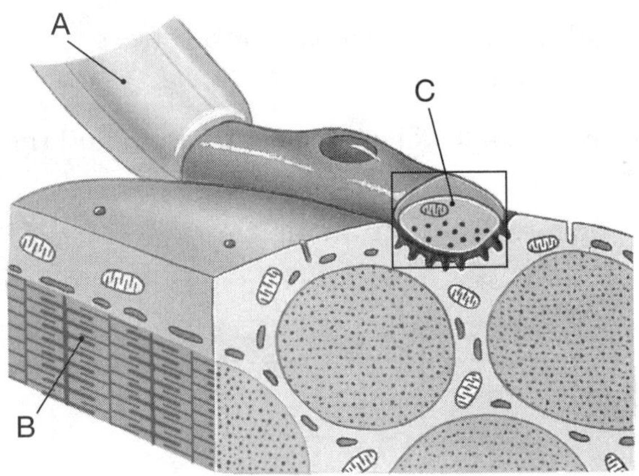

A ______________________ B ______________________

C ______________________

OBJECTIVE 6: Contrast isotonic and isometric muscle contractions (textbook p. 193).

Our day to day activities involve isotonic and isometric contractions. After studying p. 193 in the text, you should be able to distinguish between isotonic and isometric contractions and how we use each.

____ 1. An isotonic contraction _____.

a. does not generate movement
b. generates movement
c. maintains constant tension as the muscle contracts
d. both b and c

____ 2. An isometric contraction _____.

a. does not generate movement
b. generates movement
c. maintains constant tension as the muscle contracts
d. both b and c

____ 3. Which of the following is an example of an isotonic contraction?

a. lifting a chair off the floor
b. trying to lift a weight that is too heavy to budge
c. both a and b
d. neither a nor b

____ 4. Which of the following is an example of an isometric contraction?

a. lifting a chair off the floor
b. trying to lift a weight that is too heavy to budge
c. both a and b
d. neither a nor b

_____ 5. Which of the following is an example of an isotonic contraction?

a. lifting a glass of water to your mouth to drink
b. not being able to push a car down the street even though you've tried
c. steadily pushing a car down the street
d. both a and c

6. In Figure 9–5, identify the example of isotonic contraction and the example of isometric contraction.

Figure 9–5 Isotonic and Isometric Concentrations

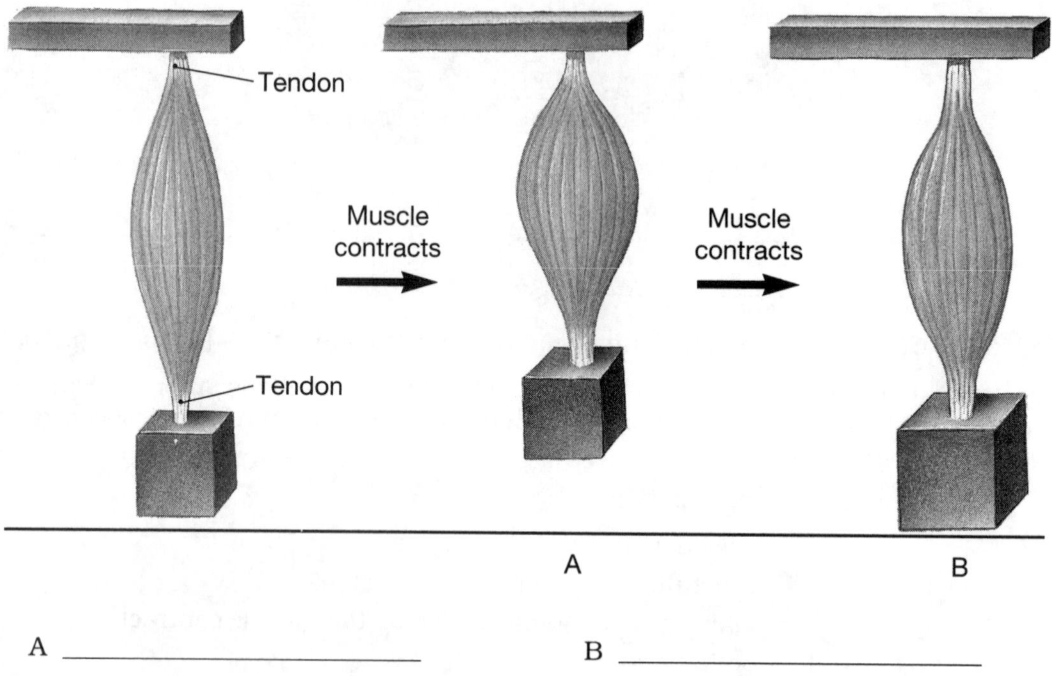

A ______________________ B ______________________

OBJECTIVE 7: Distinguish between aerobic and anaerobic exercise (textbook p. 193).

After studying p. 193 in the text, you should be able to describe the differences between aerobic and anaerobic energy production.

_____ 1. When muscles increase in size, as in the case of bodybuilders, the increase is due to _____.

a. an increase in muscle cells
b. an increase in muscle fibers
c. an increase in muscle myofibrils
d. all the above

_____ 2. _____ exercise requires oxygen and is of _____ duration than _____ exercise, which does not require oxygen.

a. Anaerobic, longer, aerobic
b. Aerobic, longer, anaerobic
c. Anaerobic, shorter, aerobic
d. Aerobic, shorter, anaerobic

_____ 3. ATP is produced for muscle usage via _____.

a. aerobic metabolism
b. anaerobic metabolism
c. glycolysis
d. all the above

_____ 4. Which of the following statements is true?

a. If you breathe properly while exercising, your muscles will be able to generate ATP via aerobic metabolism.
b. If you do not breathe properly while exercising, your muscles will not generate ATP for muscle contractions and you will develop muscle fatigue.
c. If you do not breathe properly while exercising, your muscles will rely on ATP production from anaerobic metabolism.
d. Both a and c are correct.

_____ 5. Lactic acid production causes muscle fatigue and muscle soreness. Lactic acid is produced _____.

a. during aerobic metabolism
b. during anaerobic metabolism
c. during aerobic and anaerobic metabolism
d. during neither aerobic nor anaerobic metabolism

_____ 6. Why does a runner breathe heavily after having just finished a race?

a. The body is trying to accumulate oxygen that is needed to produce ATP to convert lactic acid back to glucose.
b. The body is trying to accumulate oxygen to repay the oxygen debt created by anaerobic exercise.
c. The body is trying to accumulate oxygen in an effort to get rid of the lactic acid that caused muscle fatigue.
d. all the above

OBJECTIVE 8: Identify the major axial muscles of the body (textbook pp. 194–201).

After studying pp. 194–201 in the text, you should be able to name and identify the muscles associated with the axial skeleton.

_____ 1. The orbicularis oris muscle _____.

a. closes the lips
b. makes you smile
c. opens the lips
d. all the above

_____ 2. Which of the following muscles is attached to the zygomatic process?

a. masseter
b. orbicularis oris
c. zygomaticus
d. both a and c

_____ 3. Which of the following muscles allows straightening up from a stooped-over position?

a. rectus abdominis
b. quadratus lumborum
c. sacrospinalis
d. all the above

_____ 4. Which of the following axial muscles is (are) involved in the breathing process?

a. diaphragm
b. external intercostals
c. internal intercostals
d. all the above

_____ 5. The muscle that opposes the erector spinae is the _____.

a. external oblique
b. rectus abdominis
c. spinalis
d. sacrospinalis

6. Identify the muscles labeled in Figure 9–6.

Figure 9–6 Muscles of the Axial Skeleton

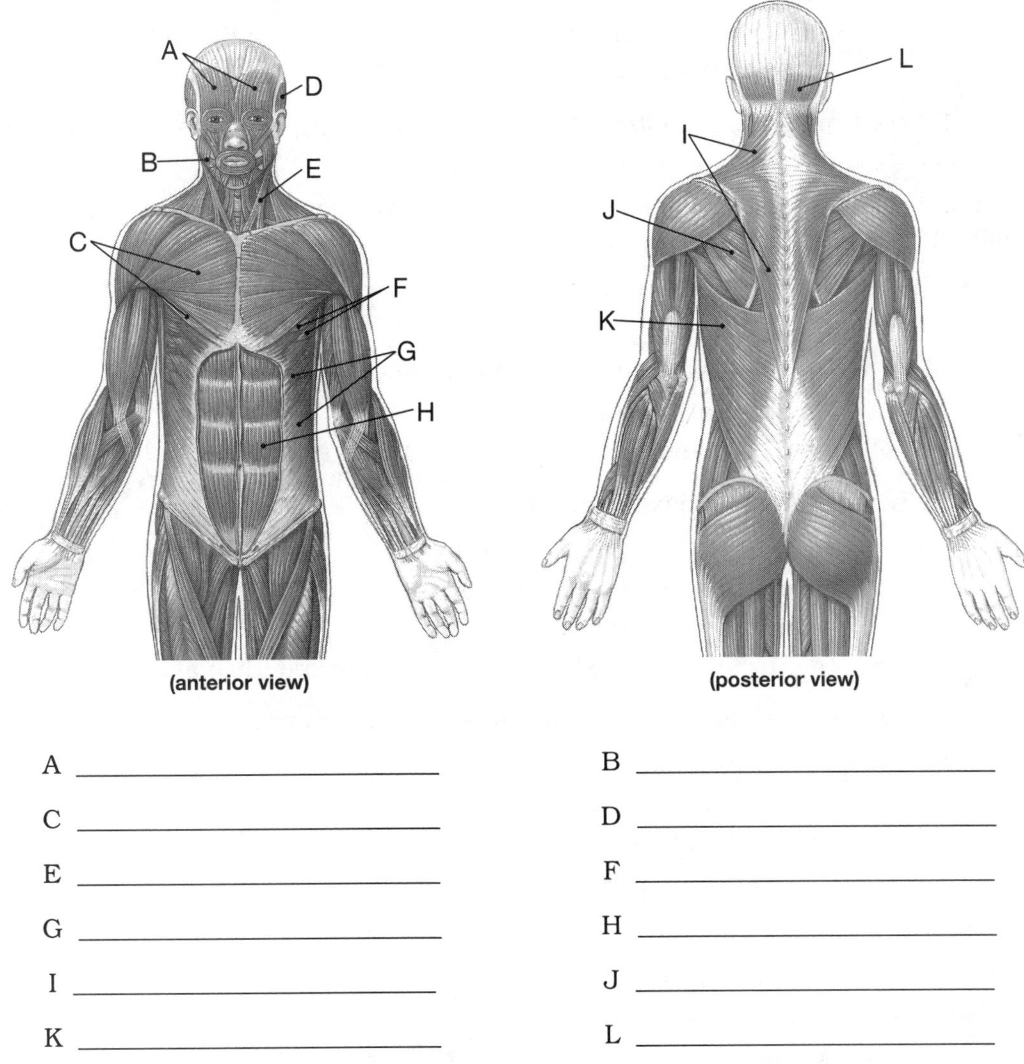

A ______________________ B ______________________

C ______________________ D ______________________

E ______________________ F ______________________

G ______________________ H ______________________

I ______________________ J ______________________

K ______________________ L ______________________

OBJECTIVE 9: Identify the major appendicular muscles of the body (textbook pp. 202–205).

After studying pp. 202–205 in the text, you should be able to name and identify the appendicular muscles, which include the muscles of the pectoral girdle and the upper limbs, the muscles of the pelvic girdle and the lower limbs, and the muscles of the torso.

_____ 1. The major chest muscle is called the _____.

a. pectoralis major
b. serratus anterior
c. teres major
d. trapezius

_____ 2. The major muscle covering the costal region is the _____.

a. latissimus dorsi
b. pectoralis major
c. rectus abdominis
d. trapezius

_____ 3. Which of the following muscles are referred to as the rotator cuff muscles?

a. deltoid and teres major
b. extensor digitorum and palmaris longus
c. flexor carpi ulnaris, flexor carpi radialis, and brachioradialis
d. infraspinatus and teres major

4. Which of these two muscles is lateral to the other: flexor carpi ulnaris; flexor carpi radialis? ______________________

5. Which of these two muscles is medial to the other: extensor digitorum; extensor carpi radialis? ______________________

_____ 6. Which of the following muscles flexes the antebrachium?

a. biceps brachii
b. flexor carpi ulnaris and flexor carpi radialis
c. flexor digitorum
d. all the above

_____ 7. When the gluteus maximus muscle contracts, it _____.

a. extends the thigh anteriorly
b. extends the thigh backward
c. moves the thigh laterally
d. The gluteus maximus does not move the thigh at all.

_____ 8. Which of the following muscles will *not* flex the lower leg?

a. biceps femoris
b. semimembranosus
c. semitendinosus
d. vastus medialis

_____ 9. The quadriceps are located on the _____ thigh.

a. anterior
b. posterior
c. lateral
d. medial

10. Which of these two muscles is lateral to the other: biceps femoris; semitendinosus? ______________________

11. Which of these two muscles is medial to the other: semitendinosus; gracilis? ______________________

12. Identify the muscles labeled in Figure 9–7.

Figure 9–7 Major Superficial Muscles of the Body (Anterior View)

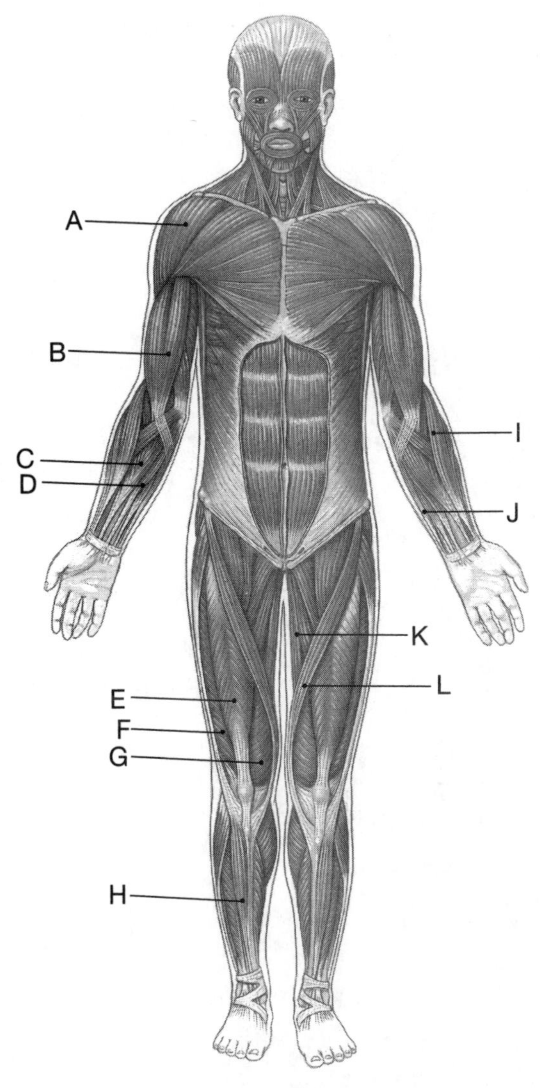

A ______________________ B ______________________

C ______________________ D ______________________

E ______________________ F ______________________

G ______________________ H ______________________

I ______________________ J ______________________

K ______________________ L ______________________

13. Identify muscles labeled in Figure 9–8.

Figure 9–8 Major Superficial Muscles of the Body (Posterior View)

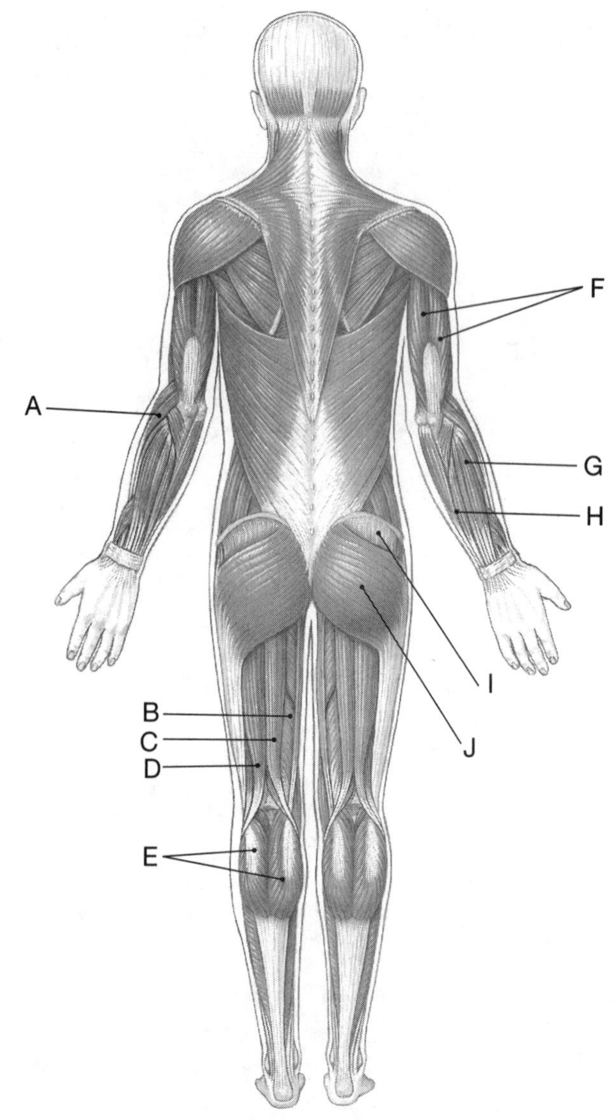

A ______________________ B ______________________

C ______________________ D ______________________

E ______________________ F ______________________

G ______________________ H ______________________

I ______________________ J ______________________

OBJECTIVE 10: Describe the effects of exercise and aging on muscle tissue (textbook p. 205).

After studying p. 205 in the text, you should be able to describe the effects of aging on the muscular system, which are almost as visible as the aging effects on the integumentary system.

_____ 1. Which of the following statements in reference to aging skeletal muscle cells (fibers) is true?

a. The diameter of the fibers decreases.
b. The fibers begin to disappear.
c. The fibers get shorter.
d. All the above.

_____ 2. As skeletal muscle fibers age they become _________________________ probably due to _____.

a. less flexible, fibrosis
b. less flexible, years of wear and tear
c. more flabby, lack of use
d. smaller in diameter, fibrosis

_____ 3. Repair of skeletal muscle tissue is limited due to the aging process. The typical result is _____.

a. a tendency for rapid muscular fatigue
b. decreased amounts of fibrous connective tissue
c. formation of scar tissue
d. overheating of muscle tissue

_____ 4. The process during which the aging skeletal muscles develop increasing amounts of fibrous connective tissue is called _____.

a. atrophy
b. fibrosis
c. hypertrophy
d. tetanus

OBJECTIVE 11: Describe the two common symptoms of muscle disorders (textbook p. 206).

There are numerous muscle disorders, but all the disorders typically have common symptoms. After studying p. 206 in the text, you should be able to describe two of the most common symptoms of any muscle disorder.

_____ 1. Muscle weakness and pain are two conditions of muscle disorders. Which of the following is true in reference to these conditions?

a. Both muscle weakness and muscle pain are signs.
b. It depends on the patient's perception as to which is a sign and which is a symptom.
c. Muscle pain is a sign, and muscle weakness is a symptom.
d. Muscle weakness is a sign, and muscle pain is a symptom.

_____ 2. Which of the following is considered a sign of a muscle disorder?

a. a muscle strain
b. muscle spasms
c. nerve damage affecting a muscle
d. viral infection of the muscle

_____ 3. Which of the following is considered a symptom of a muscle disorder?

a. a torn muscle
b. atrophy
c. ptosis
d. the presence of a muscle "knot"

_____ 4. Which of the following is classified as a symptom of a muscle disorder?

a. fibromyalgia
b. tetanus
c. trichinosis
d. all the above

_____ 5. Which of the following is classified as a sign of a muscle disorder?

a. hernia
b. bruises
c. both a and b
d. none of the above

OBJECTIVE 12: Describe the major disorders of the muscular system (textbook pp. 206–213).

There are numerous disorders associated with the various muscles of the body. After studying in the text, pp. 206–213, you should be able to describe some of the disorders and how they directly affect the actions of the body.

_____ 1. Compartment syndrome of the lateral compartment of the lower leg may cause ischemia of the _____.

a. anterior compartment
b. deep posterior compartment
c. lateral compartment
d. superficial posterior compartment

_____ 2. Jason accidentally stepped in a hole and twisted his ankle. The doctor determined that in the process of twisting his ankle, Jason actually tore his peroneus longus. This condition is known as _____.

a. a sprain
b. a strain
c. myositis
d. tendonitis

_____ 3. A viral infection that causes muscle pain but not muscle weakness is _____.

a. fibromyalgia
b. myositis
c. necrotizing fascititis
d. tetanus

_____ 4. Which of the following best describes rhabdomyosarcoma?

a. It is a benign tumor of skeletal muscles.
b. It is a benign tumor of smooth muscles.
c. It is a malignant tumor of skeletal muscles.
d. It is a malignant tumor of smooth muscles.

_____ 5. Which of the following disorders occurs more often in males than in females?

a. muscular dystrophy
b. myalgia
c. myasthenia gravis
d. sarcoma

PART II: CHAPTER-COMPREHENSIVE EXERCISES

Match the terms in column with the descriptions or statements in column B.

MATCHING I

	(A)	(B)
_____	1. actin	A. The sliding of this filament results in muscle contraction.
_____	2. aerobic	B. The moveable connection of bone to muscle
_____	3. anaerobic	C. The triceps brachii is a(n) _____ muscle to the biceps brachii.
_____	4. antagonistic	D. The type of muscle contraction that does not result in movement
_____	5. axial	E. The production and use of ATP without the presence of oxygen
_____	6. biceps brachii	F. The production and use of ATP with the presence of oxygen
_____	7. biceps femoris	G. A product of anaerobic metabolism
_____	8. insertion	H. The collective name for the muscles of the head, neck, and torso
_____	9. isometric	I. The muscle that causes flexion of the lower arm
_____	10. lactic acid	J. The muscle that causes flexion of the lower leg
_____	11. myoma	K. Benign tumor of muscles

MATCHING II

	(A)	(B)
_____	1. latissimus dorsi	A. The basic functional unit of muscle
_____	2. muscle fibers	B. The filament that contains cross-bridges
_____	3. myosin	C. Muscle cells are also called _____.
_____	4. neuromuscular	D. The junction of the nerve and the muscle
_____	5. origin	E. The stationary connection of bone to muscle
_____	6. polio	F. A type of muscle that operates under voluntary control
_____	7. rectus abdominis	G. The muscle that causes extension of the lower arm
_____	8. rectus femoris	H. The muscle that causes extension of the lower leg
_____	9. sarcoma	I. The muscle that flexes the torso
_____	10. sarcomere	J. The muscle that extends the torso
_____	11. skeletal	K. Disease resulting in inflammation of the protective membranes of the CNS
_____	12. triceps brachii	L. Malignant tumor of muscles

CONCEPT MAP

This concept map summarizes and organizes the microscopic and macroscopic features of some of the major superficial muscles of the axial and appendicular divisions. Use the following terms to fill in the boxes identified by the circled numbers, 1–13.

maintain body temperature	**gastrocnemius**	**axial**
maintain posture	**deltoid**	**myosin**
latissimus dorsi	**actin**	**torso**
triceps brachii	**ptosis**	**hernia**
carpal tunnel syndrome		

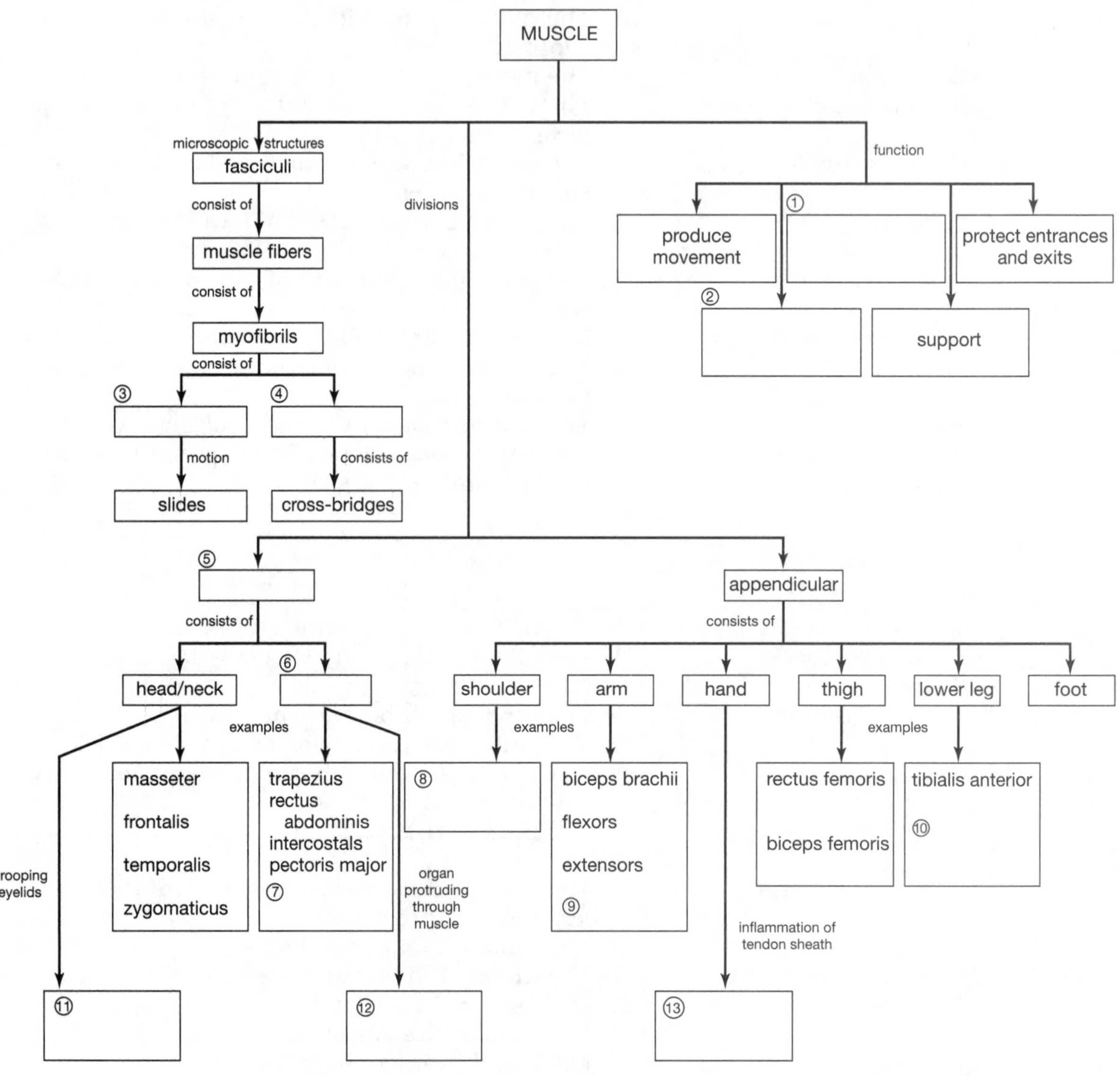

CROSSWORD PUZZLE

This crossword puzzle reviews the material in Chapter 9. To complete the puzzle, you have to know the answers to the clues given, and you must be able to spell the terms correctly.

ACROSS

5. A sheath that connects the frontalis to the occipitalis
6. This type of metabolism requires the use of oxygen.
8. A major lower back muscle
11. The muscle associated with the cheeks
13. Pain in the joints and muscles but no muscle weakness
16. The filament that slides
18. The rectus femoris will _____ the lower leg.
19. Myasthenia gravis is an _____ disease.
20. General inflammation of a skeletal muscle

DOWN

1. A voluntary type of muscle
2. One of the quadriceps muscles
3. The biceps brachii and the triceps brachii are examples of _____ muscles.
4. The thick filaments that have cross-bridges
7. The biceps brachii will _____ the antebrachium.
9. Cardiac muscle has _____ discs. No other muscle has them.
10. These connect muscle to bone
12. A neuromuscular junction is the connection between a motor neuron and a muscle _____.
14. This type of metabolism does not require the use of oxygen.
15. An early sign of myasthenia gravis
17. The deltoid will _____ arm.

FILL-IN-THE-BLANK NARRATIVE

Use the following terms to fill in the blanks of this summary of Chapter 9.

arrector pili	**abducts**	**flex**
deltoid	**biceps femoris**	**homeostasis**
hamstrings	**heat**	**gluteus maximus**
iliopsoas	**neuromuscular**	**extending**
rectus femoris	**superficial**	**isometric**
nervous	**myasthenia gravis**	**muscular dystrophy**
trichinosis	**hypocalcemia**	

Muscles do a lot more than provide movement. As muscles contract they generate (1) __________. For example, (2) __________ muscles create goosebumps. As these muscles create goosebumps, they generate heat. This is one way the body warms itself when it is cold.

Generally, when we think of muscles, we think of the (3) __________ muscles involved in movement. We take for granted many of the motions we encounter on a regular basis. Simple walking or kicking a ball requires the contraction of numerous muscles. For example, when we attempt to kick a ball with our right foot, we partially balance ourselves on our left leg. This requires the contraction of muscles in our left leg, which are (4) __________ contractions. We then move our right leg backward by contracting the (5) __________, and we (6) __________ our lower leg just a bit by contracting the (7) __________. This muscle is one of the (8) __________. When we begin to kick the ball, our (9) __________ muscles move our leg anteriorly. At just the right instant, the (10) __________ contracts, thus (11) __________ our lower leg so our foot contacts the ball. While kicking the ball, we are probably using our arms to help balance us. Our (12) __________ muscle (13) __________ the humerus, thereby causing our arms to extend laterally.

The skeletal muscles contract only after they have received an impulse from the (14) __________ system. An impulse from the nervous system contacts a muscle at the (15) __________ junction. This is another example of how one organ system works with other organ systems to accomplish a common goal, the maintenance of (16) __________.

Muscle contraction is hindered when an autoimmune disease, (17) __________, attacks the neuromuscular junctions. Muscle action is also altered when a sex-linked disorder, (18) __________, results in a progressive muscular weakness. Muscle weakness is a sign of a disease called (19) __________, which is caused by the invasion of a nematode. Another cause of muscle weakness is (20) __________. Any of the disorders will make it difficult to maintain homeostasis.

CLINICAL CONCEPT

The following clinical concept applies the information in Chapter 9. Following the application is a set of questions to help you understand the concept.

When a nerve impulse arrives at a muscle, a sequence of events occurs that results in muscle contraction. The impulse causes the sarcoplasmic reticulum to release calcium ions, which allows cross-bridge binding of myosin to the active sites of actin. The cross-bridges use ATP to "pull" the actin myofilaments. ATP is also involved in the detachment of the cross-bridges from actin. To generate the ATP necessary for cross-bridge activity, the mitochondria requires oxygen (aerobic metabolism). When an individual dies, air containing oxygen is no longer inhaled, and therefore ATP production ceases. Calcium ions diffusing into the sarcoplasm or leaking out of the sarcoplasmic reticulum trigger a sustained contraction. Without ATP, the cross-bridges cannot detach from the active sites, and the muscles lock in a contracted position.

1. What causes the release of calcium ions from the sarcoplasmic reticulum?

2. What do the cross-bridges need in order to attach to, move, or be released from actin?

3. The gas _____ is necessary to complete the formation of ATP.

4. When the body runs out of _____, it will run out of _____. At that time the muscles will develop rigor mortis.

THIS CONCLUDES CHAPTER 9 EXERCISES

CHAPTER

10

The Nervous System I: Nerve Cells and the Spinal Cord

Chapter Objectives:

1 Describe the overall organization and functions of the nervous system.

2 Describe the differences in the structure and function of neurons and neuroglia.

3 What is the relationship between neurons and nerves?

4 Describe the structure of a synapse and how an action potential (nerve impulse) passes from one neuron to another.

5 Describe the structure and function of the spinal cord.

6 Describe the essential elements of a reflex arc.

7 Describe the structures and functions of the sympathetic and parasympathetic divisions of the ANS.

8 Describe nervous system disorders associated with demyelination.

9 Describe common disorders of the spinal cord and spinal nerves.

The nervous system is the control center and communication network of the body. Its primary function is to regulate and maintain homeostasis. This complex organ system accounts for about 3 percent of the total body weight. It is vital to life as well as to our experience and appreciation of it. The nervous system provides relatively swift, but generally brief, responses to stimuli by temporarily modifying the activities of other organ systems. The modifications usually occur within a matter of milliseconds, and the effects disappear soon after neural activity ceases.

This chapter details the structure and function of neurons and neuroglia. Having an understanding of how a neuron functions will help in the understanding of how incoming information is processed and integrated with other body systems. Having an understanding of how a neuron functions also helps us to understand the abnormalities (disorders) when they occur.

Knowledge of the nervous system is an integral part of understanding many of the body's activities, all of which must be controlled and adjusted to meet changing internal and external environmental conditions. Without such control and adjustment, homeostasis could not be maintained.

PART I: OBJECTIVE-BASED QUESTIONS

OBJECTIVE 1: Describe the overall organization and functions of the nervous system (textbook pp. 219–220).

After studying pp. 219–220 in the text, you should be able to categorize the diverse components of the nervous system. You should also understand the dichotomous division of the nervous system as presented in Figure 10-1 of the text.

_____ 1. The nervous system is divided into two major categories: the _____, which consists of the brain and spinal cord, and the _____, which consists of the nerves going out to the peripheral regions of the body such as the arms and legs.

a. afferent, efferent
b. central nervous system, peripheral nervous system
c. efferent, afferent
d. parasympathetic nervous system, sympathetic nervous system

_____ 2. The peripheral nervous system can be further divided into two divisions, the _____.

a. autonomic and somatic
b. cranial nerves and spinal nerves
c. efferent and afferent
d. sympathetic and parasympathetic

_____ 3. Nerves carrying impulses from the peripheral nervous system to the central nervous system are called _____. The nerves carrying impulses from the central nervous system to the peripheral nervous system are called _____.

a. afferent, efferent
b. autonomic, somatic
c. efferent, afferent
d. somatic, autonomic

_____ 4. An impulse from the brain to the biceps brachii is _____.

a. an impulse from the CNS to the PNS. The portion of the PNS is the somatic system.
b. an impulse from the CNS to the PNS. The portion of the PNS is the autonomic system.
c. an impulse from the PNS to the CNS. The portion of the PNS is the somatic system.
d. an impulse from the PNS to the CNS. The portion of the PNS is the autonomic system.

_____ 5. Which of the following statements is true?

a. The peripheral nerves carry information to and from the central nervous system.
b. The peripheral nervous system consists of autonomic and somatic nerves.
c. The peripheral nerves lie outside the central nervous system.
d. All the above are true.

OBJECTIVE 2: Describe the differences in the structure and function of neurons and neuroglia (textbook pp. 220–225).

After studying pp. 220–225 in the text, you should be able to differentiate between neurons, the cells of the nervous system that have the ability to conduct impulses, and neuroglia, the special group of cells that support and protect the neurons.

_____ 1. Which of the following statements is true?

a. Motor neurons carry impulses to the CNS.
b. Motor neurons carry impulses to a muscle.
c. Sensory neurons are also considered to be an efferent nerve.
d. Sensory neurons carry impulses to a muscle.

_____ 2. Which of the following consists of an axon, a soma (cell body), and a dendrite?

a. astrocyte
b. neuroglial cell
c. Schwann cell
d. none of the above

_____ 3. The function of neuroglial cells is _____.

a. to conduct information to the neurons
b. to support and protect the neurons
c. both a and b
d. none of the above

_____ 4. The _____ cells form a myelin sheath around neurons in the CNS, and _____ cells form a myelin sheath around the neurons in the PNS.

a. astrocyte, glial
b. microglia, Schwann
c. oligodendrocyte, Schwann
d. none of the above

_____ 5. Which part of a neuron is typically wrapped with a myelin sheath by a glial cell?

a. axon
b. soma
c. dendrite
d. all the above

6. In Figure 10–1, identify the cell body, axon, dendrite, nucleus, and neuromuscular junction.

Figure 10–1 Structure of a Motor Neuron

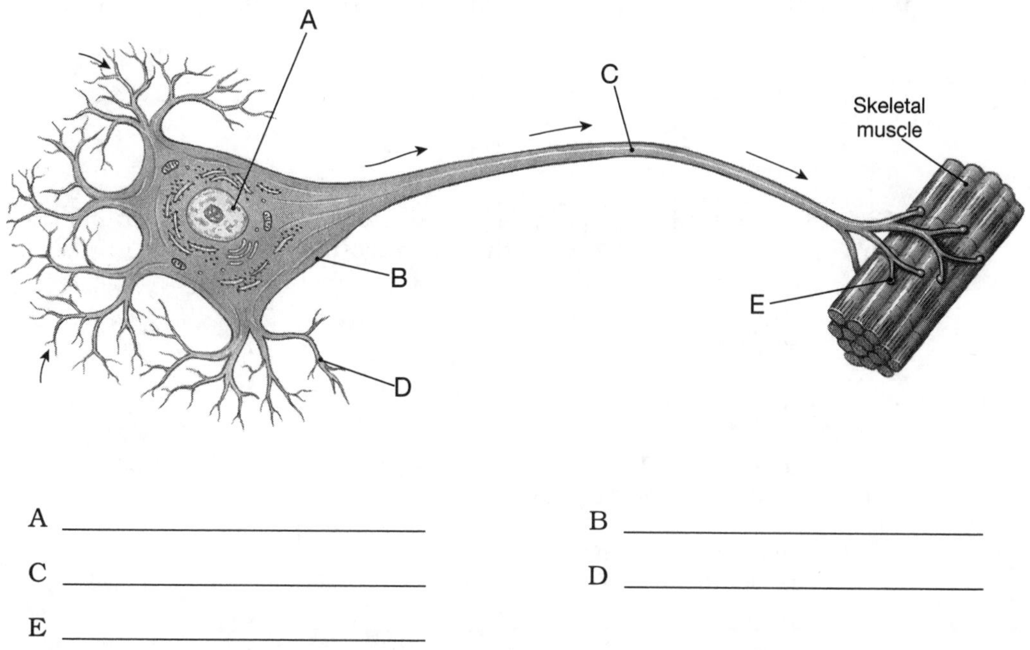

A ______________________ B ______________________

C ______________________ D ______________________

E ______________________

7. In Figure 10–2, identify the myelinated axons, astrocyte, neuron, microglial cell, and oligodendrocyte.

Figure 10–2 Neurons and Neuroglia

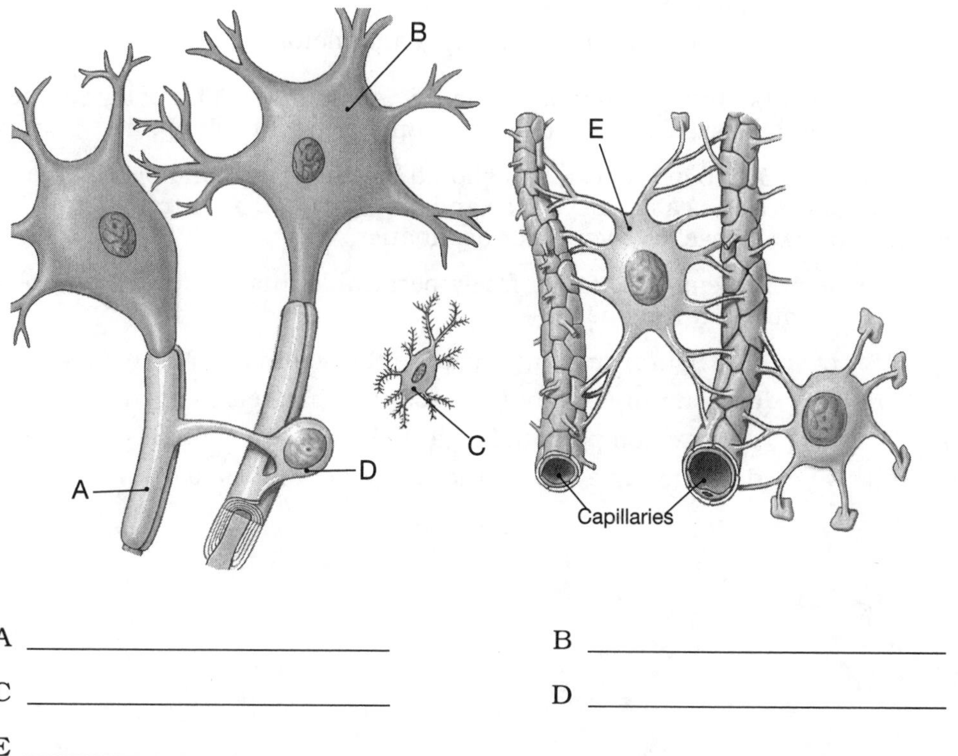

A ______________________ B ______________________

C ______________________ D ______________________

E ______________________

OBJECTIVE 3: What is the relationship between neurons and nerves? (textbook p. 223).

After studying p. 223 in the text, you should be able to differentiate between nerves, which are bundles of axons in the peripheral nervous system that appear as whitish stringlike structures, easily visible to the naked eye, and neurons, which are microscopic.

_____ 1. What part of a neuron forms bundles that are referred to as a typical nerve?

a. axon
b. soma
c. dendrite
d. all the above

_____ 2. The PNS is made of _____ pairs of nerves associated with the brain and _____ pairs of nerves associated with the spinal cord.

a. 7,12
b. 31,12
c. 12,31
d. 43,43

_____ 3. The sciatic nerve is a large nerve that goes down the leg. What part of the neuron is actually the sciatic nerve?

a. axon
b. dendrite
c. soma
d. The sciatic nerve is too large to be a neuron.

_____ 4. The bundled axons that make up the nerves are _____.

a. all sensory
b. all motor
c. neither sensory nor motor
d. a combination of sensory and motor

OBJECTIVE 4: Describe the structure of a synapse and how an action potential (nerve impulse) passes from one neuron to another (textbook pp. 223–225).

After studying pp. 223–225 in the text, you should be able to explain how a nerve impulse is transmitted across a gap called a synapse and how the impulse is transmitted in a specific direction so that it reaches the proper destination.

_____ 1. A neuron membrane is *not* freely permeable to ions. How then do positive ions enter the neuron?

a. A neurotransmitter causes the sodium channels to open.
b. A stimulus causes the sodium channels to open.
c. An action potential causes the sodium channels to open.
d. An impulse causes the sodium channels to open.

_____ 2. The gap between one neuron and another or the gap between a neuron and an effector is called a _____.

a. dendrite
b. soma
c. synapse
d. none of the above

_____ 3. Which of the following statements is true?

a. An impulse travels through the axon to the presynaptic area. The presynaptic area releases acetylcholine. Acetylcholine causes an influx of sodium ions into the postsynaptic area.
b. An impulse travels through the dendrite to the presynaptic area. The presynaptic area releases acetylcholine. Acetylcholine causes an influx of sodium ions into the postsynaptic area.
c. An impulse travels through the axon to the postsynaptic area. The postsynaptic area releases acetylcholine. Acetylcholine causes an influx of sodium ions into the presynaptic area.
d. An impulse travels through the dendrite to the postsynaptic area. The postsynaptic area releases acetylcholine. Acetylcholine causes an influx of sodium ions into the presynaptic area.

_____ 4. Which of the following statements is true?

a. The presynaptic area releases a neurotransmitter; the postsynaptic area releases a neurotransmitter deactivator.
b. The presynaptic region releases acetylcholine; the postsynaptic area releases acetylcholinesterase.
c. The presynaptic region releases acetylcholinesterase; the postsynaptic area releases acetylcholine.
d. Both a and b are true.

_____ 5. Which of the following statements is true?

a. Acetylcholine allows an impulse to go across the synapse to the next neuron.
b. Norepinephrine allows an impulse to go across the synapse to the next neuron.
c. Serotonin allows an impulse to go across the synapse to the next neuron.
d. All the above are true.

6. An impulse cannot travel from the postsynaptic area to the presynaptic area in the synapse because the postsynaptic area does not release a(n) _______________.

7. An impulse can travel from the presynaptic area to the postsynaptic area of another neuron because the presynaptic area releases a(n) _______________.

8. In Figure 10-3, identify the axon, dendrite, soma, and synapse, and indicate the direction in which the impulse will travel (E to F or F to E).

Figure 10-3 Two Neurons and a Synapse

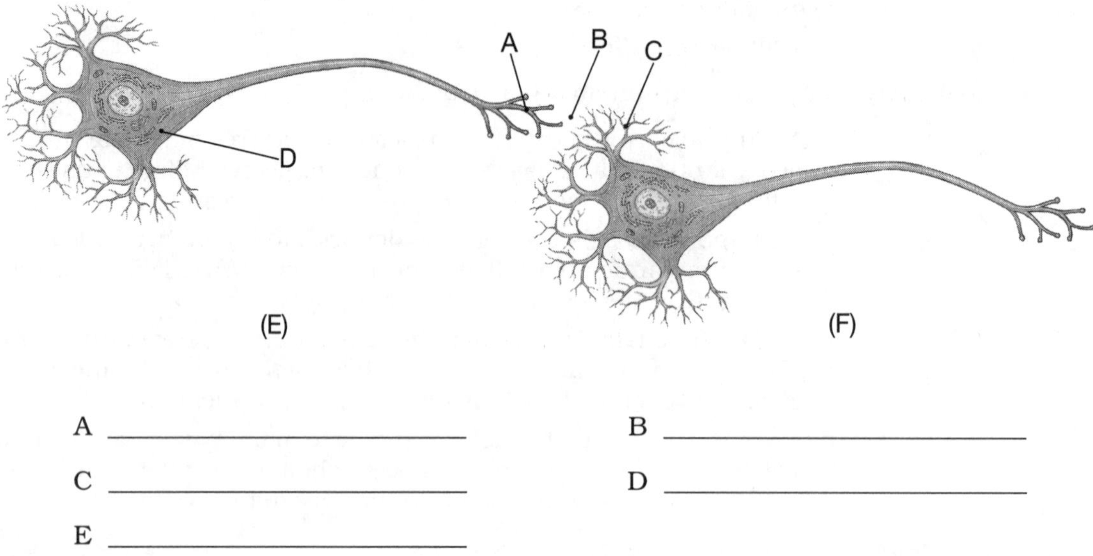

A ____________________ B ____________________

C ____________________ D ____________________

E ____________________

OBJECTIVE 5: Describe the structure and function of the spinal cord (textbook pp. 225–226).

After studying pp. 225–226 in the text, you should be familiar with the following features of the spinal cord: The spinal cord is a rather complex set of nerves consisting of ascending tracts transmitting sensory information up toward the brain and descending tracts conveying motor commands down into the spinal cord from the brain. Thirty-one pairs of spinal nerves transmit sensory and motor impulses to and from the central nervous system.

_____ 1. You have probably heard people speak of gray matter and white matter. The gray matter of the spinal cord is made of _____.

a. axons

b. dendrites

c. somas

d. myelinated axons

_____ 2. The white matter of the spinal cord is composed of _____.

a. axons

b. dendrites

c. somas

d. all the above

_____ 3. The ascending tracts of the spinal cord are _____ that carry _____ information to the brain.

a. axons, motor
b. axons, sensory
c. interneurons, motor
d. interneurons, sensory

_____ 4. The descending tracts of the spinal cord are _____ that carry _____ information to the periphery.

a. axons, motor
b. axons, sensory
c. interneurons, motor
d. interneurons, sensory

_____ 5. Damage to which of the following spinal cord regions could lead to the greatest destruction? _____

a. cervical
b. thoracic
c. lumbar
d. sacral

6. If a virus affects a nerve that causes an extensive rash in the gluteal region, according to the dermatome chart (Figure 10-7 in the text), which nerves is this virus possibly affecting? ____________________

7. In Figure 10–4, identify the cervical nerves, thoracic nerves, lumbar nerves, sacral nerves, and cauda equina.

Figure 10–4 The Spinal Cord

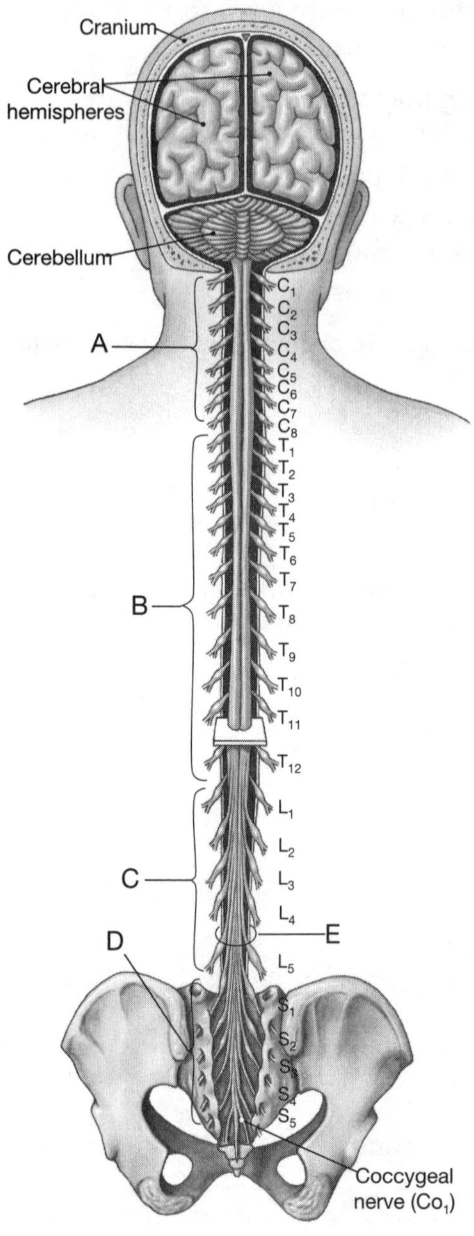

A ____________________ B ____________________

C ____________________ D ____________________

E ____________________

8. In Figure 10–5, identify the ventral root, dorsal root, anterior spinal cord, posterior spinal cord, white matter, and gray matter.

Figure 10–5 Transverse View of the Spinal Cord

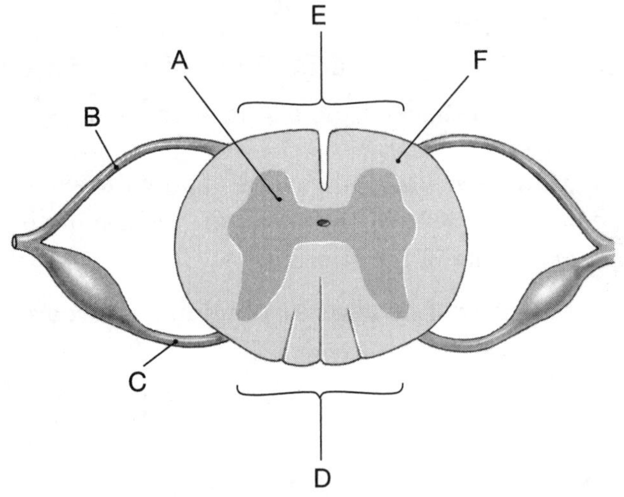

A ____________________ B ____________________

C ____________________ D ____________________

E ____________________ F ____________________

OBJECTIVE 6: Describe the essential elements of a reflex arc (textbook pp. 228–229).

After studying pp. 228–229 in the text, you should be able to describe a reflex arc as the pathway of an impulse to some part of the central nervous system (either the brain or the spinal cord) and back to an effector such as a muscle or gland.

____ 1. Which of the following describes a reflex arc?

a. A stimulus causes an impulse to travel along a sensory neuron, then the CNS sends an impulse to the effector via a motor neuron.
b. A stimulus causes an impulse to travel along a motor neuron, then the CNS sends an impulse to the effector via a sensory neuron.
c. A stimulus causes an impulse to travel along a sensory neuron, then the PNS sends an impulse to the effector via a motor neuron.
d. A stimulus causes an impulse to travel along a sensory neuron to the effector, then the CNS sends an impulse to a muscle.

____ 2. The sensory neuron associated with a reflex arc transmits the impulse ____.

a. away from an interneuron
b. away from the spinal cord
c. toward the effector
d. toward the spinal cord

_____ 3. Which of the following typically activates a motor neuron in a reflex arc?

a. the effector
b. the effector and/or interneuron
c. the sensory neuron and/or interneuron
d. the stimulus

_____ 4. In a reflex arc, where are the neurotransmitters located?

a. in the vesicles of the axons of the sensory nerve and the motor nerve
b. in the vesicles of the axons of only the sensory nerves
c. in the vesicles of the axons of only the motor nerves
d. at the site of the stimulus

5. What is the minimum number of neurons involved in a reflex arc?

____

6. In Figure 10–6, identify the effector, sensory nerve, motor nerve, spinal cord, and the site of the stimulus.

Figure 10–6 The Reflex Arc

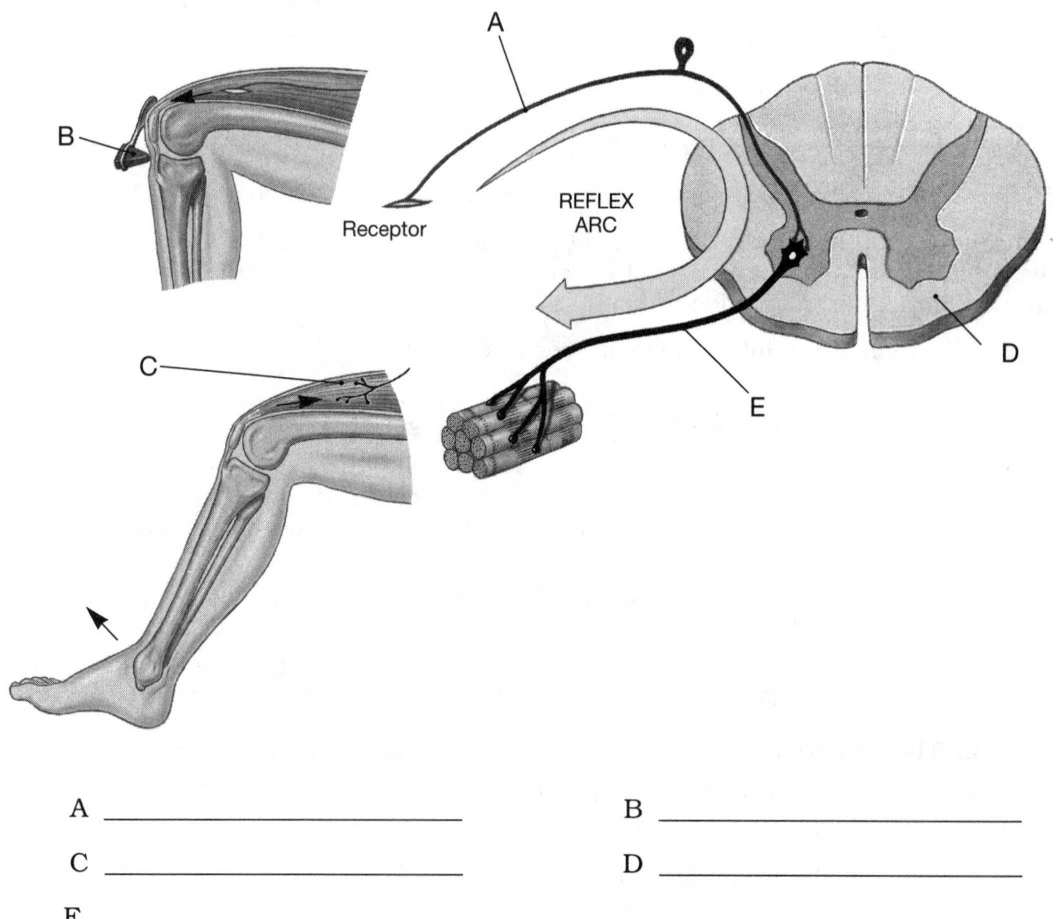

A _______________________ B _______________________

C _______________________ D _______________________

E _______________________

OBJECTIVE 7: Describe the structures and functions of the sympathetic and parasympathetic divisions of the ANS (textbook pp. 229–232).

After studying pp. 229–232 in the text, you should be able to explain the important features of the autonomic nervous system (ANS), which consists of a sympathetic and a parasympathetic division responsible for the involuntary control and regulation of smooth muscles, cardiac muscle, and glands. You should also be able to describe how these two divisions typically operate to assist the body in maintaining homeostasis.

____ 1. The autonomic nervous system consists of two divisions: the ____.

a. peripheral and parasympathetic
b. peripheral and sympathetic
c. somatic and peripheral
d. sympathetic and parasympathetic

____ 2. Usually, the parasympathetic and sympathetic nerves work in a(n) ____ manner.

a. antagonistic
b. enhancing
c. inhibitory
d. stimulatory

____ 3. The axons of the motor neurons in the somatic system transmit information from the ____ through the ____ to a ____.

a. CNS, PNS, muscle
b. CNS, preganglion, muscle
c. PNS, preganglion, muscle
d. PNS, CNS, muscle

____ 4. The axons of the motor neurons in the autonomic system transmit information from the ____ to the ____ to the ____ to a ____.

a. CNS, preganglion, autonomic ganglion muscle
b. PNS, preganglion, autonomic ganglion muscle
c. preganglion, CNS, autonomic ganglion muscle
d. preganglion, PNS, autonomic ganglion muscle

____ 5. Which of the following statements is true?

a. Preganglion axons are excitatory.
b. Preganglion axons are inhibitory.
c. Preganglion axons are known as adrenergic neurons.
d. Preganglion axons can be both excitatory and inhibitory.

____ 6. The sympathetic division is located ____, and the parasympathetic division is located ____.

a. between T_1 and T_{12}, in the cervical and sacral region
b. between T_1 and T_{12}, in the sacral region
c. in the cervical and sacral region, between T_1 and T_{12}
d. between T_1 and T_{12}, in the lumbar region

7. In Figure 10–7, identify the spinal cord, sympathetic chain, sympathetic nerves, and parasympathetic nerves.

Figure 10–7 The Sympathetic and Parasympathetic Nerves of the ANS

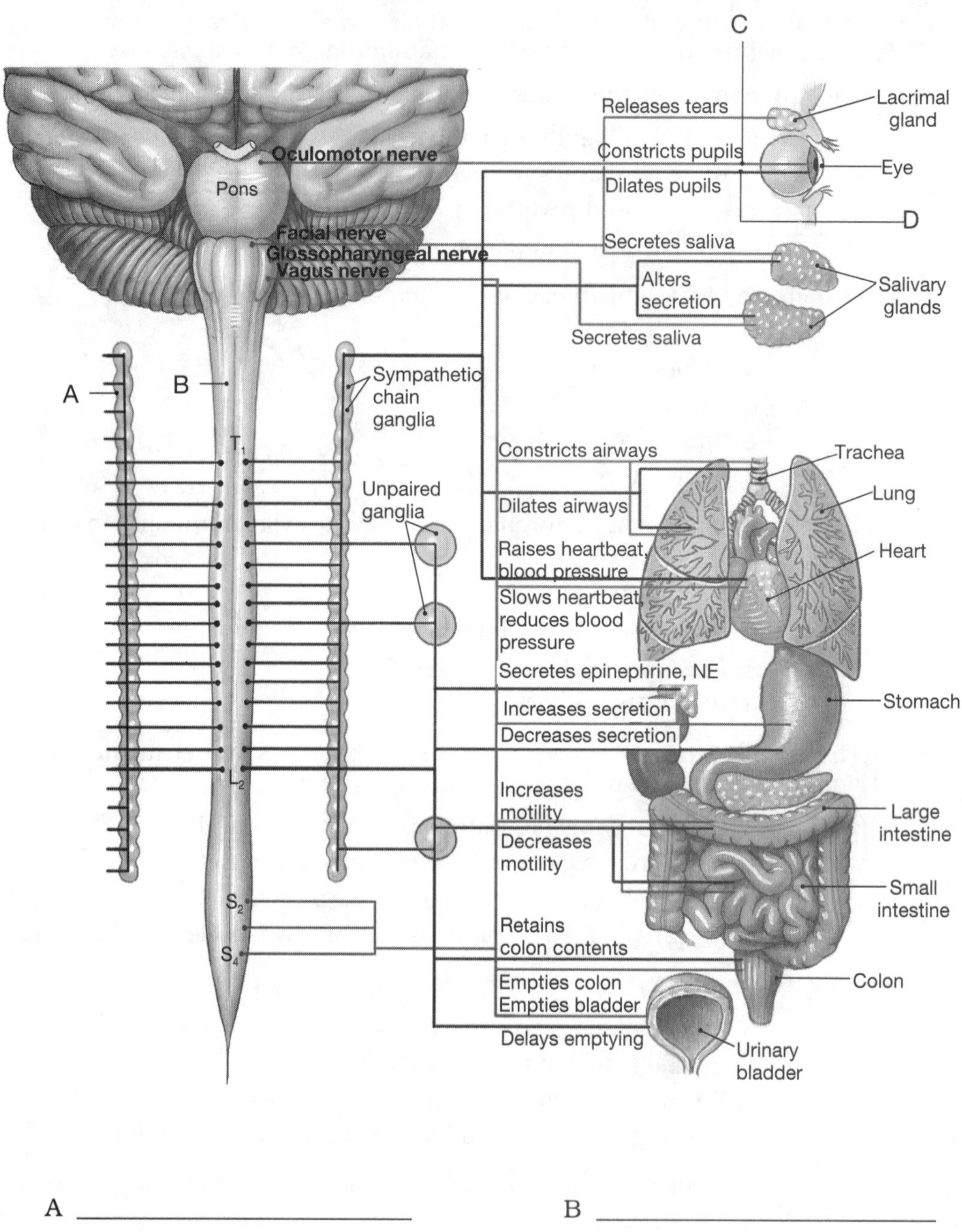

A ______________________ B ______________________

C ______________________ D ______________________

OBJECTIVE 8: Describe nervous system disorders associated with demyelination (textbook p. 233).

Many axons are protected by a myelin sheath. Some nerve disorders result in the demyelination of that sheath. After studying in the text, p. 233, you should be able to describe some of the disorders associated with myelin sheath damage.

_____ 1. Demyelination refers to _____.

a. heavy-metal poisoning of the nervous system
b. the destruction of the axon
c. the destruction of the myelin sheath
d. the formation of a neuroblastoma

_____ 2. Which of the following is not a demyelination disorder?

a. diphtheria
b. glioma
c. multiple sclerosis
d. All the above are demyelination disorders.

_____ 3. Which of the following is a condition in which the immune system attacks the myelin sheath?

a. demyelination
b. glioma
c. multiple sclerosis
d. neuroma

_____ 4. Which of the following is true in reference to brain tumors?

a. Brain tumors are a type of neuroblastoma.
b. Brain tumors are due to the demyelination of the neurons of the brain.
c. Brain tumors are gliomas that affect the glial cells of the brain.
d. Brain tumors are gliomas that destroy the myelin sheath of axons.

_____ 5. Which of the following statements is most accurate?

a. Demyelination disorders are a type of neuroma.
b. Gliomas and neuroblastomas are types of neuromas.
c. Neuroblastomas and neuromas are types of gliomas.
d. Neuromas and gliomas are types of neuroblastomas.

OBJECTIVE 9: Describe common disorders of the spinal cord and spinal nerves (textbook pp. 234–236).

There are numerous disorders associated with the spinal cord. After studying in the text, pp. 234–236, you should be able to describe some of the more common disorders and how they actually affect normal body actions.

_____ 1. Leprosy is a disease that is caused by bacteria. This bacterium invades the _____.

a. brain
b. face
c. peripheral nerves
d. spinal nerves

_____ 2. Your textbook describes several palsies. Which of the following best defines palsy?

a. a type of demyelination disorder
b. abnormal nerve formation
c. temporary or sometimes permanent loss of sensation
d. trauma to specific nerves

_____ 3. The virus that causes chickenpox can remain dormant for several years within the neurons of the ____.

a. gray matter of the spinal cord
b. nerve plexuses
c. peripheral nerves
d. white matter of the spinal cord

_____ 4. Compressing a nerve in the popliteal region may result in a condition known as ____.

a. Hansen's disease
b. paresthesia
c. peroneal palsy
d. sciatica

_____ 5. Compression of the median nerve may result in pain. This condition is known as ____.

a. carpal tunnel syndrome
b. Hansen's disease
c. paresthesia
d. ulnar palsy

PART II: CHAPTER-COMPREHENSIVE EXERCISES

Match the terms in column A with the descriptions or statements in column B.

MATCHING I

	(A)	(B)
____	1. ascending	A. One component of the central nervous system
____	2. brachial plexus	B. Axons of the autonomic nervous system synapse with these structures.
____	3. brain	C. A type of cell that provides protection for the neurons
____	4. cauda equina	D. Tracts of nerves in the spinal cord that carry impulses to the brain
____	5. dermatomes	E. Contains the axons that bring information to the spinal cord
____	6. dorsal root	F. Specific body surface areas innervated by a specific spinal segment
____	7. ganglia	G. The distal end of the spinal cord
____	8. glial	H. A group of nerves that innervate the arms
____	9. glioma	I. A tumor in the brain region

MATCHING II

	(A)	(B)
____	1. axon	A. A bundle of axons
____	2. lumbosacral plexus	B. The portion of a neuron that dominates gray matter
____	3. myelin	C. A chemical released by the axons
____	4. nerve	D. A system of nerves that transmits information from the CNS to the muscles
____	5. neurotransmitter	E. The gap between the axon of one neuron and the dendrite of the next neuron
____	6. soma	F. A group of nerves that innervate the legs
____	7. somatic	G. Contains the axons that carry information from the spinal cord to the muscles and glands
____	8. synapse	H. A protective sheath found around some axons
____	9. ventral root	I. Demyelination typically affects the ____ of a nerve.

CONCEPT MAP

This concept map summarizes and organizes the information in Chapter 10 concerning the nervous system. Use the following terms to complete the map by filling in the boxes identified by the circled numbers, 1–10.

glands	**gray matter**	**neurotransmitter**
PNS	**soma**	**spinal cord**
sympathetic	**somatic nerves**	**palsy**
demyelination		

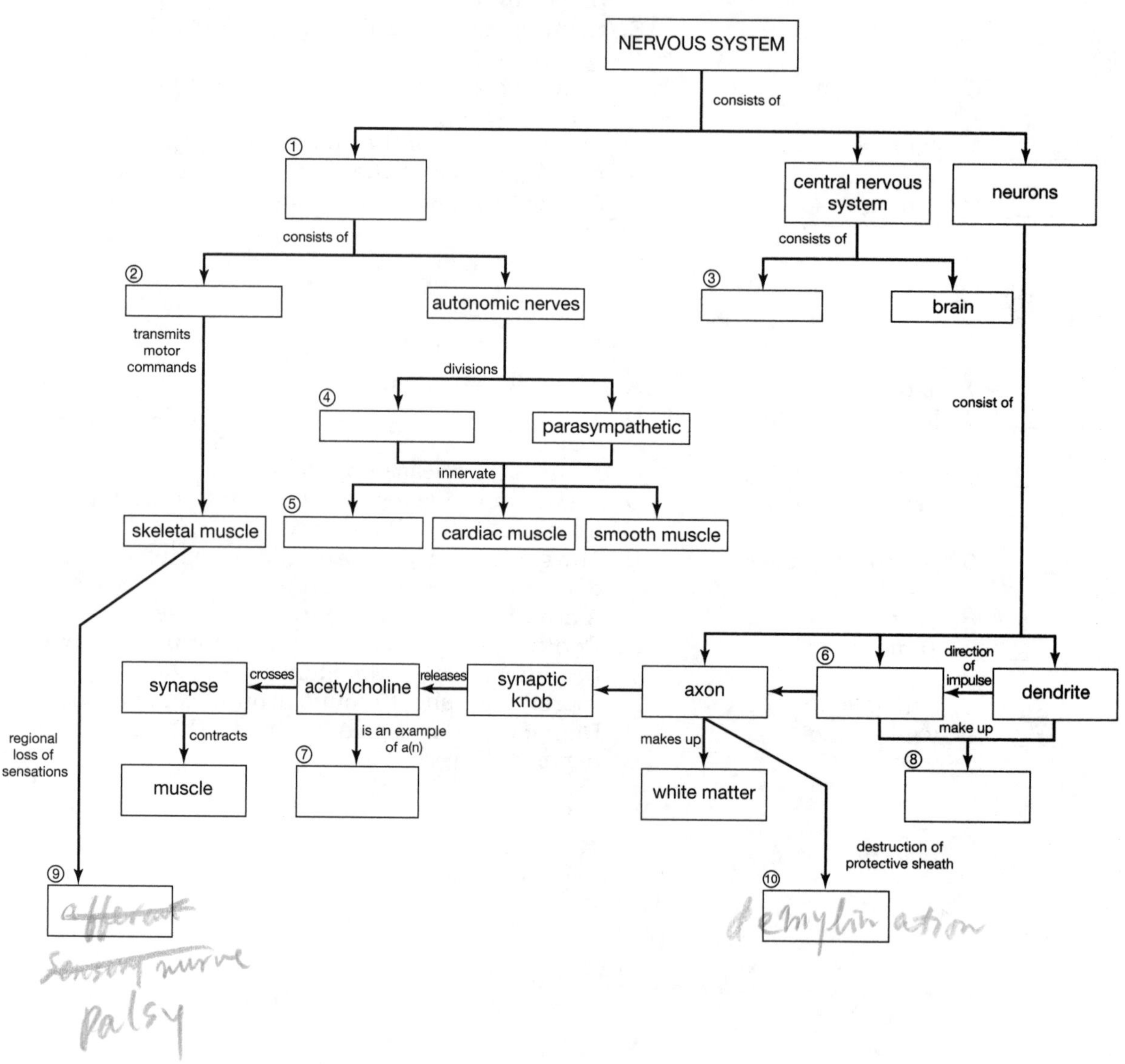

CROSSWORD PUZZLE

The following crossword puzzle reviews the material in Chapter 10. To complete the puzzle, you have to know the answers to the clues given, and you must be able to spell the terms correctly.

ACROSS

1. Nerves antagonistic to the parasympathetic nervous system
4. The same as a motor nerve
6. An arm or leg "falling asleep"
8. One of the many types of neurotransmitters
10. Inflammation of a nerve
11. The descending tracts in the spinal cord are _____ nerves.

DOWN

1. The portion of the neuron that contains organelles
2. The genus name of the organism that causes leprosy
3. The same as a sensory nerve
5. One of the two components of the CNS
7. A major nerve of the lumbosacral plexus
9. The abbreviation for acetylcholine

FILL-IN-THE-BLANK NARRATIVE

Use the following terms to fill in the blanks of this summary of Chapter 10.

acetylcholine	**actin**	**autonomic**
homeostasis	**motor**	**negative**
myosin	**permeable**	**reflex arc**
sensory	**stimulus**	**synapse**
parasympathetic	**neurotransmitter**	**ATP**

Imagine that you have touched a hot stove. The heat from the stove is the (1) __________. The nerve membrane becomes (2) __________ to positive ions. The (3) __________ impulse travels to the CNS, which sends (4) __________ impulses to effectors such as the biceps brachii muscle. As the impulse reaches the axon end, (5) __________ is released. This (6) __________ travels across the (7) __________ to the muscle and causes the release of calcium ions from the muscle cell's sarcoplasmic reticulum. The released calcium ions cause a change in the orientation of the troponin-tropomyosin complex that exposes the binding sites on (8) __________. Cross-bridges from (9) __________ bond to the binding sites, and (10) __________ causes the cross-bridges to pivot. The biceps brachii contracts and you move your hand from the hot stove.

In any reflex, a sensory nerve is activated, a motor nerve is activated, and an effector is stimulated. This sequence is known as a (11) __________. One very crucial division of the nervous system is the (12) __________ nervous system (ANS). This system consists of sympathetic and (13) __________ nerves. The sympathetic nerves typically speed up many processes, and the parasympathetic nerves typically slow down activity. These two sets of nerves usually help maintain (14) __________ via (15) __________ feedback mechanisms.

CLINICAL CONCEPTS

The following clinical concepts apply the information in Chapter 10. Following the application is a set of questions to help you understand the concepts.

1. Novocain is typically used by dentists to "deaden the nerve" associated with the teeth. Novocain causes the sensory nerve to the brain to become impermeable to sodium ions. Because sodium ions cannot enter the membrane (in this case), there will be no impulse. If no impulse arrives at the brain from the tooth, the brain cannot interpret pain. Therefore, you will not feel anything until the Novocain wears off.
2. Various diseases affect nerve function. An example is multiple sclerosis, which is a degeneration of the myelin sheath around the peripheral nerves associated with skeletal muscles. This disorder may result in general paralysis.

1. What happens when a nerve is "deadened" or becomes "numb?"

2. How does Novocain prevent patients from feeling what the dentist is doing to their teeth and gums?

3. Occasionally, the Novocain begins to wear off while the dentist is still working on a patient's teeth. Suddenly, the patient feels pain. Explain what has happened.

4. What does the myelin sheath do for the axons of the nerves?

5. How does multiple sclerosis affect the nerves?

THIS CONCLUDES CHAPTER 10 EXERCISES

CHAPTER 11

The Nervous System II: The Brain and Cranial Nerves

The brain is probably the most fascinating organ in the body, yet little is known about its structural and functional complexities. This large, complex organ in the central nervous system is located in the cranial cavity, where it is completely surrounded by cerebrospinal fluid, the cranial meninges, and the bony structures of the cranium. Activities such as memory, conscious and subconscious thoughts, sight, hearing, taste, touch, and many others are the result of brain activity. This chapter discusses numerous abnormalities of the brain and its surrounding tissues that can affect many activities.

Many activities are carried out by the cranial nerves. Many disorders that are linked to cranial nerve malfunctions have been identified. Cranial nerve malfunctions also affect numerous activities. There is much that we still do not understand about these activities.

The brain consists of billions of neurons that are organized with extensive interconnections that provide great versatility and variability. Because the interconnections are complex, variability of responses is possible.

The exercises and questions in Chapter 10 examine the major regions and functional roles of the brain and their relationships to the 12 pairs of cranial nerves, which are located on the inferior surface of the brain.

Chapter Objectives:

1. Name the major regions of the brain.
2. Describe the three membrane layers that cover the brain.
3. Describe the relationship of cerebrospinal fluid with the ventricles of the brain.
4. Locate the motor, sensory, and association areas of the cerebral cortex and discuss their functions.
5. Describe the location and functions of the limbic system.
6. Describe the functions of the diencephalon.
7. Describe the functions of the brain stem components.
8. Describe the functions of the cerebellum.
9. Identify the cranial nerves and relate each pair to its major functions.
10. Summarize the effects of aging on the nervous system.
11. Describe the three symptoms characteristic of many nervous system disorders.
12. Describe common disorders of the brain and cranial nerves.

PART I: OBJECTIVE-BASED QUESTIONS

OBJECTIVE 1: Name the major regions of the brain (textbook pp. 241–242).

It is common knowledge that the brain is a very complex organ. We know very little about its functional complexities. To help understand the brain a little better, scientists have divided the brain into various regions. After studying pp. 241–242 in the text, you should be able to identify the major regions of the brain and their functions.

____ 1. The brain stem comprises ____.

a. the diencephalon, thalamus, and hypothalamus
b. the midbrain, pons, and hypothalamus
c. the midbrain, pons, and medulla oblongata
d. the pituitary gland, thalamus, and midbrain

____ 2. The diencephalon consists of ____.

a. the brain stem and midbrain
b. the cerebellum and pons
c. the midbrain, pons, and medulla oblongata
d. the thalamus and hypothalamus

____ 3. The structure that helps coordinate skeletal muscle movement is the ____.

a. cerebellum
b. cerebrum
c. medulla oblongata
d. pons

____ 4. The pons is ____ to the medulla oblongata.

a. inferior
b. lateral
c. medial
d. superior

____ 5. The thalamus is responsible for ____.

a. controlling emotions
b. controlling involuntary responses
c. relaying sensory information
d. all the above

____ 6. The hypothalamus contains centers involved with the ____.

a. control of emotions
b. control of involuntary responses
c. relay of sensory information
d. all the above

7. In Figure 11–1, identify the pons, medulla oblongata, cerebrum, cerebellum, and thalamus.

Figure 11–1 The Brain

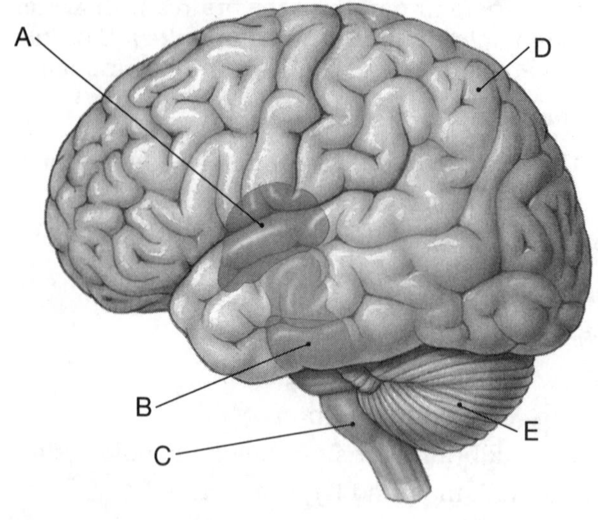

A ___thalamus___ B ___pons___

C ___medulla oblongata___ D ~~cerebellum~~ Cerebrum

E ___crebellum___

OBJECTIVE 2: Describe the three membrane layers that cover the brain (textbook pp. 242–243).

After studying pp. 242–243 in the text, you should be able to name the three layers of membranes collectively called the meninges that lie underneath the skull and help the brain remain suspended in a cushioning fluid substance.

____ 1. The term *meninges* refers to ____ membranes surrounding the CNS structures.

a. 1
b. 2
c. 3
d. 4

____ 2. The meningeal layer closest to the brain tissue is the ____.

a. arachnoid
b. blood–brain barrier
c. dura mater
d. pia mater

____ 3. The cerebrospinal fluid lies between the ____ and the ____.

a. arachnoid, pia mater
b. dura mater, arachnoid
c. dura mater, skull
d. pia mater, brain

_____ 4. The meningeal layer that is innervated by blood vessels to supply the brain is the _____.

a. arachnoid
b. blood–brain barrier
c. dura mater
d. pia mater

_____ 5. The two-layered _____ consists of the large veins called dura sinuses.

a. arachnoid
b. dura mater
c. pia mater
d. blood–brain barrier

6. In Figure 11–2, identify the dura mater, arachnoid mater, pia mater, skull, brain tissue, and cerebrospinal fluid location.

Figure 11–2 The Meninges of the Brain

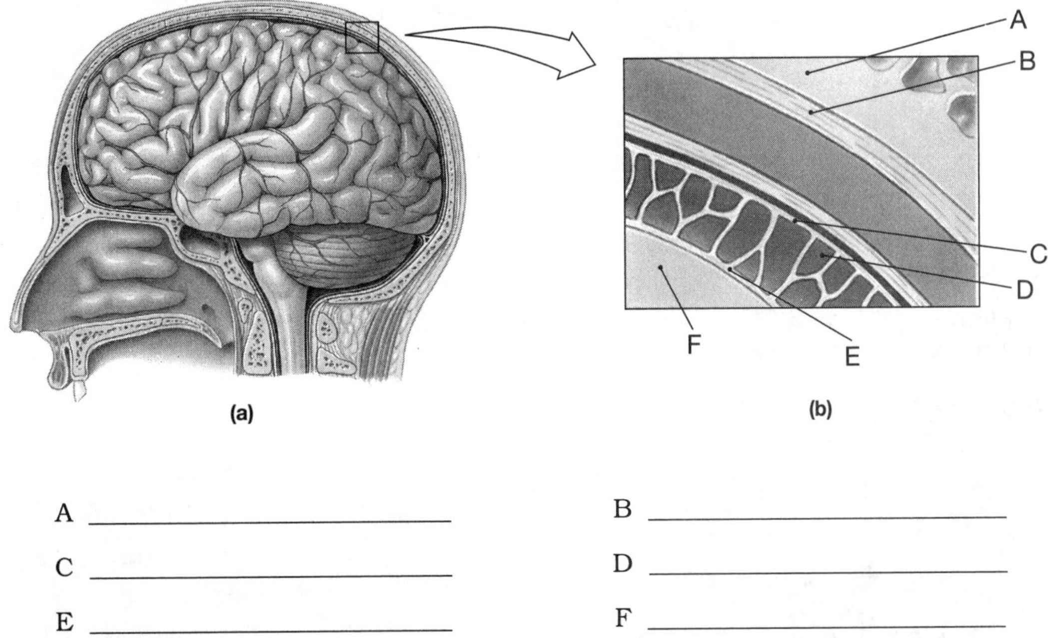

A _______________ B _______________

C _______________ D _______________

E _______________ F _______________

OBJECTIVE 3: Describe the relationship of cerebrospinal fluid with the ventricles of the brain (textbook p. 244).

After studying p. 244 in the text, you should be able to explain the roles of the cerebrospinal fluid and ventricles in the brain.

_____ 1. Cerebrospinal fluid passes through the central canal associated with the central nervous system. Where is the central canal found?

a. in the brain
b. in the choroid plexus
c. in the spinal cord
d. in the ventricles

_____ 2. The left and right cerebral hemispheres consist of which of the following ventricles?

a. second
b. third
c. fourth
d. lateral

_____ 3. How many ventricles are there in the brain?

a. 1
b. 2
c. 3
d. 4

_____ 4. The fourth ventricle is located in the _____.

a. cerebellum
b. diencephalon
c. midbrain
d. pons and medulla oblongata

_____ 5. What is the function of the cerebrospinal fluid?

a. It provides cushioning for the brain.
b. It transports nutrients to the CNS.
c. It transports waste away from the CNS.
d. all the above

6. Cerebrospinal fluid is produced by the ______________________.

7. In Figure 11–3, identify the pons, medulla oblongata, third ventricle, fourth ventricle, and the lateral ventricles.

Figure 11–3 Ventricles of the Brain

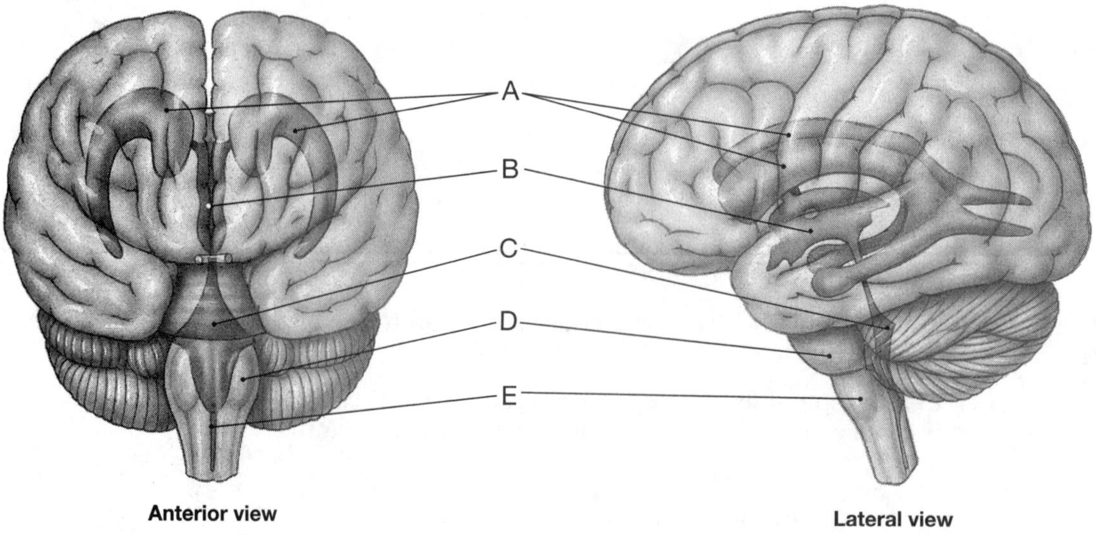

A
B
C
D
E

Anterior view

Lateral view

A lateral ventricle

B third ventricle

C fourth ventricle

D pons

E medulla oblongata

8. In Figure 11–4, identify the choroid plexus, fourth ventricle, central canal, third ventricle area, cerebellum, pons, and medulla oblongata.

Figure 11–4 Cerebrospinal Fluid Circulation through the Ventricles

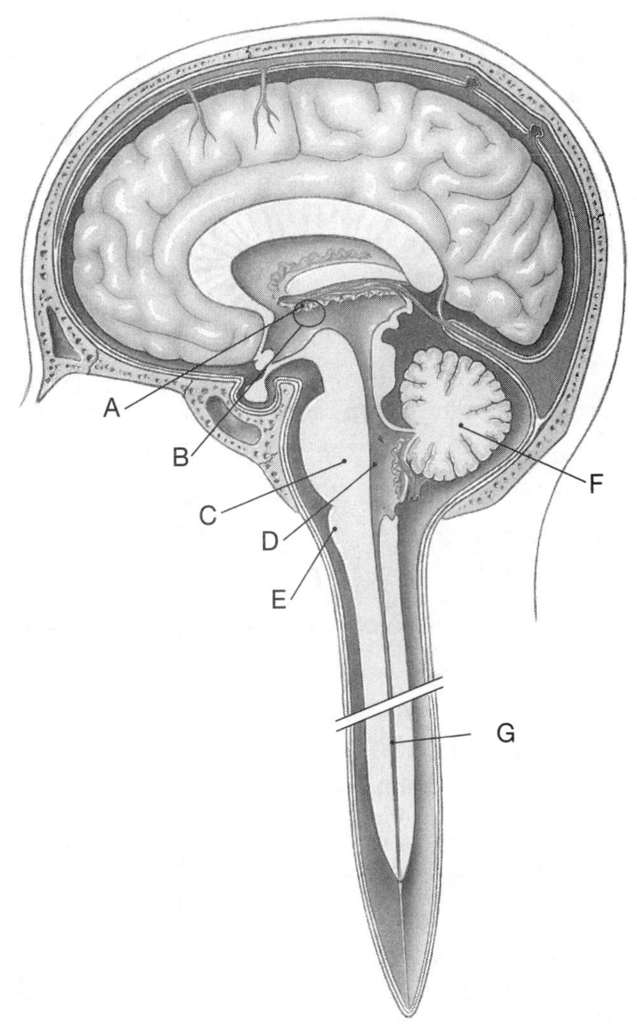

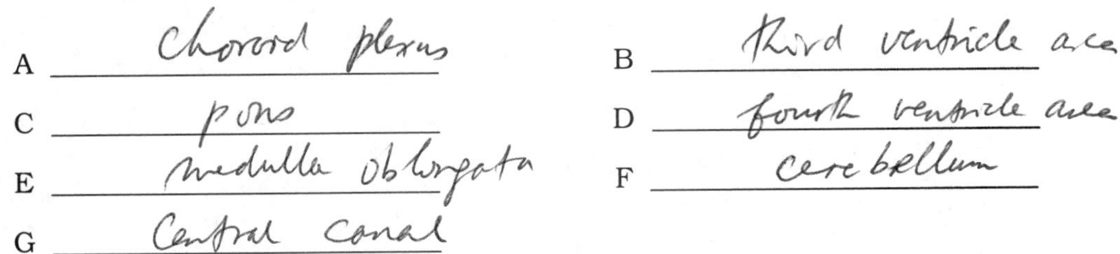

OBJECTIVE 4: Locate the motor, sensory, and association areas of the cerebral cortex and discuss their functions (textbook pp. 245–247).

After studying pp. 245–247 in the text, you should be able to identify the elevated ridges called gyri and the valleys or depressions called sulci of the cerebral cortex.

_____ 1. Where are the primary motor areas located?

a. in the precentral gyrus area
b. in the postcentral gyrus area
c. in the corpus callosum region
d. all the above

_____ 2. Where are the primary sensory areas located?

a. in the precentral gyrus area
b. in the postcentral gyrus area
c. in the corpus callosum region
d. all the above

_____ 3. The interpretive area for vision is located in the _____.

a. frontal lobe
b. occipital lobe
c. temporal lobe
d. none of the above

_____ 4. Neurons in the primary motor cortex area are responsible for _____.

a. auditory responses
b. involuntary responses
c. visual responses
d. voluntary responses

_____ 5. Damage to the _____ may cause a person to have difficulty in speaking and/or interpreting what they are hearing.

a. left hemisphere
b. right hemisphere
c. occipital lobe
d. frontal lobe

6. In Figure 11–5, identify the frontal lobe region, occipital lobe region, temporal lobe region, parietal lobe region, and central sulcus.

Figure 11–5 The Left Cerebral Hemisphere

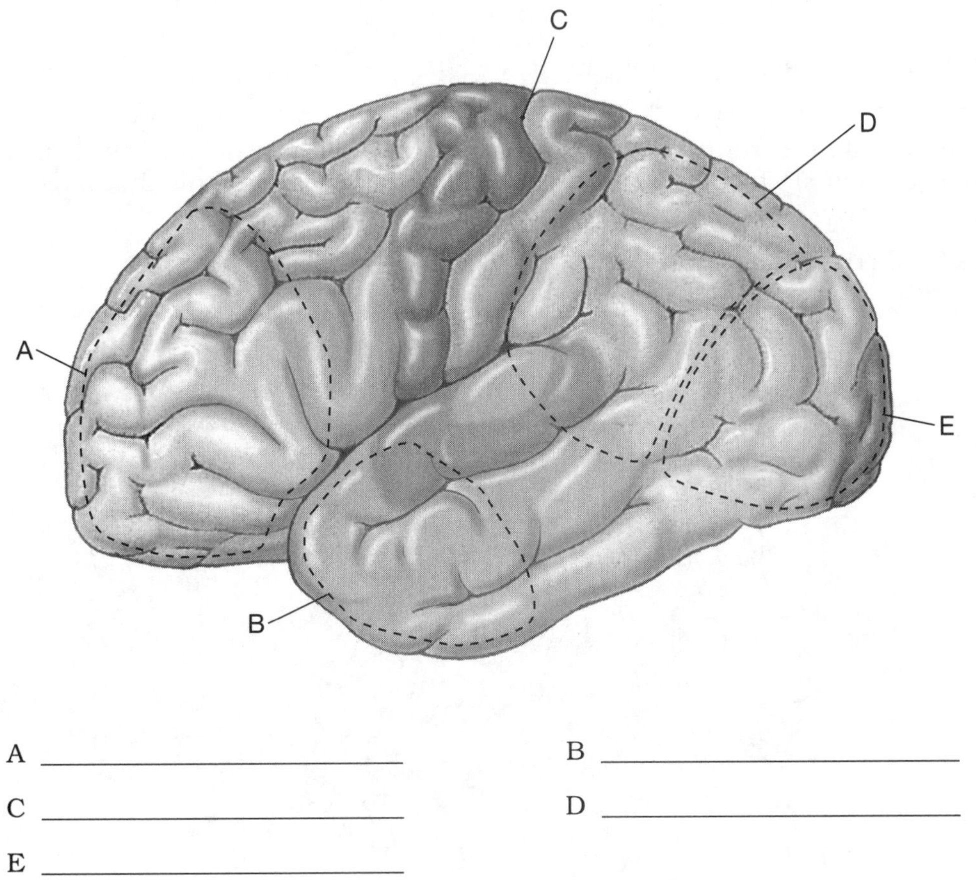

A ______________________ B ______________________

C ______________________ D ______________________

E ______________________

OBJECTIVE 5: Describe the location and functions of the limbic system (textbook p. 247).

With an increased incidence of Alzheimer's disease, the limbic system of the brain has received a great deal of attention. Damage to the hippocampus, a part of the limbic system, interferes with memory storage and retrieval that is characteristic of an Alzheimer patient. After studying p. 247 in the text, you should be able to locate the limbic system of the brain and explain the role of the hippocampus.

____ 1. The limbic system is located ____.

a. between the cerebrum and the cerebellum
b. between the cerebrum and the diencephalon
c. between the left and right hemispheres
d. between the thalamus and the hypothalamus

____ 2. The function of the limbic system includes ____.

a. conscious reflexes and emotion
b. long-term storage and conscious reflexes
c. sense of smell and long-term storage
d. all the above

_____ 3. Automatic reflexes such as chewing are controlled by the _____.

a. cerebellum
b. cerebrum
c. limbic system
d. medulla oblongata

4. Alzheimer's disease is probably due to a malfunction of the _____ of the limbic system.
_____ hippo campus

5. According to the description of the location of the limbic system (from the text and also from question 1) which letter, A, B, or C, in Figure 11–6 represents the limbic system?
_____

Figure 11–6 The Limbic System

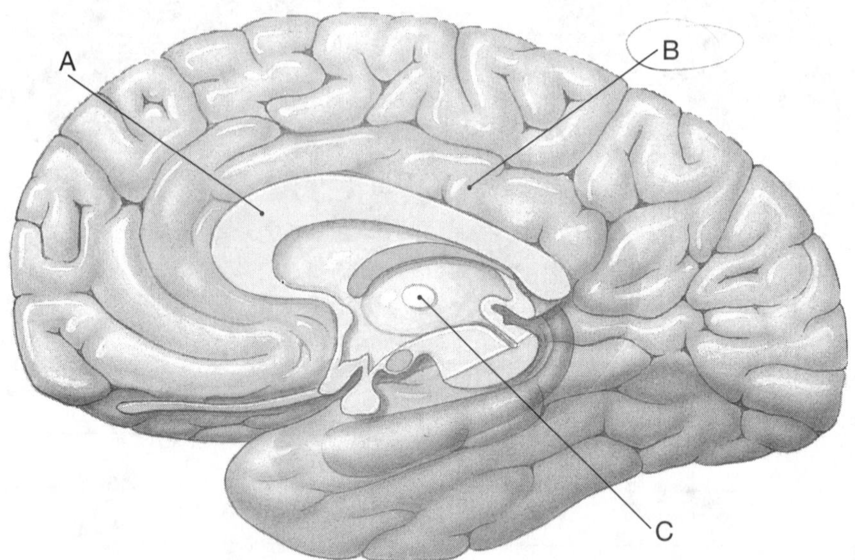

OBJECTIVE 6: Describe the functions of the diencephalon (textbook p. 248).

After studying p. 248 in the text, you should be able to identify the parts of the diencephalon, especially the thalamus and the hypothalamus.

_____ 1. The hypothalamus portion of the diencephalon does which of the following?

a. controls thirst
b. maintains body temperature
c. provides some hormones
d. all the above

_____ 2. The thalamus portion of the diencephalon does which of the following?

a. coordinates autonomic centers in the pons
b. coordinates voluntary functions
c. relays sensory information to the cerebrum
d. none of the above

____ 3. The sides of the diencephalon make up the ____, and the top consists of the ____.

a. hypothalamus, choroid plexus
b. hypothalamus, pineal gland
c. thalamus, choroid plexus
d. thalamus, pineal gland

____ 4. Which of the following produces hormones?

a. choroid plexus
b. hypothalamus
c. thalamus
d. all the above

5. The diencephalon comprises the thalamus and the hypothalamus. The thalamus is the area inferior to the corpus callosum, and the hypothalamus is inferior to the thalamus. Based on that information, identify the thalamus and the hypothalamus in Figure 11–7.

Figure 11–7 The Diencephalon

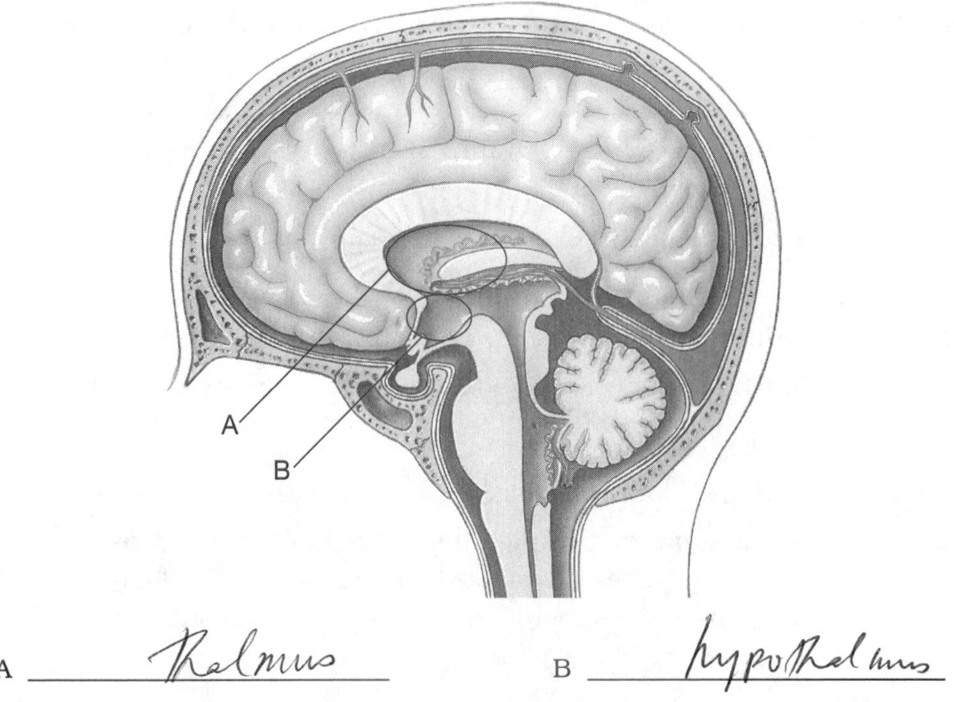

A thalmus B hypothalmus

OBJECTIVE 7: Describe the functions of the brain stem components (textbook p. 248).

After studying p. 248 in the text, you should be able to locate the brain stem and discuss its functions.

____ 1. The brain stem comprises the ____.

a. cerebellum and medulla oblongata
b. midbrain, pons, and medulla oblongata
c. pons and the cerebellum
d. pons, medulla oblongata, and spinal cord

____ 2. Which area of the brain stem is primarily responsible for keeping us alert?

a. the medulla oblongata
b. the midbrain
c. the pons
d. none of the above

____ 3. Which of the brain stem components has nuclei that are involved with one or more cranial nerves?

a. the medulla oblongata
b. the midbrain
c. the pons
d. none of the above

____ 4. Rhythmical breathing is controlled by ____.

a. the medulla oblongata
b. the midbrain
c. the pons
d. none of the above

____ 5. Which of the following structures connects the cerebellum to the cerebrum?

a. the medulla oblongata
b. the pons
c. the reticular activating system
d. the thalamus

6. The structure within the midbrain involved in keeping us awake is the reticular activating system

OBJECTIVE 8: Describe the functions of the cerebellum (textbook p. 249).

After studying p. 249 in the text, you should be able to explain the role of the cerebellum, which is sometimes referred to as the "little brain" because it coordinates a lot of information coming from the cerebrum itself.

____ 1. Which of the following statements about the cerebellum is true?

a. Ataxia may result if the cerebellum malfunctions.
b. It provides coordinated movement of skeletal muscles.
c. It provides coordinated movement of smooth muscles.
d. All the above are true.

____ 2. Which of the following is a manifestation of ataxia?

a. a lack of balance
b. jerky motions of the skeletal muscles
c. malfunctioning of the smooth muscles of the stomach
d. all the above

_____ 3. The cerebellum coordinates movement via _____ neurons.

a. parasympathetic
b. sensory
c. motor
d. sympathetic

_____ 4. The initial impulse to move a skeletal muscle originates in the _____ and is "fine tuned" in the _____.

a. cerebellum, cerebrum
b. cerebrum, cerebellum
c. midbrain, cerebellum
d. pons, cerebellum

5. The cerebellum is located at the base of the _____ lobe of the brain.

OBJECTIVE 9: Identify the cranial nerves and relate each pair to its major functions (textbook pp. 249–252).

All the nerves of the body are specialized according to their specific functions. However, there are 12 pairs of nerves that are associated with the inferior portion of the brain that have caught the attention of anatomists for many years. We identify these nerves as the **"12 pairs of cranial nerves."** After studying pp. 249–252 in the text, you should be able to describe the location and function of each of the 12 pairs of cranial nerves.

_____ 1. The cranial nerves are numbered _____.

a. from the inferior side of the brain, anterior to posterior
b. from the inferior side of the brain, posterior to anterior
c. from the superior side of the brain, anterior to posterior
d. None of the above is correct.

_____ 2. Which pair of cranial nerves is responsible for the sense of smell?

a. oculomotor
b. olfactory
c. optic
d. none of the above

_____ 3. What is the number of the pair of cranial nerves involved in vision?

a. I
b. II
c. III
d. IV

_____ 4. Which pair of cranial nerves controls the pupil of the eye?

a. abducens
b. oculomotor
c. optic
d. trochlear

_____ 5. Which pairs of cranial nerves control the eye muscles?

a. oculomotor, facial, abducens
b. oculomotor, trochlear, abducens
c. oculomotor, trochlear, trigeminal
d. optic, oculomotor, trochlear

_____ 6. When a person feels pain from sinus headaches, the _____ nerve is probably involved.

a. facial
b. ophthalmic
c. trochlear
d. none of the above

_____ 7. What is the number of the pair of cranial nerves involved in taste?

a. 5
b. 7
c. 11
d. 12

_____ 8. Which pair of cranial nerves is involved in balance?

a. the accessory
b. the cochlear portion of CN VIII
c. the ophthalmic portion of CN V
d. the vestibular portion of CN VIII

_____ 9. Which pair of cranial nerves controls the breathing muscle?

a. accessory
b. hypoglossal
c. vagus
d. none of the above

_____ 10. Which pair of cranial nerves controls tongue movement?

a. accessory
b. facial
c. glossopharyngeal
d. hypoglossal

11. Which pair of cranial nerves controls the sternocleidomastoid muscles?

_____________________ accessary nerve

12. In Figure 11–8, identify and label the 12 pairs of cranial nerves, including their numbers.

Figure 11–8 The 12 Pairs of Cranial Nerves

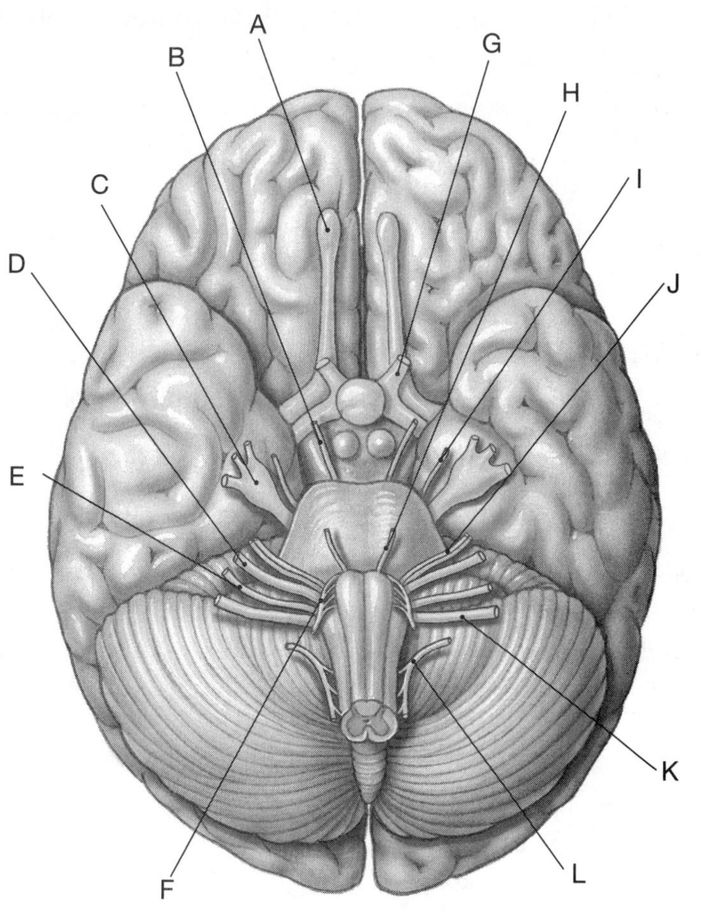

A olfactory CN I
B oculomotor CN III
C trigeminal CN V
D vestibulocochlear CN VIII
E glossopharyngeal CN IX
F hypoglossal CN XII
G optic CN II
H abducens CN VI
I trochlear CN IV
J facial CN VII
K vagus CN X
L accessary CN XI

OBJECTIVE 10: Summarize the effects of aging on the nervous system (textbook p. 252).

After studying p. 252 in the text, you should be able to explain what happens to the nervous system with age.

_____ 1. With age, it appears that the _____ get narrower and the _____ get wider in the cerebral cortex.

a. grooves, ridges
b. gyri, meninges
c. ridges, grooves
d. none of the above

_____ 2. A reduction in brain size associated with aging is primarily due to _____.

a. a decrease in synaptic connections
b. a decreased flow of blood to the brain
c. narrowing of the gyrus, thus a reduction of cerebral volume
d. all the above

_____ 3. Anatomical changes can occur in neurons as they age due to _____.

a. abnormal deposits of protein
b. extracellular deposits
c. intracellular deposits
d. all the above

_____ 4. With age, short-term memory fades but long-term memory endures, probably due to _____.

a. anatomical changes in the neurons
b. decomposition of the meninges
c. decreased flow of blood to the brain
d. diminished size of the overall brain itself

OBJECTIVE 11: Describe the three symptoms characteristic of many nervous system disorders (textbook p. 254).

Most neurological disorders are associated with symptoms such as headaches, abnormal muscle function, and paresthesia. After studying p. 254 in the text, you should be able to describe these symptoms.

_____ 1. Demyelination disorders are a ____.

a. form of paresthesia
b. muscle disease
c. nerve disease
d. neuromuscular disease

_____ 2. Myopathies is a general term for ____.

a. muscle diseases
b. neuromuscular diseases
c. paresthetic diseases
d. peripheral nerve damage

_____ 3. Muscle palsy causes ____.

a. loss of feeling
b. muscle weakness
c. myopathy
d. spasms

_____ 4. Paresthesia causes _____.

a. muscle weakness
b. myopathy
c. spasms
d. tingling sensations

OBJECTIVE 12: Describe common disorders of the brain and cranial nerves (textbook pp. 254–261).

There are numerous brain and cranial disorders. After studying pp. 254–261 in the text, you should be able to describe these disorders and how these they affect the human body.

_____ 1. Which of the following may cause meningitis?

a. *Haemophilus influenzae*
b. mumps virus
c. *Streptococcus pneumoniae*
d. all the above

_____ 2. An infection of the membranes associated with the brain and spinal cord is known as ____.

a. Huntington's disease
b. meningitis
c. Parkinson's disease
d. rabies

_____ 3. Generally, a benign tumor of the brain can be removed because ____.

a. it forms between the arachnoid and dura mater
b. it forms between the arachnoid and the pia mater
c. it forms between the dura mater and the skull
d. it forms on the sulci and not on the gyri of the brain

_____ 4. Identify the disease being described: A patient wants a specific muscle to contract. Before the patient can get that muscle to contract, the antagonistic muscles must be overcome with a relatively forceful contraction.

a. Alzheimer's disease
b. cerebral palsy
c. Huntington's disease
d. Parkinson's disease

_____ 5. Bell's palsy is an infliction of cranial nerve number ____.

a. III
b. V
c. VII
d. XI

PART II: CHAPTER-COMPREHENSIVE EXERCISES

Match the descriptions or statements in column B with the terms in column A.

MATCHING

	(A)	(B)
_____	1. abducens	A. This structure relays information from the surroundings to the cerebrum.
_____	2. Bell's palsy	B. This structure helps regulate body temperature.
_____	3. cerebellum	C. A group of nerves connecting the two hemispheres
_____	4. choroid plexus	D. This structure receives information from the cerebrum and alters the information to provide smooth coordinated activity.
_____	5. corpus callosum	E. This lobe is involved in the interpretation of vision.
_____	6. frontal	F. This lobe is involved in fine-motor dexterity.
_____	7. Hutchinson's disease	G. This structure provides rhythmic breathing.
_____	8. hypothalamus	H. The membranes that surround the structures of the CNS
_____	9. medulla oblongata	I. This structure produces cerebrospinal fluid.
_____	10. meninges	J. This nerve controls the lateral rectus muscle of the eye.
_____	11. occipital	K. This nerve controls the superior oblique muscle of the eye.
_____	12. oculomotor	L. This nerve controls the lens of the eye.
_____	13. Parkinson's disease	M. An inherited disease that causes a gradual decline in intellectual abilities
_____	14. thalamus	N. A disease that causes tremors
_____	15. trochlea	O. An inflammation of the facial nerve

CONCEPT MAP

This concept map summarizes and organizes the information in Chapter 11. Use the following terms to complete the map by filling in the boxes identified by the circled numbers, 1–13.

blood pressure	**cerebellum**	**diencephalon**
pons	**body temperature**	**thalamus**
meninges	**meningitis**	**Parkinson's disease**
Huntington's disease	**N V**	**N VII**
N XI		

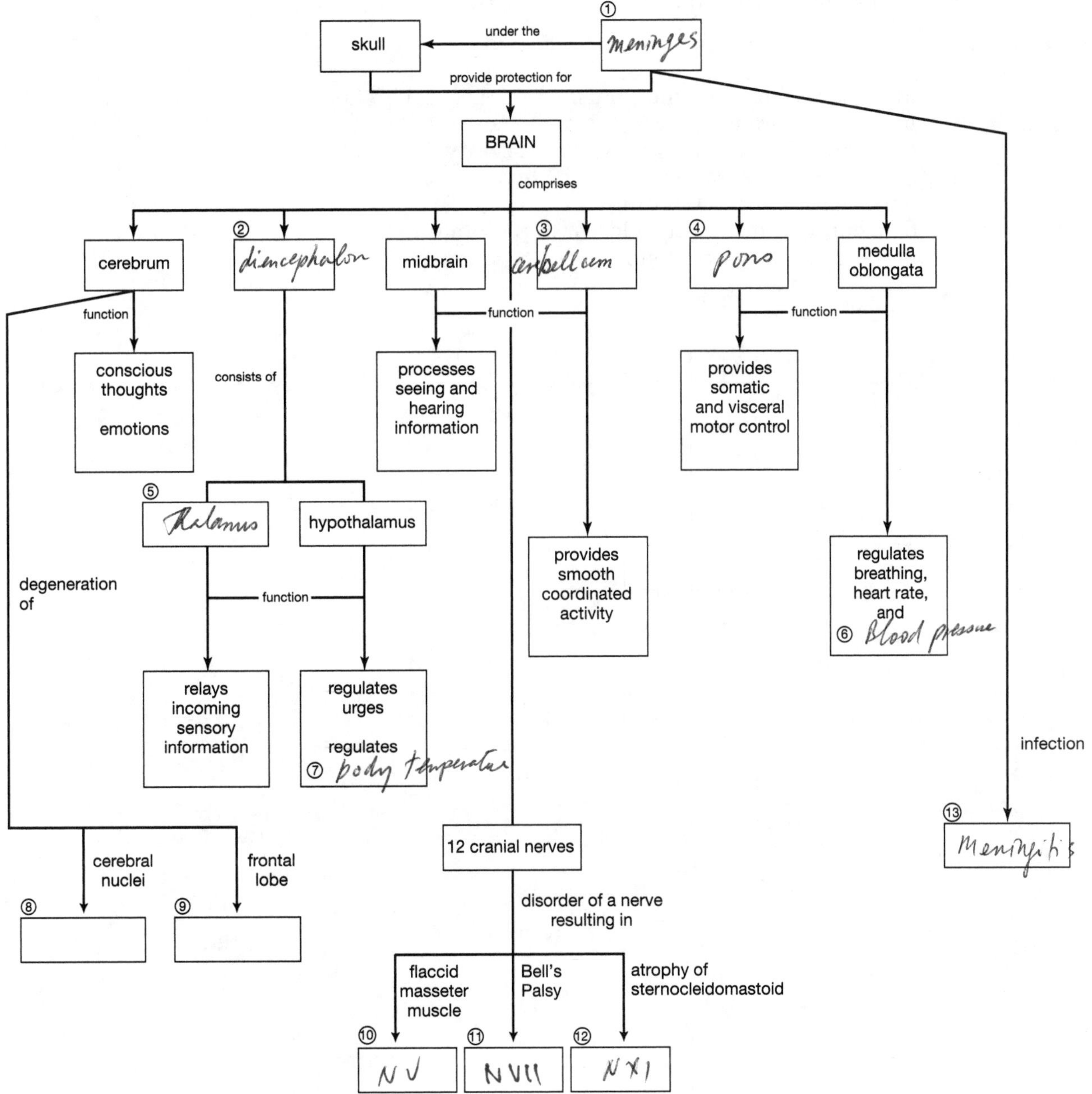

CROSSWORD PUZZLE

The following crossword puzzle reviews the material in Chapter 11. To complete the puzzle, you have to know the answers to the clues given, and you must be able to spell the terms correctly.

ACROSS

1. A disease that results in a deterioration of mental organization
4. A disease that results in tremors
8. This part of the brain produces oxytocin.
11. The mandibular portion of the _____ nerve is involved in chewing.
12. Smell is interpreted in the _____ lobe of the brain.

DOWN

2. A disease that results in diminished intellectual abilities
3. Ridges of the brain
5. Structural changes in cerebral _____ occur with aging.
6. The _____ gland is connected to the hypothalamus.
7. Shallow grooves of the brain
9. The meningeal membrane closest to the brain
10. Speech centers are located in the _____ hemisphere.

FILL-IN-THE-BLANK NARRATIVE

Use the following terms to fill in the blanks of this summary of Chapter 11.

accessory	**alert**	**cerebellum**
hemisphere	**hypoglossal**	**masseter**
occipital	**oculomotor**	**olfactory**
optic	**sleep**	**synaptic**
temporal	**trigeminal**	**Parkinson's disease**
dopamine	**levodopa**	**Bell's palsy**

While walking through a general store, Adam detects the odor of perfume. The (1) ___________ nerves send signals to the (2) ___________ lobe of his brain to interpret the odor. Adam's brain interprets this to be a very pleasant smell; however, Adam is at the store to buy a candy bar. He purchases the candy bar and eats it. Adam is able to find the candy bar he likes because the (3) ___________ nerve allows his eyes to focus properly. The image of the candy bar is sent to the (4) ___________ lobe via the (5) ___________ nerve. His cerebrum sends a signal indicating "eat the candy bar," and his (6) ___________ provides that activity with smooth, efficient, coordinated muscle control. The mandibular portion of the (7) ___________ nerve stimulates the (8) ___________ muscle, allowing Adam to chew the candy bar. The (9) ___________ nerve controls his tongue so he can move the chewed food to the back of his mouth and begin to swallow. Swallowing is controlled by the (10) ___________ nerves.

As Adam chews and swallows, he comments that the candy bar tastes good. He can speak about the candy bar because the speech center in his left (11) ___________ is functioning, allowing him to articulate his thoughts into words.

Adam decides to purchase another candy bar to take to class with him. The action of eating a candy bar may activate the reticular activating system (RAS) of his midbrain, which will keep him (12) ___________, but soon, Adam's RAS becomes inactive and he begins to drift off to (13) ___________.

With age, (14) ___________ changes will occur in Adam's brain, and his memory of recent events will begin to diminish even though he might still remember things from the past. As Adam ages he notices that the simple task of eating candy bars becomes a chore. His hand and arm muscles exhibit a tremor-like action. Adam apparently has developed (15) ___________. Adam has been seen by a physician and has been prescribed (16) ___________, which crosses the blood-brain barrier and is converted to (17) ___________, which reduces, to a certain degree, the tremor action. In addition, Adam notices that he has lost his sense of taste, and one corner of his mouth is drooping. The physician has also diagnosed Adam as having (18) ___________, which is an affliction of cranial nerve number 7.

CLINICAL CONCEPTS

The following clinical concepts apply the information in Chapter 11. Following the applications is a set of questions to help you understand the concept.

1. Bell's palsy is a condition that affects the facial cranial nerve, making it difficult for the patient to produce tears. The palsy also makes it difficult to close the eyelids or to be out in bright sunlight, as the pupils of the eyes will not dilate and constrict appropriately. Often, the patient has to use eye drops to continually lubricate the eye and also has to wear an eye patch.
2. The "job" of the cerebellum is to receive "commands" from the cerebrum and then execute them. The cerebrum acts as the "boss." The cerebrum "tells" the cerebellum what to do, and the cerebellum carries out the instructions in a smooth, coordinated manner. For example, the cerebrum may give the command to walk from point A to point B. The cerebellum will allow you to walk from point A to point B in a very smooth, coordinated fashion. Without the cerebellum, the cerebrum would still allow you to walk from point A to point B but perhaps not in a smooth, coordinated fashion.

1. What is the number for the facial nerve?

2. Why does a patient with Bell's palsy need to use eye drops continually?

3. Which structure, the cerebrum or the cerebellum, initiates the command to begin walking?

4. Which structure, the cerebrum or cerebellum, carries out the motor commands and permits walking in a coordinated manner?

5. What symptoms might you expect if a patient who had cancer of the cerebellum was getting progressively worse?

THIS CONCLUDES CHAPTER 11 EXERCISES

CHAPTER

12

The Senses

Our awareness of the world within and around us is based on information provided by our senses. Sensory receptors are specialized cells or cell processes that react to stimuli within the body or to stimuli in the environment outside the body. The sensory receptors receive stimuli, and neurons transmit action potentials to the CNS, where the sensations are processed and translated. The result is motor responses that serve to maintain homeostasis.

Chapter 12 considers the **general senses** of temperature, pain, touch, pressure, vibration, and proprioception as well as the **special senses** of smell, taste, balance, hearing, and vision. The activities in this chapter examine receptor function and basic concepts in sensory processing.

Chapter 12 also discusses a few select disorders of the senses and how these can upset homeostasis.

Chapter Objectives:

- Distinguish between the general and special senses.
- Identify the receptors for the general senses and describe how they function.
- Describe the receptors and processes involved in the sense of smell.
- Discuss the receptors and processes involved in the sense of taste.
- Identify the parts of the eye and their functions.
- Explain how we are able to see objects and distinguish colors.
- Discuss the receptors and processes involved in the sense of equilibrium.
- Describe the parts of the ear and their roles in the process of hearing.
- Describe the assessment of the general senses, and pain relief by non-narcotic and narcotic drugs.
- Describe examples of disorders in the senses of smell, taste, vision, equilibrium, and hearing.

PART I: OBJECTIVE-BASED QUESTIONS

OBJECTIVE 1: Distinguish between the general and special senses (textbook p. 267).

After studying p. 267 in the text, you should be able to describe the two categories of senses of the nervous system based on function: general senses and special senses.

_____ 1. Which of the following is considered a general sense?

a. balance
b. pain
c. smell
d. taste

_____ 2 Which of the following is considered a special sense?

a. balance
b. pain
c. pressure
d. temperature

_____ 3. The receptors for special senses are located _____.

a. all over the body
b. in specific sense organs
c. in the cerebral area
d. all the above

4. Why are some senses considered "special" and others "general?"

____

OBJECTIVE 2: Identify the receptors for the general senses and describe how they function (textbook pp. 268–269).

The general senses have receptors located all over the body. After studying pp. 268–269 in the text, you should be able to describe the locations of the receptors for the general senses and the precise function of each.

_____ 1. Which of the following statements is false?

a. Pain receptors consist of free nerve endings.
b. There are different types of pain receptors.
c. There are pain receptors in the brain.
d. There are pain receptors scattered throughout our skin.

_____ 2. Which of the following statements is true?

a. Temperature receptors are free nerve endings, whereas pain receptors do not exist as free nerve endings.
b. The receptors for detecting cold are the same as the receptors for detecting hot.
c. The temperature receptors are typically located in deep tissue.
d. There are separate temperature receptors for detecting cold and hot.

_____ 3. Proprioceptors are located _____.

a. in muscles
b. in tendons
c. within joints
d. all the above

_____ 4. If you walk into an air-conditioned room, the temperature receptors send signals to your brain and you interpret cold. After a while, you no longer feel that it is cold but rather that the temperature is comfortable. This is because of _____.

a. a process called adaptation
b. an increase of sensory responses to the constant stimuli of cold
c. an increase in other stimuli that "masks" the temperature stimuli
d. Both a and b are correct.

5. Touch and pressure sensory receptors are called ________________________ receptors.

OBJECTIVE 3: Describe the receptors and processes involved in the sense of smell (textbook pp. 269–270).

The sense of smell is not nearly as acute in humans as it is in many other animals, however, we rely on it constantly to detect and identify the aromas around us. After studying pp. 269–270 in the text, you should be able to explain the role the olfactory receptors play in relaying the sensory information of smell to the olfactory regions of the brain.

_____ 1. What is the function of the mucus produced by the olfactory glands?

a. It keeps the nasal area moist.
b. It prevents the buildup of overpowering stimuli.
c. It provides a medium for molecules to dissolve in.
d. All the above are correct.

_____ 2. An impulse is sent to the temporal lobe of the brain via cranial nerve I when _____.

a. a dissolved chemical contacts the olfactory cilia
b. a dissolved chemical contacts the olfactory nerve
c. a dissolved chemical contacts the olfactory organ
d. a dissolved chemical contacts the olfactory gland

3. The olfactory axons must pass through the ________________________ before reaching the olfactory bulb portion of cranial nerve I.

4. Before odorous compounds can stimulate olfactory receptors, they must be ________________________ in the mucus.

5. Why do dogs have a greater sensitivity to odors than do humans ?

____

6. In Figure 12–1, identify the cribriform plate area, olfactory nerve, olfactory cilia, and olfactory receptor cells.

Figure 12–1 The Olfactory Organs

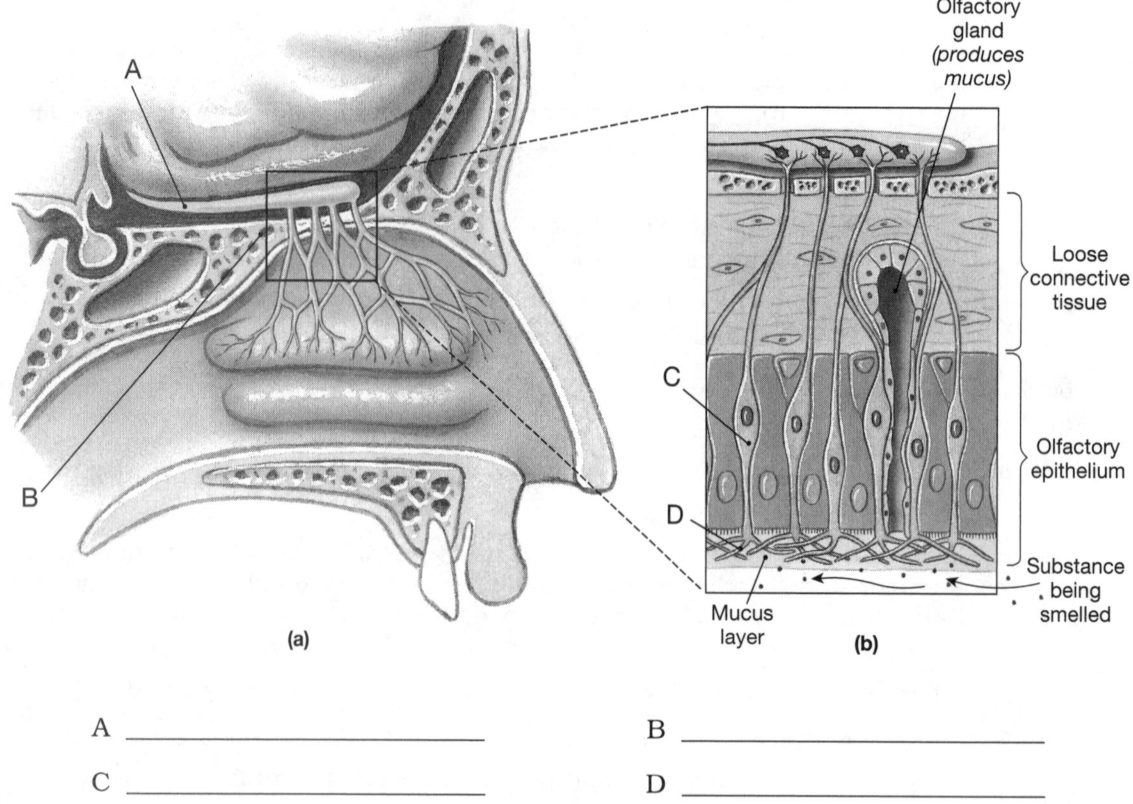

A ____________________ B ____________________

C ____________________ D ____________________

OBJECTIVE 4: Discuss the receptors and processes involved in the sense of taste (textbook p. 270).

After studying p. 270 in the text, you should be able to explain the differences among papillae, taste buds, and gustatory cells, which are responsible for detecting flavor.

_____ 1. Which of the following statements is true?

a. Papillae of the tongue consist of gustatory cells, which consists of taste buds.

b. Papillae of the tongue consist of taste buds, which consist of gustatory cells.

c. Taste buds of the tongue consist of gustatory cells, which consist of papillae.

d. Taste buds of the tongue consist of papillae, which consist of gustatory cells.

_____ 2. The sensation of taste is due to receptors called _____.

a. gustatory cells

b. papillae

c. taste buds

d. none of the above

____ 3. In order to initiate a nerve impulse along the facial nerve, ____.

a. dissolved chemicals contact taste hairs, which stimulate a change in the facial nerve
b. dissolved chemicals contact taste hairs, which stimulate a change in the gustatory cells
c. dissolved chemicals contact taste hairs, which stimulate a change in the papillae
d. dissolved chemicals contact the gustatory cells, which activate the facial nerve

____ 4. The function of the gustatory cells is very similar to that of ____.

a. olfactory cells
b. pain receptors
c. proprioceptors
d. tactile receptors

5. If your tongue were completely dry, you would not be able to detect the flavor of food because molecules must ______________________ before contacting the taste hairs.

6. In Figure 12–2, identify the gustatory cells, papillae, and taste buds.

Figure 12–2 Taste Buds and Gustatory Cells

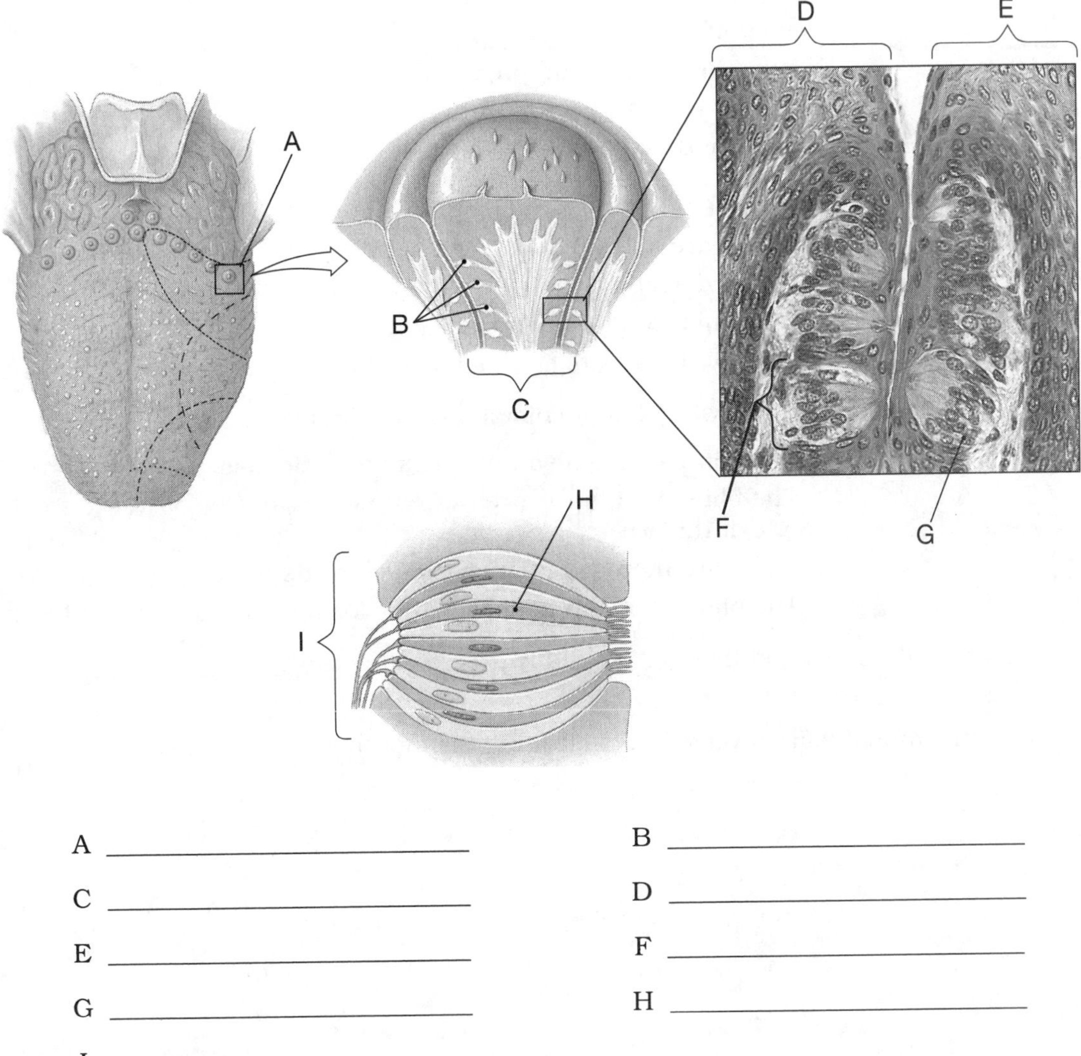

A ______________________ B ______________________

C ______________________ D ______________________

E ______________________ F ______________________

G ______________________ H ______________________

I ______________________

OBJECTIVE 5: Identify the parts of the eye and their functions (textbook pp. 271–275).

After studying pp. 271–275 in the text, you should be able to describe the anatomy of the eye and explain how the eyes facilitate the sense of sight by sending messages to the brain for interpretation.

_____ 1. Which of the following statements is true?

a. The choroid is the outer layer of the eye. The cornea is the transparent portion of the choroid.
b. The retina develops into the choroid, which develops into the sclera.
c. The retina is the outer layer of the eye. The cornea is the transparent portion of the retina.
d. The sclera is the outer layer of the eye. The cornea is the transparent portion of the sclera.

_____ 2. The pupil is an opening through the _____.

a. cornea
b. iris
c. lens
d. retina

_____ 3. The inner layer of the eye is called the _____. It is made of _____.

a. choroid, retinal cells
b. choroid, the iris and pupil
c. retina, rods and cones
d. none of the above

_____ 4. Which of the following statements is false?

a. Cones are activated with bright light.
b. Cones detect color.
c. Rods allow us to see in dim light.
d. The fovea consists of a high concentration of rods.

_____ 5. Which of the following statements about the blind spot is false?

a. The blind spot is also known as the optic disc.
b. The blind spot is the area where nerves and blood vessels enter or exit the eye.
c. There are no rods or cones located in the blind spot.
d. The blind spot can be detected by looking straight at an object.

6. The eye is lubricated and therefore partially protected by the ____________________ gland.

7. How many muscles are involved in controlling the movement of each eye?

____

8. In Figure 12–3, identify the optic disc, optic nerve, retina, sclera, cornea, pupil, lens, ciliary bodies, and vitreous chamber.

Figure 12–3 Anatomy of the Eye

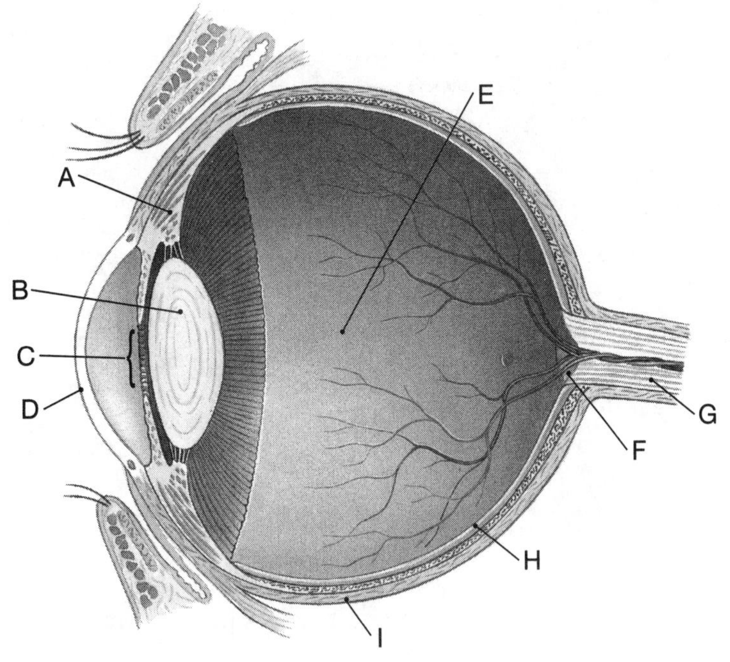

A ______________________ B ______________________

C ______________________ D ______________________

E ______________________ F ______________________

G ______________________ H ______________________

I ______________________

9. In Figure 12–4, identify and label the six different eye muscles. For letter F name the muscle that cannot be seen in this view.

Figure 12–4 The Extrinsic Eye Muscles (lateral view of the left eye)

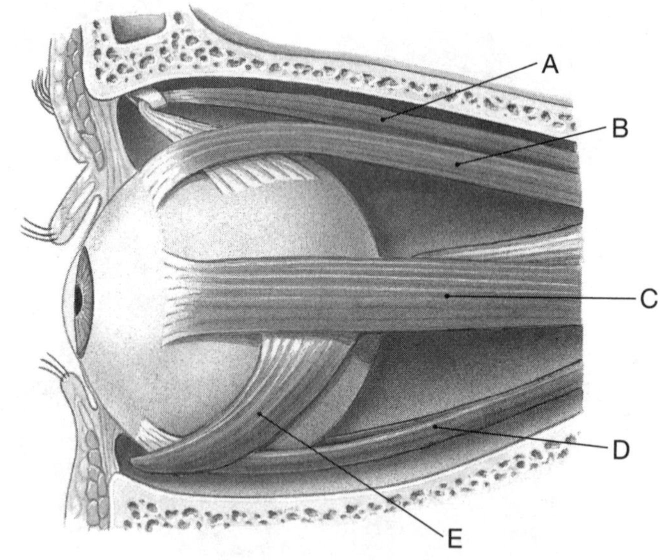

A ______________________ B ______________________

C ______________________ D ______________________

E ______________________

OBJECTIVE 6: Explain how we are able to see objects and distinguish colors (textbook pp. 273–274).

The sense of vision is initiated in the retina of the eye due to the presence of photoreceptor cells called rods and cones. When the cones are activated, we see color. At night or in dim light, only the rods are activated. Therefore, we cannot see color at night. After studying pp. 273–274 in the text, you should be able to describe the process of vision and the role of rods and cones in differentiating colors.

_____ 1. The _____ changes in diameter in response to the intensity of light entering the eye. Light is necessary to activate the cells of the _____.

a. lens, retina

b. pupil, optic disc

c. pupil, retina

d. retina, pupil

_____ 2. Color vision is interpreted when the _____ is (are) stimulated.

a. cones

b. optic disc

c. rods

d. none of the above

_____ 3. Axons from the _____ converge on the _____ and proceed toward the brain as cranial nerve _____.

a. ganglion cells, optic disc, II
b. cornea, fovea, I
c. fovea, optic disc, I
d. optic disc, fovea, II

_____ 4. Nearsightedness, or _____, is a condition in which you can see _____.

a. hyperopia, close to you
b. myopia, close to you
c. hyperopia, far away from you
d. myopia, far away from you

5. There are hundreds of different colors. How many different types of cones are there to detect all those colors? ______________________________

6. We typically cannot see color at night because the ______________________________ are not activated in dim light.

7. In Figure 12–5, identify the rods, cones, outer layer of the retina, and inner layer of the retina.

Figure 12–5 Structure of the Retina

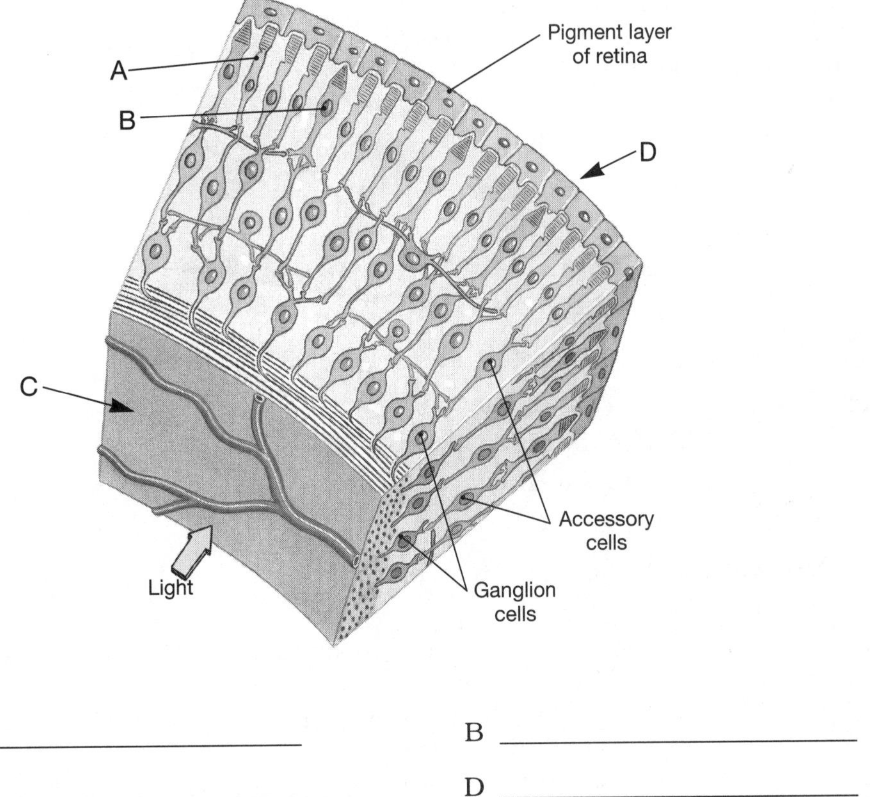

A ______________________________ B ______________________________

C ______________________________ D ______________________________

8. In Figure 12–6, identify myopia, hyperopia, the retina, cornea, and lens.

Figure 12–6 Two Common Visual Problems

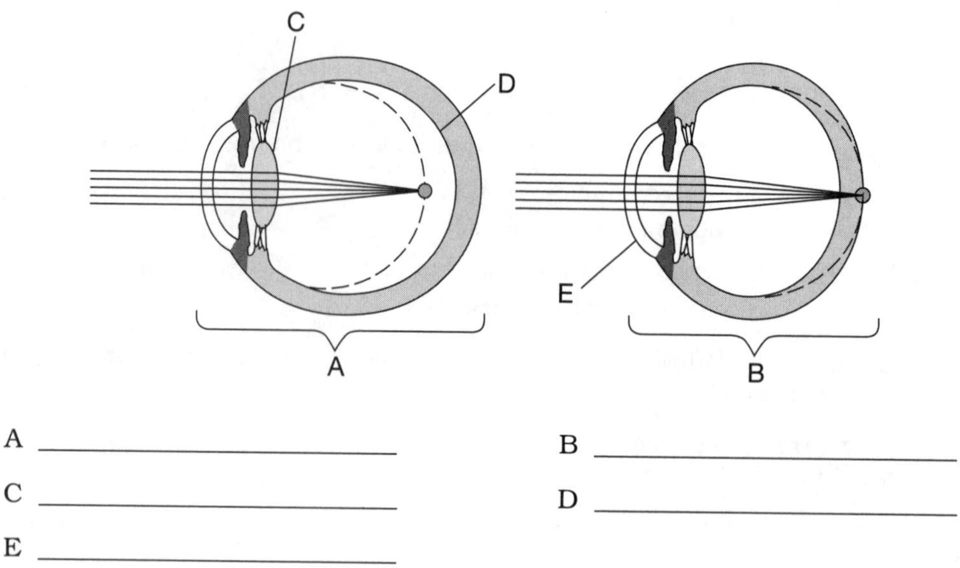

A ______________________ B ______________________

C ______________________ D ______________________

E ______________________

OBJECTIVE 7: Discuss the receptors and processes involved in the sense of equilibrium (textbook pp. 275–276).

It may surprise you to learn that the ear is involved not only in hearing but also in maintaining balance. After studying pp. 275–276 in the text, you should be able to explain the role of the ears, and especially the fluid in special structures inside the ears in helping maintain balance.

_____ 1. Which portion of the ear is involved in balance?

a. the inner ear
b. the middle ear
c. the outer ear
d. all the above

_____ 2. Where are the receptors involved in balance found?

a. in the cochlea
b. in the organ of Corti
c. in the semicircular canals
d. none of the above

_____ 3. Which of the following statements best explains how balance is achieved by the ears?

a. When the fluid in the ears moves, a signal is sent to the brain via CN VIII.
b. When the fluid in the cochlea moves, a signal is sent to the brain via CN VIII.
c. When the fluid in the cochlea moves, a signal is sent to the brain via the vestibular portion of CN VIII.
d. When the fluid in the semicircular canals moves, a signal is sent to the brain via CN VIII.

_____ 4. Often, a(n) _____ infection causes dizziness.

a. cochlear
b. inner ear
c. middle ear
d. outer ear

5. What is the name of the fluid that can be found in the semicircular canals of the inner ear? ______________________

6. In Figure 12–7, identify the outer ear area, inner ear area, middle ear area, semicircular canals, and CN VIII.

Figure 12–7 Structure of the Ear

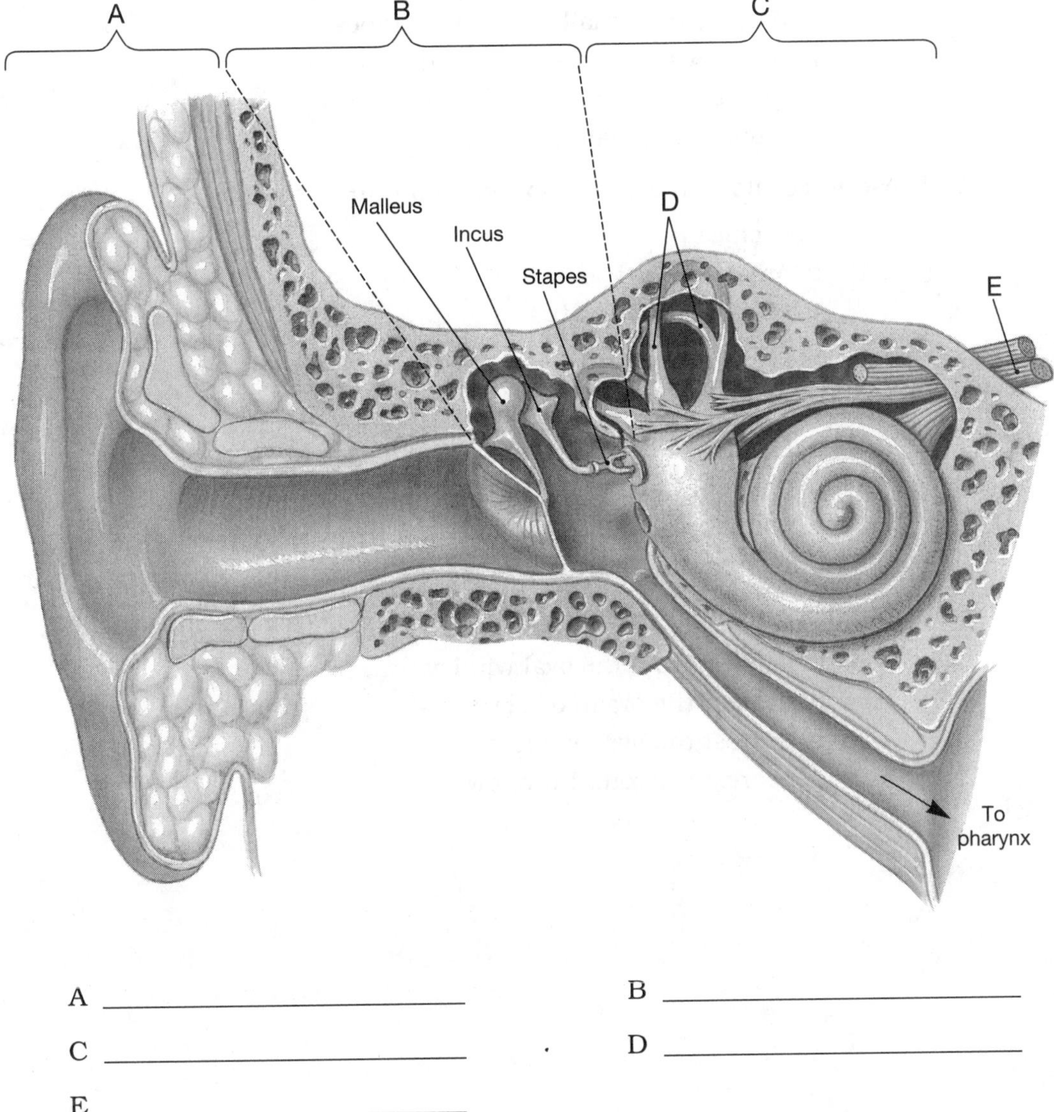

A ______________________ B ______________________

C ______________________ D ______________________

E ______________________

OBJECTIVE 8: Describe the parts of the ear and their roles in the process of hearing (textbook pp. 276–279).

After studying pp. 276–279 in the text, you should be able to describe the anatomy of the ear and explain the role of each part including the three smallest bones of the body in the process of hearing.

____ 1. The ____, located in the ____ ear, is involved in hearing.

a. cochlea, inner
b. cochlea, middle
c. vestibule, inner
d. vestibule, middle

____ 2. Sound waves vibrate the ____, which sets the ossicles in motion. The sequence of ossicles is ____.

a. oval window, malleus; incus, stapes
b. round window, malleus; incus, stapes
c. tympanum, incus; stapes, malleus
d. tympanum, malleus; incus, stapes

____ 3. Sensory receptors that detect sound are located in the ____.

a. cochlea
b. organ of Corti
c. vestibular apparatus
d. none of the above

____ 4. The ossicles connect the ____.

a. cochlea to the organ of Corti
b. organ of Corti to CN VIII
c. tympanum to the oval window
d. tympanum to the round window

____ 5. High-pitched sounds are detected by movement of cochlear fluid that is ____.

a. farthest from the oval window
b. nearest the organ of Corti
c. nearest the oval window
d. nearest the round window

6. In Figure 12–8, identify the malleus, incus, stapes, tympanum, oval window, endolymph area, perilymph area, area of high-pitched sounds, and the area of low-pitched sounds.

Figure 12–8 The Structures of Hearing

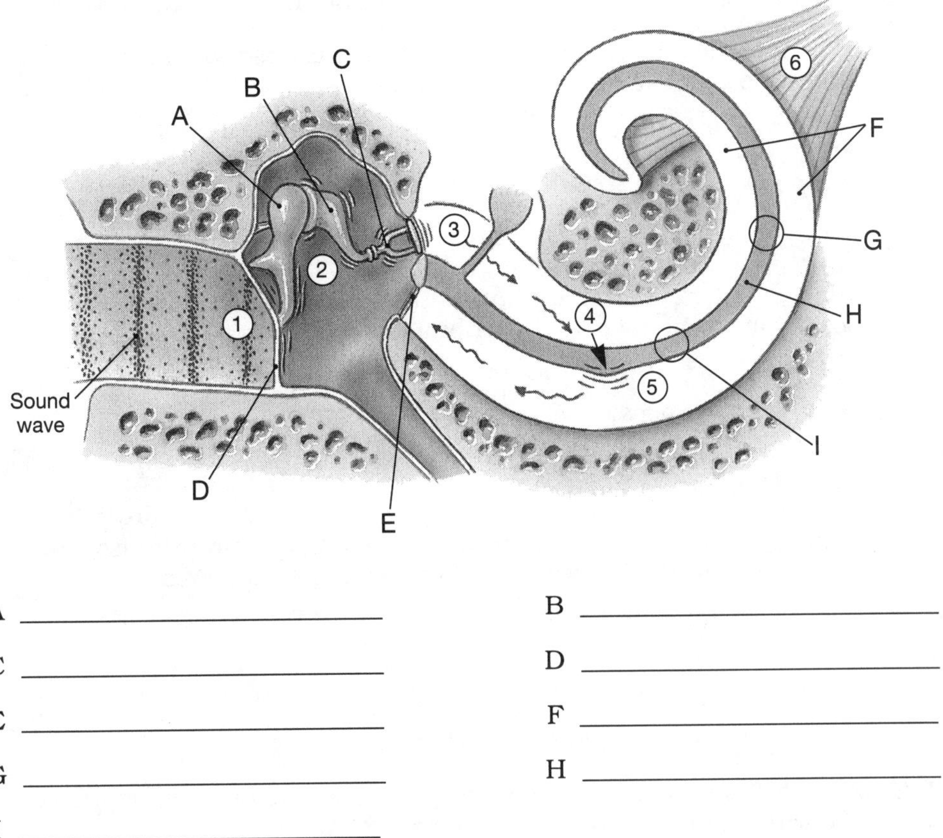

A ______________________ B ______________________

C ______________________ D ______________________

E ______________________ F ______________________

G ______________________ H ______________________

I ______________________

OBJECTIVE 9: Describe the assessment of the general senses and pain relief by non-narcotic and narcotic drugs (textbook p. 281).

A series of assessments is used to determine the source of the problem causing a patient to be out of homeostasis. Once the assessment indicates the source of the pain, for example, pain relievers may be given. After studying p. 281 in the text, you should be able to describe some of the assessment procedures and be able to define pain reliever terminology.

_____ 1. Which of the following statements is true in reference to the two-point discrimination test?

a. On the back of the hand, a patient normally cannot feel two points that are closer together than 4 cm.

b. On the back of the hand, a patient normally cannot feel two points that are farther apart than 4 cm.

c. On the back of the hand, a patient normally can feel two points only if the two points are 4 - 7 cm apart.

_____ 2. Analgesics reduce pain by _____.

a. reducing inflammation

b. reducing stimulation of neurons associated with the pain area

c. Both a and b are correct.

d. None of the above is correct.

_____ 3. Analgesics are _____.

a. non-narcotic
b. narcotic
c. Both a and b are correct.
d. None of the above is correct because they are anesthetics.

_____ 4. Which of the following types of pain is being described? A person stubs a toe and feels the pain.

a. acute pain
b. analgesic
c. chronic pain
d. paresthesia

_____ 5. A(n) _____ is a drug that reduces the feeling of pain but does not cause the loss of any other sensitivity within a specified area.

a. anesthetic
b. analgesic
c. narcotic
d. prostaglandin

_____ 6. A(n) _____ is a drug that reduces the feeling of pain but also causes temporary loss of any other sensitivity within a specified area.

a. anesthetic
b. analgesic
c. narcotic
d. prostaglandin

OBJECTIVE 10: Describe examples of disorders in the senses of smell, taste, vision, equilibrium, and hearing (textbook pp. 281-286).

There are numerous disorders associated with the five main senses. After studying pp. 281-286 in the text, you should be able to describe a few select disorders.

_____ 1. Damage to the cribriform plate (Chapter 8, The Skeletal System) could cause problems with _____.

a. the ability to feel
b. the ability to hear
c. the ability to see
d. the ability to smell

_____ 2. Damage to cranial nerve VII affects _____.

a. the ability to feel
b. the ability to taste
c. the ability to see
d. the ability to smell

____ 3. The presence of a whitish area in the pupil is characteristic of _____.

a. bacterial conjunctivitis
b. cataracts
c. glaucoma
d. retinoblastoma

____ 4. Steve wears glasses. He can read without his glasses but needs to put his glasses back on when he looks at distant objects. Steve has a case of _____.

a. astigmatism
b. hyperopia
c. myopia
d. strabismus

____ 5. Glaucoma is a condition in which _____.

a. pressure in the eye distorts the retina
b. pressure on the cornea causes vision problems
c. the lens of the eye becomes less elastic
d. the retinal artery is blocked resulting in blindness

____ 6. When the endolymph within the semicircular canals begins to move, even though you are standing still, you begin to think you are moving. This condition is known as _____.

a. onchocerciasis
b. otitis media
c. presbycusis
d. vertigo

PART II: CHAPTER-COMPREHENSIVE EXERCISES

Match the terms or phrases in column A with the descriptions or statements in column B.

MATCHING I

	(A)	(B)
____	1. adaptation	A. Special cells located in the taste buds
____	2. cerumen	B. The transparent portion of the sclera
____	3. ceruminous	C. The layer of eye tissue between the sclera and the retina
____	4. choroid	D. The inner ear portion involved with hearing
____	5. ciliary muscles	E. The gland that produces earwax
____	6. cochlea	F. The central portion of the retina that is 100% cones
____	7. cornea	G. Structures that change the shape of the lens
____	8. fovea	H. We typically don't "feel" our clothes on our body all day long because of _____.
____	9. gustatory cells	I. The name for earwax
____	10. hyperopia	J. Farsightedness due to abnormal shape of the eye
____	11. presbyopia	K. Farsightedness due to age

MATCHING II

	(A)	(B)
_____	1. astigmatism	A. The process of smelling
_____	2. malleus	B. The innermost layer of the eye consisting of rods and cones
_____	3. olfaction	C. The inner ear portion involved with balance
_____	4. optic disc	D. The ossicle attached to the oval window
_____	5. presbyopia	E. The ossicle attached to the tympanum
_____	6. proprioceptors	F. A form of hyperopia found in the elderly
_____	7. retina	G. The portion of the retina that does not contain rods and cones
_____	8. semicircular canals	H. Special structures that monitor tendons and ligaments
_____	9. stapes	I. Results in the "unwanted" movement of endolymph
_____	10. vertigo	J. Abnormal-shaped cornea

CONCEPT MAP

This concept map summarizes and organizes the information concerning the senses. Use the following terms to complete the map by filling in the boxes identified by the circled numbers, 1–12.

organ of Corti	**gustatory receptors**	**taste**
vision	**proprioceptors**	**tactile receptors**
pain	**vertigo**	**glaucoma**
affliction of N VII	**affliction of N I**	**otitis media**

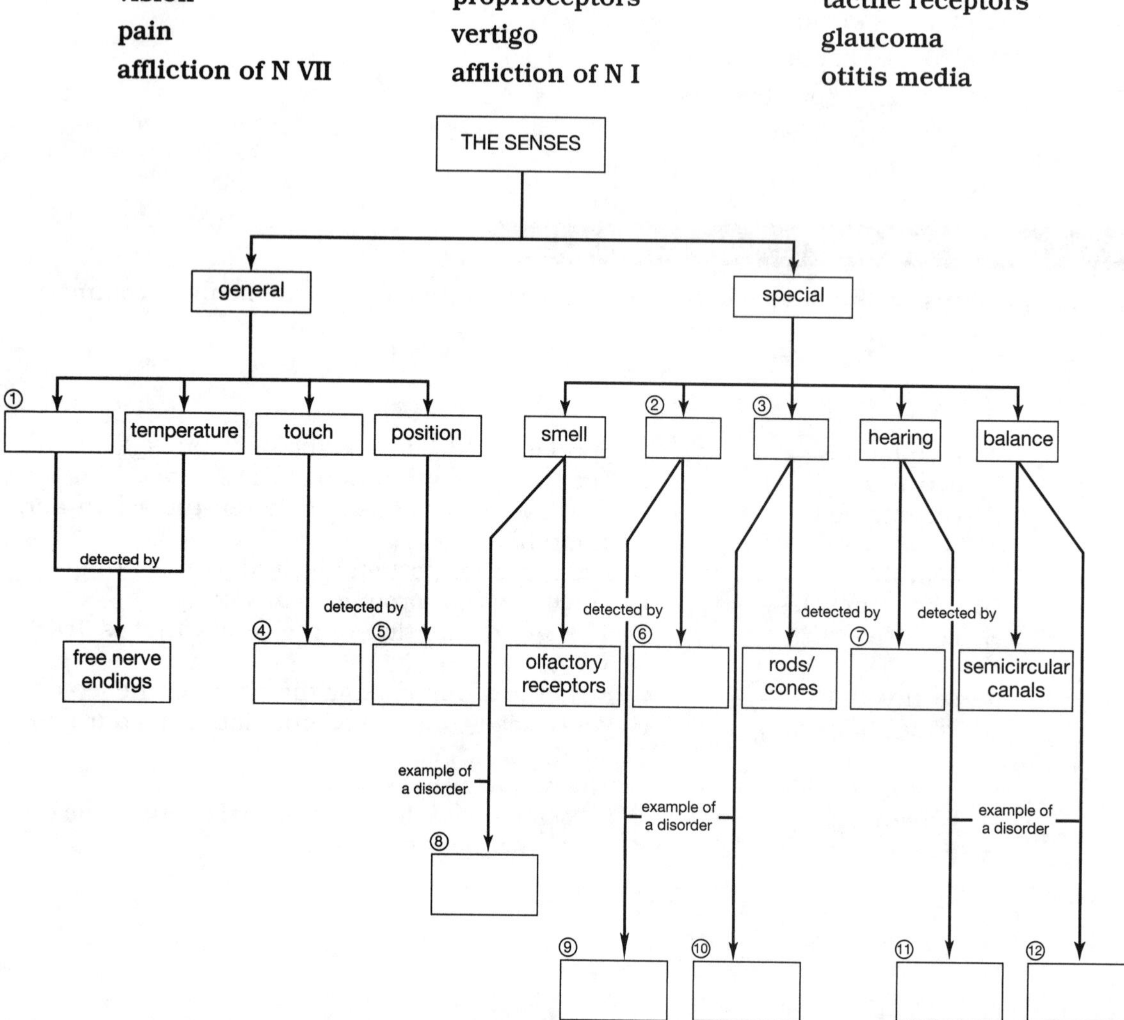

CROSSWORD PUZZLE

The following crossword puzzle reviews the material in Chapter 12. In order to solve the puzzle, you have to know the answers to the clues given, and you must be able to spell the terms correctly.

ACROSS

2. The location of the organ of Corti
7. Irregularities in the shape of the lens
9. The second ossicle involved in the sequence of transmitting sound waves
10. The transparent portion of the sclera
11. The portion of the retina where the optic nerve exits the eye
13. Deafness due to the fusion of the ossicles is called _____ deafness.
17. Deafness due to problems with the cochlea is called _____ deafness.
19. Ability to see far but not near (farsightedness)

DOWN

1. Rods and cones are a type of sensory receptor called a _____.
3. The process of tasting food
4. An infection of the lacrimal gland
5. Sensory receptors involved with touch
6. The three smallest bones of the body, which are located in the middle ear
8. Fluid in the semicircular canals
12. The name for earwax
14. _____ receptors are sensory nerve cells called free nerve endings
15. Ability to see near but not far (nearsightedness)
16. A cranial nerve involved in taste
18. The cells of the retina that cannot detect color

FILL-IN-THE-BLANK NARRATIVE

Use the following terms to fill in the blanks of this summary of Chapter 12.

cochlea	**dissolved**	**efferent**
cones	**emmetropia**	**myopia**
olfactory	**semicircular canals**	**facial**
tactile	**vestibular**	**vestibulocochlear**
Corti	**rods**	**gustatory**

Let's pretend we are going to the movies. We arrive a bit late and walk into the theater and see that the movie is already in progress. We can see the colored images because the light is bright enough to activate our (1) ___________. Now, we must find a seat. As soon as we divert our eyes from the screen to the seats, we experience total darkness. Our cones "shut off" instantly. Our (2) ___________ need to be activated so we can see in the dark. But they are slow to activate, so for a minute or two, we are technically blind. Nonetheless, we find a seat. After several minutes we look around and find we can see every seat in the house. Our cones are at work when we look at the screen, and our rods are now working when we look away from the screen.

Some people have a condition known as (3) ___________, which makes it difficult for them to see the screen, without their glasses. Other people may have (4) ___________, which indicates that the lenses of their eyes focus images correctly without glasses or contact lenses.

During the movie, our (5) ___________ nerves are stimulated, and we immediately interpret the familiar smell of hot buttered popcorn. As we eat the popcorn, the food molecules are partially (6) ___________ and activate the (7) ___________ cells. These cells send signals to our brain via the (8) ___________ nerve. We can hear the movie because the organ of (9) ___________, located in our (10) ___________, is activated and sends signals to our brain via the (11) ___________ nerve.

Now, the movie is over. We stand up to leave, and the slope of the floor makes us feel as if we are going to topple over. We get that feeling because the fluid in our (12) ___________ moves when we stand up. This movement activates the sensory cells that send signals to our brain via the (13) ___________ portion of cranial nerve VIII. Once our brain has received that signal, it sends (14) ___________ messages to our leg muscles to help us maintain proper position. As we exit the theater, someone steps on our toes. Our (15) ___________ receptors are activated and we feel the pain.

Thus, as simple an act as going to a movie requires many interactions of various organ systems of the body, in this case, interactions of the various general and specialized senses of the nervous system.

CLINICAL CONCEPT

The following clinical concept applies the information in Chapter 12 regarding the ears. Following the clinical concept is a set of questions to help you understand the concept.

Brady walks into the doctor's office and gives the following information. "When I step outside to pick up the newspaper off the step, I feel dizzy. When I step outside, without stooping over, on a cold winter day, I feel dizzy. When I'm riding in a car I feel dizzy. When I get up from a lying position too quickly, I feel dizzy. Why do I always feel dizzy?"

We know from Chapter 12 that any movement of the fluid within the semicircular canals will send signals to the brain, and the brain will interpret movement. When Brady stoops over to pick up the newspaper, the fluid starts moving. When he stands up, the fluid is still moving. Even though Brady is standing still at that point, the fluid is still moving just enough to send signals to the brain indicating he is moving.

A change in temperature will also cause fluids to move. The fluid is at body temperature while Brady is inside in the house. As soon as he steps outside, the fluid is exposed to colder temperatures. Because the semicircular canals are located close to the surface of the temporal region of the skull, the fluid is easily affected by temperature changes.

While Brady is riding in a car, it probably rocks back and forth just enough to set the semicircular canal fluid in motion. Brady's brain is interpreting even more motion than is actually occurring.

Getting up too quickly from a prone position also will create fluid movement. However, in this case, the blood must flow against gravity. If Brady's blood pressure is too low, standing up suddenly will result in a momentary decreased blood flow to the brain. This, too, causes dizziness.

The doctor may run tests to be sure Brady does not have a bacterial or viral infection of the inner ear. If that is ruled out, perhaps the drug Dramamine will help his motion sickness.

1. How does fluid in the semicircular canals send signals to the brain for the interpretation of movement?

2. After a person stops moving, the brain sometimes continues to receive signals indicating the person is still moving. Why is this so?

3. How does a simple change in temperature sometimes cause the brain to interpret motion?

4. How might blood pressure be involved in creating dizziness?

5. Why might bacterial infection of the inner ear create dizzy feelings?

THIS CONCLUDES CHAPTER 12 EXERCISES

CHAPTER

13

The Endocrine System

The nervous system and the endocrine system are the control systems in the body. Together, they monitor and adjust the physiological activities throughout the body to maintain homeostasis. The effects of nervous system regulation are usually rapid and short term, whereas the effects of endocrine regulation are ongoing and long term. The endocrine system includes all the endocrine cells and tissues of the body. Endocrine cells are glandular secretory cells that release chemicals, called **hormones**, into the bloodstream for distribution to **target** organs and tissues throughout the body. These target organs and tissues may be adjacent to the hormone-releasing endocrine cells, or they may be located throughout the body, perhaps even at a significant distance from the cells that have released the hormones. The influence of these chemical "messengers" results in facilitating processes that include growth and development, reproduction and sexual maturation, and the maintenance of homeostasis within other systems. This chapter introduces the components of the endocrine system and explores the interactions between the nervous and endocrine systems and their integrative activities with the other organ systems of the body.

The endocrine hormones exert action on all organs of the body. Any abnormality existing within the endocrine system can exert a negative response in other organs. When this happens, the body is out of homeostasis.

To make the study of the endocrine system a little simpler, ask yourself and answer the following questions:

- What is the name of the gland producing the hormone?
- Where is the gland located?
- What hormone(s) does the gland secrete?
- What is the hormone's target organ in the body?
- What is the action of the hormone?

Chapter Objectives:

1. Contrast the response times to changing conditions by the nervous and endocrine systems.
2. List the types of molecules that form the two main groups of hormones .
3. Explain the general action of hormones.
4. Describe how endocrine organs are controlled by negative feedback.
5. Discuss the location, hormones, and functions of the following endocrine glands and tissues: pituitary, thyroid, parathyroids, thymus, adrenals, pancreas, testes, ovaries, and pineal gland.
6. Briefly describe the functions of the hormones secreted by the kidneys, heart, digestive system, and adipose tissue.
7. Explain how the endocrine system responds to stress.
8. Describe the causes of symptoms of most endocrine disorders.
9. Describe examples of disorders of the major endocrine glands.

PART I: OBJECTIVE-BASED QUESTIONS

OBJECTIVE 1: Contrast the response times to changing conditions by the nervous and endocrine systems (textbook p. 291).

After studying p. 291 in the text, you should be able to explain how the body maintains homeostasis in responding to various internal and external stimuli through the activities of the nervous and the endocrine systems.

_____ 1. Which of the following responds to and creates changes to maintain homeostasis?

a. endocrine system
b. nervous system
c. both a and b
d. neither a nor b

_____ 2. Which of the following systems responds to external stimuli?

a. endocrine system
b. nervous system
c. both a and b
d. neither a nor b

3. Which system referred to in the objective creates long-term responses?

____

4. Which system referred to in the objective creates rapid responses?

____

5. Which system referred to in the objective creates slower responses?

____

OBJECTIVE 2: List the types of molecules that form the two main groups of hormones (textbook pp. 291–293).

After studying pp. 291–293 in the text, you should be able to identify the two main groups of hormones, which are categorized according to their molecular makeup.

_____ 1. What are the two main groups of hormones discussed in this chapter?

a. amino acid and protein
b. lipid-based and prostaglandin
c. protein and lipid
d. steroid and lipid-based

_____ 2. What types of molecules make up the protein hormones?

a. amino acids
b. cholesterol
c. fatty acids
d. lipids

_____ 3. What types of molecules make up the steroid hormones?

a. amino acids
b. cholesterol
c. prostaglandin
d. protein

_____ 4. Protein hormones are produced by the _____.

a. adrenal medulla
b. ovaries
c. pituitary gland
d. testes

_____ 5. Steroid hormones are produced by the _____.

a. adrenal cortex
b. ovaries
c. pituitary gland
d. both a and b

OBJECTIVE 3: Explain the general action of hormones (textbook p. 293).

When a hormone leaves a gland, it travels through the bloodstream and targets a specific organ or tissue. After studying p. 293 in the text, you should be able to explain how hormones exert their effects on specific organs or tissues.

_____ 1. The target organ of a hormone is _____.

a. the stimulus that has caused the release of the hormone
b. where the hormone is coming from
c. where the hormone is going
d. where the hormone is produced

_____ 2. Hormones affect a cell by _____.

a. changing the cell's activity
b. changing the cell's enzymes
c. changing the cell's identity
d. all the above

_____ 3. Testosterone targets _____ and causes the production of _____, which results in _____.

a. muscle cells, enzymes and proteins, increased muscle size
b. muscle fibers, more hormones, increased muscle size
c. proteins, enzymes, increased muscle size
d. testes, enzymes and proteins, sperm production

_____ 4. Which of the following statements is true?

a. Hormones are produced by the target cells.
b. Hormones can target only one organ.
c. When hormones target a cell, the activities of the cell change.
d. When hormones target a cell, the hormone changes.

5. What must be present on a target cell in order for any hormone to have an effect on that target? ______________________

6. In Figure 13–1, identify the receptor sites, target organ, endocrine gland, and bloodstream.

Figure 13–1 Hormones and the Target Cells

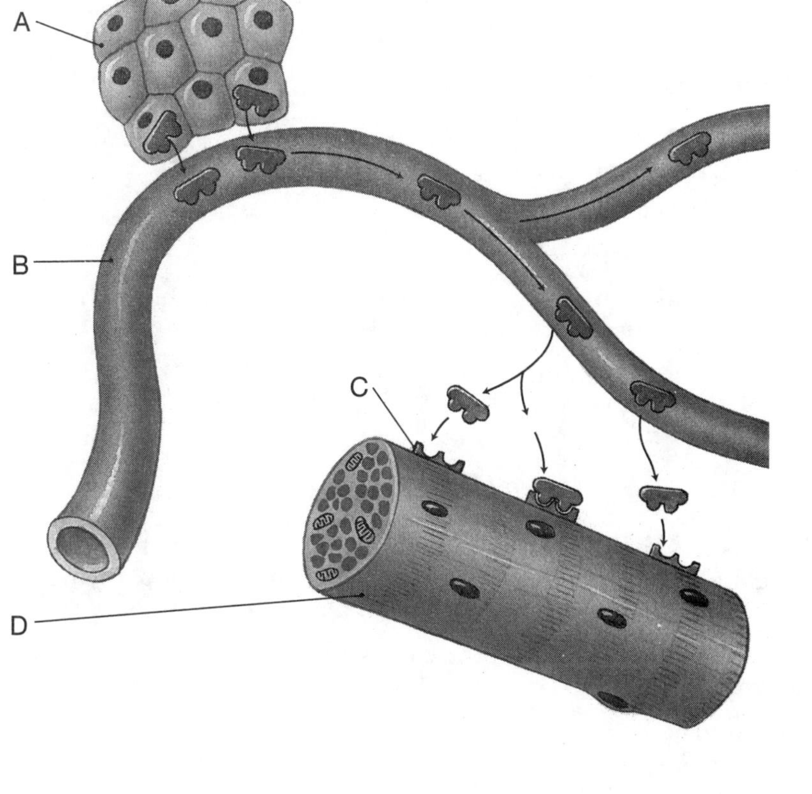

A ____________________ B ____________________

C ____________________ D ____________________

OBJECTIVE 4: Describe how endocrine organs are controlled by negative feedback (textbook pp. 293–294).

A hormone targets the cells of an organ and causes the cells to respond a specific way. To make sure the cells do not overrespond, a second hormone may be involved in order to maintain homeostasis. This is typical of the negative feedback mechanisms. After studying pp. 293–294 in the text, you should be able to explain how endocrine activity is controlled by negative feedback mechanisms.

_____ 1. Which of the following pairs of hormones produce opposite effects?

a. oxytocin, ADH
b. parathyroid hormone, calcitonin
c. testosterone, estrogen
d. thyroid hormone, parathyroid hormone

_____ 2. The parathyroid hormone is produced by the _____ gland and _____.

a. parathyroid, lowers blood calcium ion levels
b. parathyroid, raises blood calcium ion levels
c. thyroid, lowers blood calcium ion levels
d. thyroid, raises blood calcium ion levels

_____ 3. The thyroid gland releases _____ which _____.

a. calcitonin, lowers blood calcium ion levels
b. calcitonin, raises blood calcium ion levels
c. epinephrine, raises blood calcium ion levels
d. parathyroid hormone, lowers blood calcium ion levels

_____ 4. If blood calcium ion levels become too high, which gland will be activated?

a. hypothalamus
b. parathyroid
c. pituitary
d. thyroid

_____ 5. If blood calcium ion levels become too low, which hormone is going to appear in higher concentrations in the blood?

a. calcitonin
b. epinephrine
c. oxytocin
d. parathyroid hormone

OBJECTIVE 5: Discuss the location, hormones, and functions of the following endocrine glands and tissues: pituitary, thyroid, parathyroids, thymus, adrenals, pancreas, testes, ovaries, and pineal gland (textbook pp. 294–301).

When studying the endocrine system, the easiest way to learn the basic concepts is always to do the following:

1. Identify the gland.
2. Identify the hormone secreted by that gland.
3. Identify the destination for that hormone (the target).
4. Identify that hormone's action on its target.

_____ 1. Which of the following is *not* produced by the pituitary gland?

a. adrenocorticotropic hormone
b. antidiuretic hormone
c. epinephrine
d. prolactin

_____ 2. The _____ is secreted by the _____ and targets the _____.

a. adrenocorticotropic hormone, adrenal medulla, general cells of the body
b. antidiuretic hormone, anterior pituitary gland, kidney
c. thyroid stimulating hormone, anterior pituitary gland, thyroid gland
d. thyroid stimulating hormone, posterior pituitary gland, thyroid gland

_____ 3. _____ is secreted by the _____ and causes _____.

a. Oxytocin, posterior pituitary gland, milk production
b. Oxytocin, posterior pituitary gland, milk release
c. Prolactin, anterior pituitary gland, milk release
d. Prolactin, posterior pituitary gland, milk production

_____ 4. The _____ hormone causes egg development, whereas the _____ hormone causes ovulation.

a. follicle stimulating, luteinizing
b. gonadotropin, progestins
c. luteinizing, follicle stimulating
d. progesterone, estrogen

_____ 5. The thyroid gland is located _____.

a. on the trachea, superior to the sternum
b. posterior to the "Adam's apple"
c. posterior to the sternum
d. superior to the heart

_____ 6. The parathyroid glands are located _____.

a. inferior to the thyroid gland
b. posterior to the sternum
c. on the posterior side of the thyroid gland
d. None of the above is correct.

_____ 7. The thymus gland is located _____.

a. inferior to the heart
b. posterior to the body of the sternum
c. posterior to the manubrium of the sternum
d. superior to the "Adam's apple"

_____ 8. Glucocorticoids are a group of hormones produced by the _____ that speed up glucose formation.

a. adrenal cortex
b. adrenal medulla
c. anterior pituitary
d. kidney

_____ 9. The _____ gland may be responsible for regulating biological rhythms.

a. adrenal
b. pineal
c. reproductive
d. thymus

10. In Figure 13–2 identify the various glands of the endocrine system.

Figure 13–2 The Endocrine System

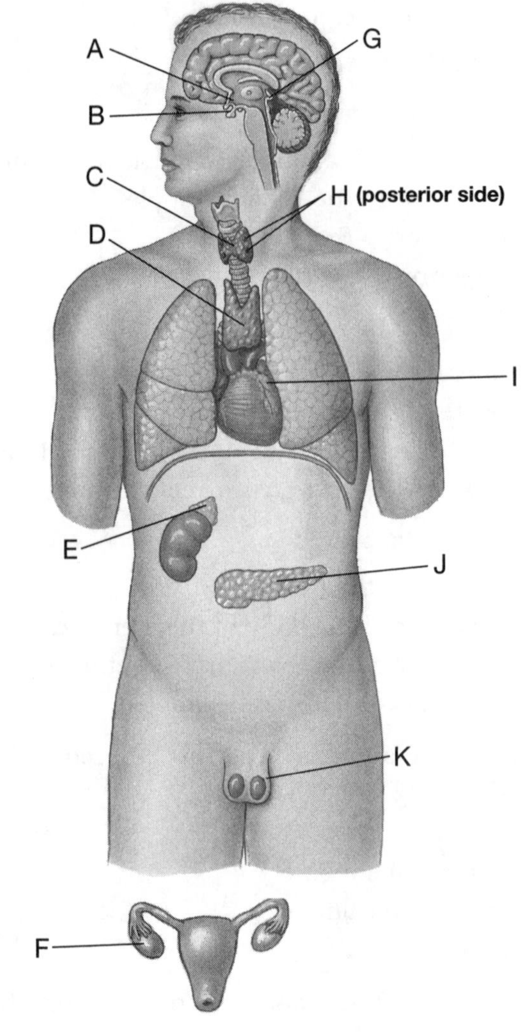

A ______________________ B ______________________

C ______________________ D ______________________

E ______________________ F ______________________

G ______________________ H ______________________

I ______________________ J ______________________

K ______________________

In Figure 13–3, identify the pituitary secretions in column A, and identify the actions of those pituitary secretions in column B.

Figure 13–3 Pituitary Secretions, Hormones, Their Targets, and Actions

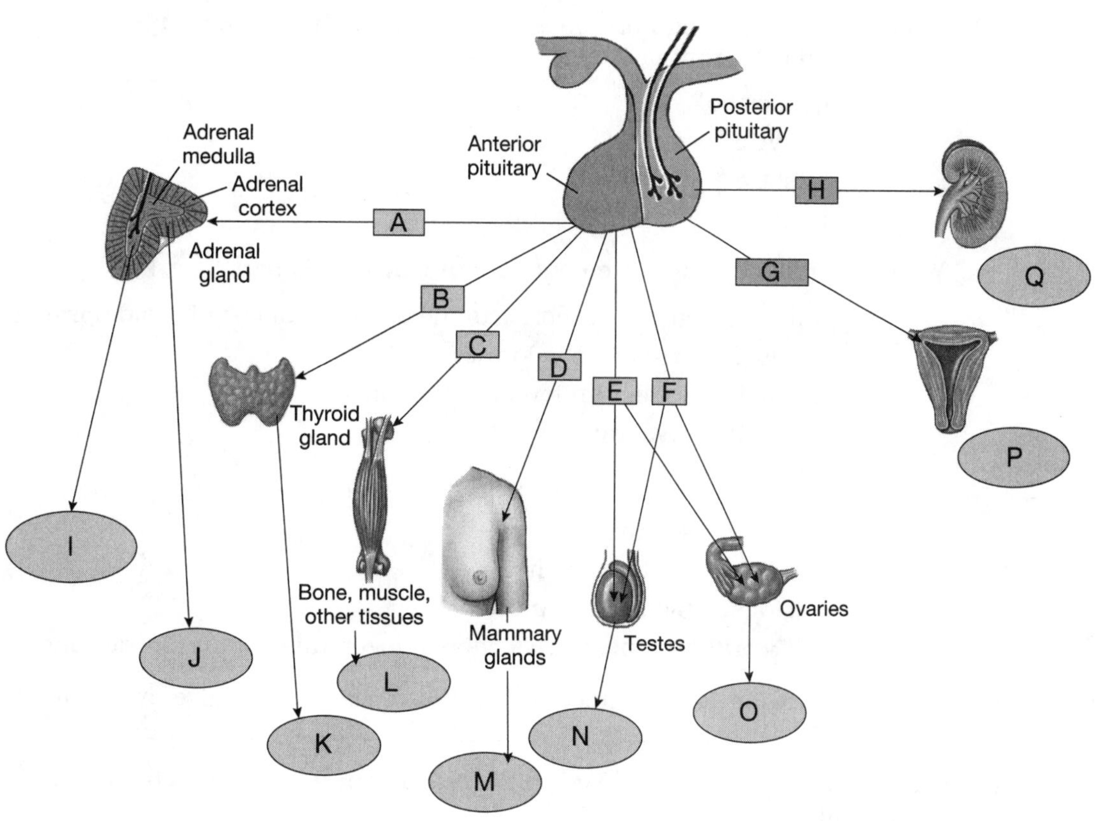

(A) Hormones released from the pituitary gland (letters A through H)	(B) Actions due to the pituitary secretions (letters I through Q)
____ 11. adrenocorticotropic hormone	____ 19. causes bone growth
____ 12. antidiuretic hormone	____ 20. causes milk production
____ 13. follicle stimulating hormone	____ 21. causes the release of thyroid hormones
____ 14. growth hormone	____ 22. causes the release of epinephrine
____ 15. luteinizing hormone	____ 23. causes the release of glucocorticoids
____ 16. oxytocin	____ 24. causes the release of testosterone
____ 17. prolactin	____ 25. causes the retention of water
____ 18. thyroid stimulating hormone	____ 26. causes the release of progesterone
	____ 27. causes uterine contractions

OBJECTIVE 6: Briefly describe the functions of the hormones secreted by the kidneys, heart, digestive system, and adipose tissue (textbook p. 301).

The organs discussed in this section typically are not referred to as glands. After studying p. 301 in the text, you should be able to describe their glandular functions.

____ 1. Which of the following hormones is secreted by the kidneys and is involved in red blood cell production?

a. calcitriol
b. erythropoietin
c. leptin
d. renin

____ 2. Which of the following statements about calcitriol is true?

a. It is secreted in response to the presence of parathyroid hormone.
b. It is produced by the kidneys.
c. It reduces calcium levels in the blood.
d. All the above are true.

____ 3. Which of the following statements about ANP is true?

a. It decreases urinary output.
b. It increases blood volume.
c. It raises blood pressure.
d. It works in a negative feedback mechanism with aldosterone.

4. Name the hormone that is produced by adipose cells, and give its function.

____

5. Who would have a higher amount of EPO in the blood: a person living at high altitudes or a person living at low altitudes? ____________________

6. In Figure 13–4, identify the heart, the kidneys, the thymus gland, and area of the small intestine.

Figure 13–4 Organs with Glandular Function

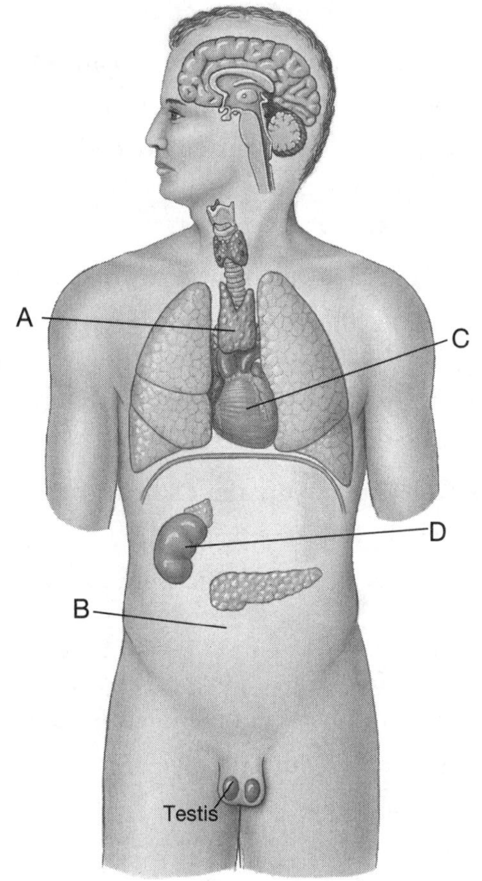

A ______________________ B ______________________

C ______________________ D ______________________

OBJECTIVE 7: Explain how the endocrine system responds to stress (textbook pp. 301–302).

After studying pp. 301–302 in the text, you should be able to describe how the nervous system and the endocrine system respond to stress and how they react and interact to restore homeostasis.

_____ 1. Which of the following correctly describes the body's response to stress?

a. Activation of the autonomic system, rapid use of glucose, the release of glucocorticoids, increase in glucose synthesis

b. Activation of the autonomic system, the release of glucocorticoids, rapid use of glucose, increase in glucose synthesis

c. The rapid use of glucose, activation of the autonomic system, the release of glucocorticoids, increase in glucose synthesis

d. The release of glucocorticoids, rapid use of glucose, activation of the autonomic system, increase in glucose synthesis

_____ 2. An abnormal secretion of aldosterone caused by stress may result in _____.

a. high blood pressure
b. imbalance of ions
c. low blood pressure
d. Both a and b are correct.

_____ 3. An abnormal secretion of glucocorticoids caused by stress may result in _____.

a. excessively high levels of glucose in the blood
b. excessively low levels of glucose in the blood
c. sluggish blood flow
d. Both a and c are correct.

_____ 4. An abnormal secretion of glucocorticoids caused by stress may result in _____.

a. decreased immune system activity
b. excessively high levels of glucose in the blood
c. excessively low levels of glucose in the blood
d. Both a and b are correct.

5. If someone says that stress caused a patient's heart to fail, what probably caused the heart failure? _______________________

OBJECTIVE 8: Describe the causes of symptoms of most endocrine disorders (textbook pp. 302–305).

The symptoms associated with endocrine disorders are generally due to hypersecretion or hyposecretion of hormones. After studying pp. 302–305 in the text, you should be able to describe how abnormal hormonal secretion can cause specific symptoms.

_____ 1. The _____ of ADH may result in polyuria.

a. hypersecretion
b. hyposecretion

_____ 2. The _____ of thyroid hormones may result in myxedema.

a. hypersecretion
b. hyposecretion

_____ 3. The _____ of glucocorticoids may result in Cushing's disease.

a. hypersecretion
b. hyposecretion

_____ 4. The _____ of the growth hormone may result in acromegaly.

a. hypersecretion
b. hyposecretion

_____ 5. When the endocrine system malfunctions, it generally results in the _____ (underproduction) of a hormone or the _____ (overproduction) of a hormone.

a. hyposecretion, hypersecretion
b. hypersecretion, hyposecretion

OBJECTIVE 9: Describe examples of disorders of the major endocrine glands (textbook pp. 305–311).

After studying pp. 305–311 in the text, you should be able to describe select hormonal abnormalities including some signs, symptoms, and treatments.

____ 1. Diabetes insipidus is ____, which results in ____.

a. lack of ADH, loss of water
b. lack of ADH, retention of water
c. excess ADH, loss of water
d. excess ADH, retention of water

____ 2. Goiter is ____, which is caused by ____.

a. an enlargement of the thyroid gland, a lack of iodine in the diet
b. an enlargement of the thyroid gland, an excess of iodine in the diet
c. the shrinkage of the thyroid gland, a lack of iodine in the diet
d. the shrinkage of the thyroid gland, an excess of iodine in the diet

____ 3. Hypothyroidism in infants is called ____, and in adults it is called ____.

a. Addison's disease, Graves' disease
b. cretinism, Graves' disease
c. cretinism, myxedema
d. Graves' disease, myxedema

____ 4. Tetany (muscle spasm) is due to ____, which results in ____.

a. hyperparathyroidism, a decrease in calcium ions
b. hyperparathyroidism, an increase in calcium ions
c. hypoparathyroidism, a decrease in calcium ions
d. hypoparathyroidism, an increase in calcium ions

____ 5. Excess ACTH production results in ____, and decreased ACTH production results in ____.

a. Addison's disease, Cushing's disease
b. Cushing's disease, Addison's disease
c. Grave's disease, Addison's disease
d. hyperaldosteronism, hypoaldosteronism

____ 6. In which of the following conditions is there normal production of insulin but the cells are not responsive?

a. Addison's disease
b. diabetes insipidus
c. type I diabetes mellitus
d. type II diabetes mellitus

7. Hyperparathyroidism results in ____________________.

PART II: CHAPTER-COMPREHENSIVE EXERCISES

Match the terms in column A with the appropriate descriptions or statements in column B.

MATCHING I

	(A)	(B)
_____	1. Addison's disease	A. A chemical whose name means "to set in motion"
_____	2. antidiuretic hormone	B. The hormone that lowers calcium ion levels in the blood
_____	3. calcitonin	C. The anatomical name for the pituitary gland
_____	4. circadian	D. The hormone that causes eggs to mature in the female and causes sperm production in males
_____	5. cortex	E. The hormone that causes a reduction of urinary output
_____	6. diabetes insipidus	F. Inadequate iodine in the diet causes a reduction in thyroxine and may result in _____.
_____	7. follicle stimulating hormone	G. The part of the adrenal gland that produces the glucocorticoids
_____	8. goiter	H. This term refers to daily cycles
_____	9. hormone	I. Hyposecretion of glucocorticoids
_____	10. hypophysis	J. Hyposecretion of ADH

MATCHING II

	(A)	(B)
_____	1. hypothalamus	A. A group of hormones that are derived from cholesterol
_____	2. luteinizing hormone	B. The part of the brain that produces oxytocin
_____	3. osteoblast	C. The hormone that triggers the release of thyroid hormones
_____	4. osteoclast	D. The hormone that stimulates the production of testosterone in males and ovulation in females
_____	5. oxytocin	E. The hormone that causes milk ejection and uterine contractions
_____	6. prolactin	F. The parathyroid hormone stimulates the _____ cells.
_____	7. steroid	G. Calcitonin stimulates the _____ cells.
_____	8. thyroid-stimulating hormone	H. The hormone that causes milk production
_____	9. Type I diabetes	I. Normal insulin levels but the insulin receptor sites are nonresponsive.
_____	10. Type II diabetes	J. Hyposecretion of insulin

CONCEPT MAP

This concept map summarizes and organizes the information in Chapter 13. Use the following terms to complete the map by filling in the boxes identified by the circled numbers, 1–9.

ACTH	**calcitonin**	**kidneys**
mammary glands	**LH**	**skeleton**
uterus	**posterior**	**TSH**

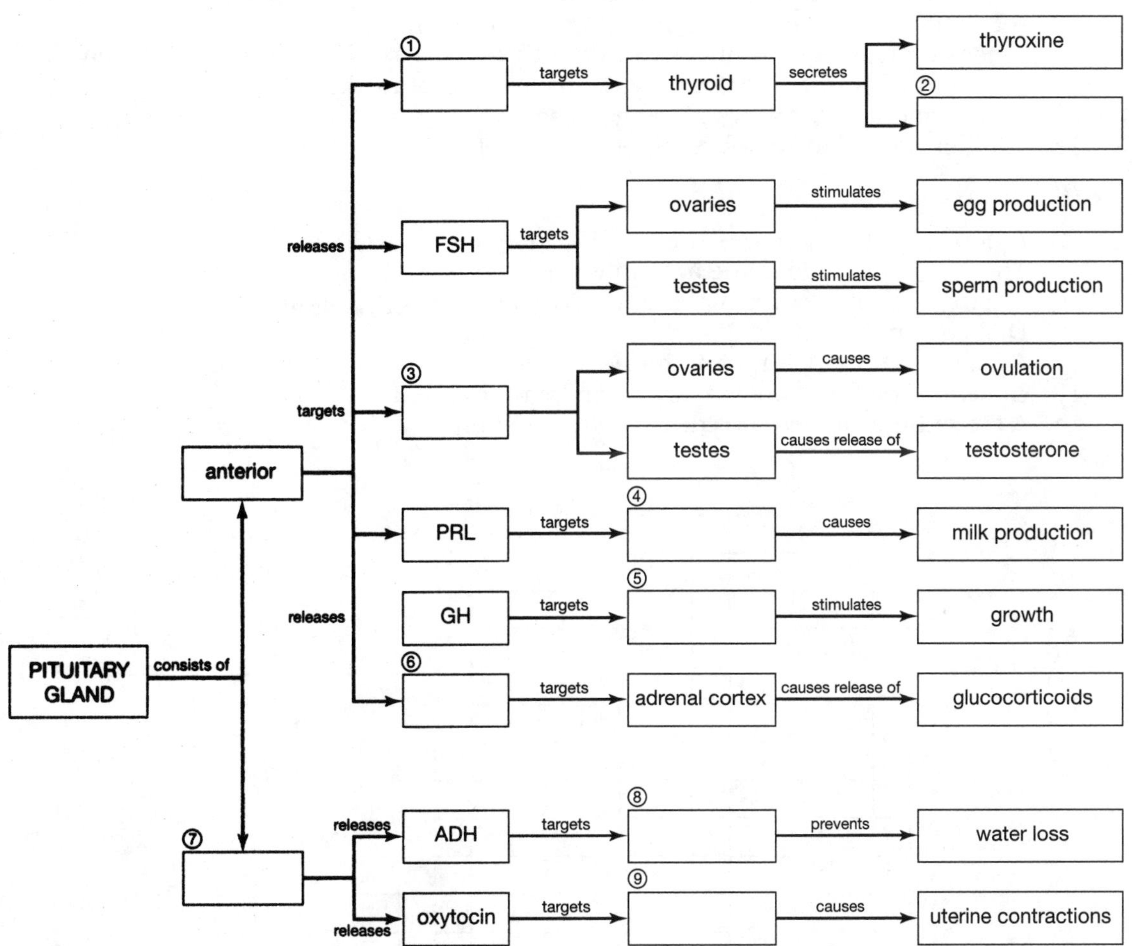

CROSSWORD PUZZLE

The following crossword puzzle reviews the material in Chapter 13. To complete the puzzle, you have to know the answers to the clues given, and must be able to spell the terms correctly.

ACROSS

1. A hormone produced by adipose cells
4. A hormone involved in secondary sex characteristics in males
5. A hormone that prevents the loss of sodium ions
7. A hormone that causes the loss of sodium ions
8. Hypersecretion of the growth hormone
10. A type of gland that secretes hormones into the blood
13. Progesterone is a hormone involved in the _____ sex characteristics in females.
15. _____ of the pituitary gland may result in hyperpituitarism.
16. The hormone that lowers blood glucose levels
17. The hormone that raises blood glucose levels

DOWN

2. A hormone that prepares the body for pregnancy
3. The _____ is an endocrine and an exocrine organ.
6. After a hormone leaves a gland, it travels to a(n) _____ organ.
9. Diabetes insipidus results in _____.
11. A gland involved with the immune system
12. A dietary ingredient necessary to make thyroxine
14. A steroid hormone that reduces inflammation

FILL-IN-THE-BLANK NARRATIVE

Use the following terms to fill in the blanks of this summary of Chapter 13.

ACTH	**blood pressure**	**osteoblast**
calcitonin	**sodium**	**thyroid hormones**
EPO	**bone marrow**	**parathyroid**
thyroxine	**thymus**	**FSH**
hypothalamus	**osteoclasts**	**oxytocin**
target	**ANP**	**testosterone**
epinephrine	**thymosin**	**oxygen**
LH	**glucocorticoids**	**progesterone**
cortisol	**insulin**	

The pituitary gland has distinct anterior and posterior portions. The posterior portion releases (1) ___________, which targets the myometrium of the uterus to produce contractions. This hormone is actually produced in the (2) ___________. The anterior portion of the pituitary produces and releases numerous hormones. Some of those hormones (3) ___________ other glands and cause them to release their hormones. For example, thyroid-stimulating hormone (TSH) targets the thyroid gland, causing it to release (4) ___________. (5) ___________ from the anterior pituitary targets the adrenal cortex, causing it to release (6) ___________, which target the liver cells, causing them to manufacture glycogen. Other hormones from the anterior pituitary target structures other than glands. For example, in the male, (7) ___________ targets the testes and causes sperm production. In the female, (8) ___________ targets the ovaries and causes ovulation of the egg.

Inferior to the "Adam's apple" is the thyroid gland. Due to pituitary secretions of TSH, it releases (9) ___________. C cells in the thyroid gland produce (10) ___________, which targets (11) ___________ cells, causing them to remove calcium ions from the blood in an effort to make bone. On the posterior side of the thyroid are the (12) ___________ glands. These glands produce hormones that target the (13) ___________, which break down bone to increase the concentration of calcium ions in the bloodstream.

Inferior to the thyroid gland is the (14) ___________ gland. This gland produces (15) ___________, which causes the development of white blood cells called lymphocytes.

Inferior to the thymus are the atria of the heart. The atria release (16) ___________, which targets the kidneys and causes the release of (17) ___________ ions. As these ions are lost, water is also lost, which ultimately reduces (18) ___________.

Superior to each kidney are the adrenal glands. Each gland has a medulla and a cortex. The adrenal cortex is activated by pituitary secretions and releases glucocorticoids such as (19) ___________. Pituitary secretions do not control the adrenal medulla. The adrenal medulla releases (20) ___________, which targets muscle cells and causes them to utilize glucose rapidly.

Inferior to the adrenal glands are the kidneys. Some kidney cells release a hormone called (21) ___________ in response to low (22) ___________ levels. Erythropoietin targets the (23) ___________ to increase development of red blood cells in an effort to transport more oxygen to the kidney cells.

Posterior to the stomach is the pancreas. The pancreas releases (24) ___________, which stimulates cells to take in glucose for metabolism. This lowers blood glucose levels.

The reproductive organs are glands that are under pituitary control. The testes release (25) ___________, which determines male characteristics, and (26) ___________ is released from ovaries, which determines female characteristics and prepares the body for pregnancy.

CLINICAL CONCEPT

The following clinical concept applies the information in Chapter 13. Following the application is a set of questions to help you understand the concept.

Kelsey took her 7-year-old son, Lionel, to the pediatrician because Lionel was wetting the bed. Kelsey told the doctor she has tried several things to prevent the bed-wetting episodes, including having him go to the bathroom just prior to bedtime and restricting his fluid intake prior to bed, but they hadn't helped.

The doctor ordered a urine test to rule out infection. It came back negative. The doctor explained that the body produces antidiuretic hormone (ADH). With normal levels of ADH, a person's kidneys excrete about 1% of the water they process into the urinary bladder, and they recirculate about 99% of the water into the bloodstream. With low levels of ADH, the kidneys put less water back into the bloodstream and more water into the urinary bladder, causing the urinary bladder to fill up during the night and increasing the potential for bed-wetting accidents.

The doctor prescribed desmopressin, a synthetic version of the hormone that is administered as a spray in each nostril at bedtime, and restricted fluids after 6:00 P.M. The pediatrician reassured Kelsey that Lionel would grow up to be a normal healthy adult.

1. Where is the antidiuretic hormone produced?

2. Which organ secretes the antidiuretic hormone?

3. What is the target of antidiuretic hormone?

4. What is the action of antidiuretic hormone?

5. In the presence of ADH, if 2000 ml of water passed through the kidneys, how many milliliters would go to the urinary bladder?

THIS CONCLUDES CHAPTER 13 EXERCISES

CHAPTER 14

The Blood

Blood is a specialized **fluid connective tissue** consisting of two basic components: the formed elements and cell fragments, and the fluid plasma in which they are carried. Every living cell relies on a continuous supply of circulating blood to bring to each cell the oxygen, nutrients, and chemical substances necessary for its proper functioning, and, at the same time, to remove waste products.

The blood, the heart, and a network of blood vessels, constitute the cardiovascular system, which, acting in concert with other systems, plays an important role in the maintenance of homeostasis.

Many times we go into the doctor's office and tell the doctor we are not feeling well. To help determine what might be our problem, the doctor proceeds to take a blood sample. By examining blood components, such as white blood cells, and comparing the blood sample with normal values, the doctor can narrow down what might possibly be wrong with us. Once it is narrowed down, the doctor can, if necessary, proceed with precise tests to pinpoint the exact problem. Whatever is happening to our body is going to have some effect on our blood components. In order to understand if there is something wrong with our blood components, we must first begin by examining the composition and functions of normal circulating blood.

The exercises in this chapter help you identify and study the components of blood and review the functional roles of the components in meeting the cellular needs of the body. Additional activities involve blood types, blood clotting, blood cell formation, and blood abnormalities.

Chapter Objectives:

1 Describe the components of the cardiovascular system.
2 Describe the three major functions of blood.
3 Describe the important components of blood.
4 Discuss the composition and functions of plasma.
5 Discuss the characteristics and functions of red blood cells.
6 Describe the various kinds of white blood cells and their functions.
7 Describe the formation of the formed elements in blood.
8 Describe the mechanisms that control blood loss after an injury.
9 Explain what determines blood type and why blood types are important.
10 Give an example of each of the primary blood disorders.
11 Distinguish among the different types of anemia.

PART I: OBJECTIVE-BASED QUESTIONS

OBJECTIVE 1: Describe the components of the cardiovascular system (textbook p. 317).

After studying p. 317 in the text, you should be able to identify components of the cardiovascular system and their general functions.

_____ 1. Which of the following is not a component of the cardiovascular system?

a. blood
b. fluid connective tissue
c. plasma
d. thymus

_____ 2. The cardiovascular system consists of vessels for the transport of white blood cells that are involved in fighting infections. Which other system also consists of vessels that transport material for fighting infections?

a. endocrine
b. lymphatic
c. respiratory
d. none of the above

OBJECTIVE 2: Describe the three major functions of blood (textbook p. 317).

After studying p. 317 in the text, you should be able to name the major functions of blood in maintaining life.

_____ 1. Blood transports which of the following?

a. It transports hormones to cells and away from glands.
b. It transports nutrients to cells and wastes away from cells.
c. It transports oxygen to cells and carbon dioxide away from cells.
d. All the above are correct.

_____ 2. Blood helps to regulate which of the following?

a. pH
b. body temperature
c. both a and b
d. neither a nor b

_____ 3. Blood provides protection against pathogens via the activity of _____.

a. plasma
b. red blood cells
c. white blood cells
d. all the above

_____ 4. The skin acts as the first line of defense against foreign invaders. If invaders get past the skin, which cells, found in blood, act as the second line of defense?

a. erythrocytes
b. leukocytes
c. platelets
d. all the above

5. How does blood help regulate body temperature?

OBJECTIVE 3: Describe the important components of blood (textbook pp. 318–319).

Our blood consists of many components essential for maintaining homeostasis. While all the components of the blood are important and essential in regard to survival, this objective will concentrate on components such as: plasma, proteins, ions, and cells. After studying pp. 318–319 in the text, you should be able to list the major components of blood and their important features.

_____ 1. Which of the following statements about whole blood is true?

a. 45 percent of the blood is made of cells, and 55 percent is plasma.
b. 45 percent of the blood is made of red blood cells, and 55 percent is plasma.
c. 60 percent of the blood is plasma, and 40 percent is made of cells.
d. 92 percent of the blood is plasma, and 8 percent is made of cells.

_____ 2. Which of the following statements about the makeup of plasma is true?

a. Plasma is 45 percent water and 55 percent protein and ions.
b. Plasma is 55 percent water and 45 percent protein and ions.
c. Plasma is 60 percent protein and 40 percent water.
d. Plasma is 92 percent water, 7 percent protein, and 1 percent ions.

_____ 3. The most abundant plasma protein is _____.

a. albumin
b. fibrinogen
c. immunoglobulin
d. transport protein

_____ 4. If a sample of whole blood was placed into a test tube, how would the blood settle in the tube?

a. Plasma would be on top, and the cellular components would settle to the bottom.
b. Plasma would be on top followed by a small layer of white blood cells, and the red blood cells would be on the bottom.
c. The cellular components would be on top, and the plasma would settle to the bottom.
d. The plasma and cellular components would be mixed.

5. What are the two main cellular components of whole blood?

____

6. In Figure 14–1, identify the plasma, white blood cells, and red blood cells.

Figure 14–1 A Sample of Centrifuged Blood

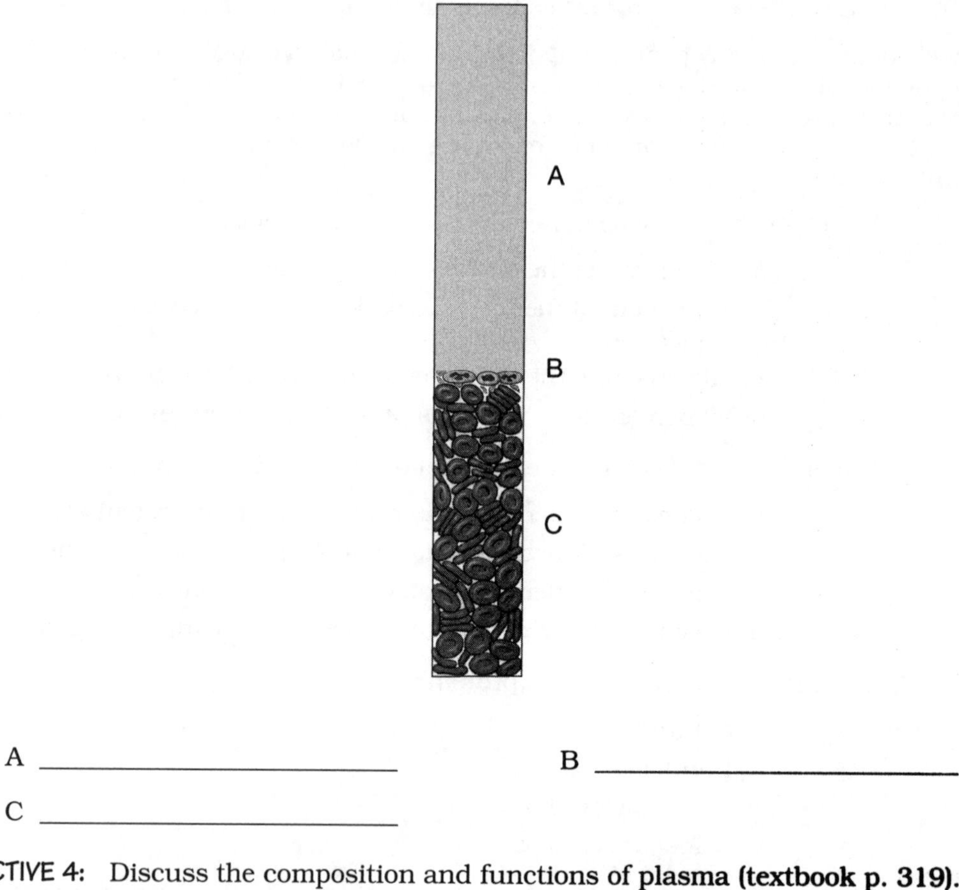

A ____________________ B ____________________

C ____________________

OBJECTIVE 4: Discuss the composition and functions of plasma (textbook p. 319).

When studying the blood, it is necessary to distinguish between the solid and liquid portion. After studying p. 319 in the text, you will learn that plasma is more than just a liquid circulating in our blood stream. It plays a vital role in helping to maintain homeostasis.

_____ 1. Plasma is made mostly of _____.

a. cells
b. ions
c. protein
d. water

_____ 2. Which of the following is a function of plasma?

a. It absorbs and releases heat as needed by the body.
b. It transports ions.
c. It transports red blood cells.
d. All the above

_____ 3. Which of the following make up the majority of plasma?

a. red blood cells
b. white blood cells
c. platelets
d. none of the above

_____ 4. Which of the following plasma proteins is involved in fighting infections?

a. albumin
b. fibrinogen
c. immunoglobulin
d. lipoprotein

5. What is the name of the plasma protein involved in blood clotting?

____

OBJECTIVE 5: Discuss the characteristics and functions of red blood cells (textbook pp. 320–321).

Red blood cells have unique characteristics, which make them different from other cells you have previously studied. Mature red blood cells do not have a nucleus or organelles. Mature red blood cells consist mainly of hemoglobin. After studying pp. 320–321 in the text, you should be able to describe the structure of red blood cells—hemoglobin surrounded by a cell membrane—and the cells' primary function.

_____ 1. Erythrocytes contain _____, which binds to and transports _____.

a. bilirubin, carbon monoxide
b. hemoglobin, oxygen
c. mitochondria, oxygen
d. protein, oxygen

_____ 2. Which of the following statements best describes the role of red blood cells?

a. Oxygen binds to the hemoglobin and is transported to body tissues.
b. Oxygen binds to the iron portion of hemoglobin and is transported to body tissues.
c. Oxygen binds to the nucleus of the erythrocyte and is transported to body tissues.
d. Oxygen binds to the erythrocyte and is transported to body tissues.

_____ 3. Carbon monoxide (CO) is a deadly gas because _____.

a. CO binds to hemoglobin and is transported to the tissues. CO then kills the tissue.
b. CO binds to hemoglobin, so oxygen cannot. Therefore, oxygen is not transported to the tissues.
c. CO binds to oxygen, which prevents the transport of oxygen to the tissues.
d. Both a and c are correct.

4. Why does a deficiency of iron result in anemia?

5. What is another name for red blood cells? _______________

6. In Figure 14–2, identify which letter, A or B, represents the location where oxygen binds to the hemoglobin molecule. ___________________________

Figure 14–2 The Hemoglobin Molecule

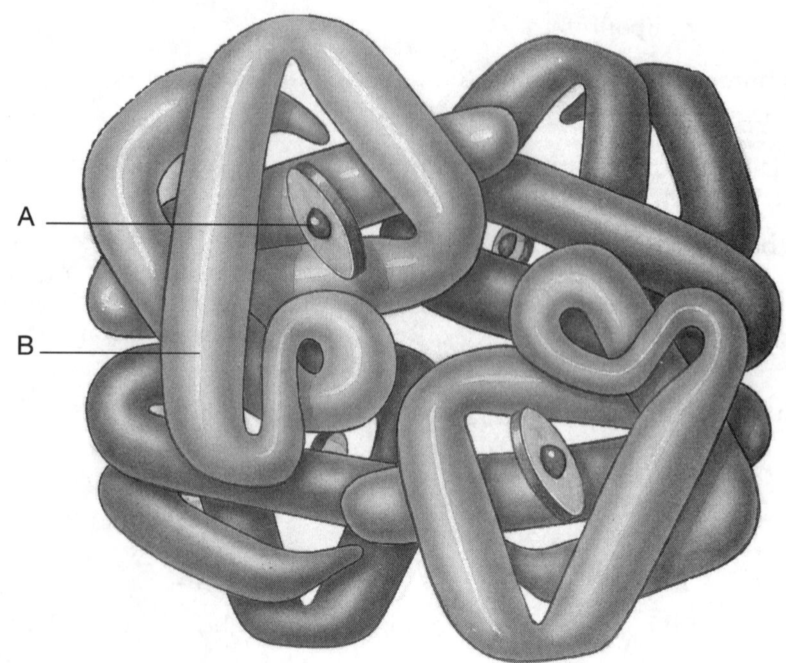

OBJECTIVE 6: Describe the various kinds of white blood cells and their functions (textbook pp. 321–322).

White blood cells will respond to the foreign invaders that might cause us to become ill. A certain number of white blood cells are necessary to fight off the "general, everyday" invaders. However, if the invaders become too great, the white blood cells will increase in number. After studying pp. 321–322 in the text, you should be able to differentiate among the different types of white blood cells and describe the function of each. White blood cell count is used as a diagnostic tool in the treatment of disease.

_____ 1. A normal number of white blood cells in a healthy person is ___________________________ /mm^3.

a. 6000–9000
b. 100,000
c. 350,000
d. none of the above

_____ 2. Which multilobed white blood cell typically fights bacteria?

a. basophil
b. lymphocyte
c. monocyte
d. neutrophil

_____ 3. Which of the following leukocytes respond to allergens?

a. eosinophils
b. lymphocytes
c. monocytes
d. neutrophils

_____ 4. Inflamed tissue, like a pimple, typically is red and swollen. Which leukocyte may be responsible for this red swollen condition?

a. basophil
b. lymphocyte
c. monocyte
d. neutrophil

_____ 5. Which leukocyte can fuse with another of its kind to create a giant phagocytic cell?

a. lymphocyte
b. monocyte
c. neutrophil
d. none of the above

_____ 6. Leukemia is a type of _____.

a. leukocytosis
b. leukopenia
c. lymphocytosis
d. lymphopenia

7. How can a doctor conclude that you have a viral infection by simply looking at a differential count of your white blood cells?

8. T cells and B cells are actually which type of leukocyte? ______________________

9. For each of the cells in Figure 14–3, answer the following questions:

    i. What is the name of the white blood cell?
    ii. What is the normal percentage of that white blood cell in the blood?
    iii. What is the general function of that white blood cell?

Figure 14–3 White Blood Cells

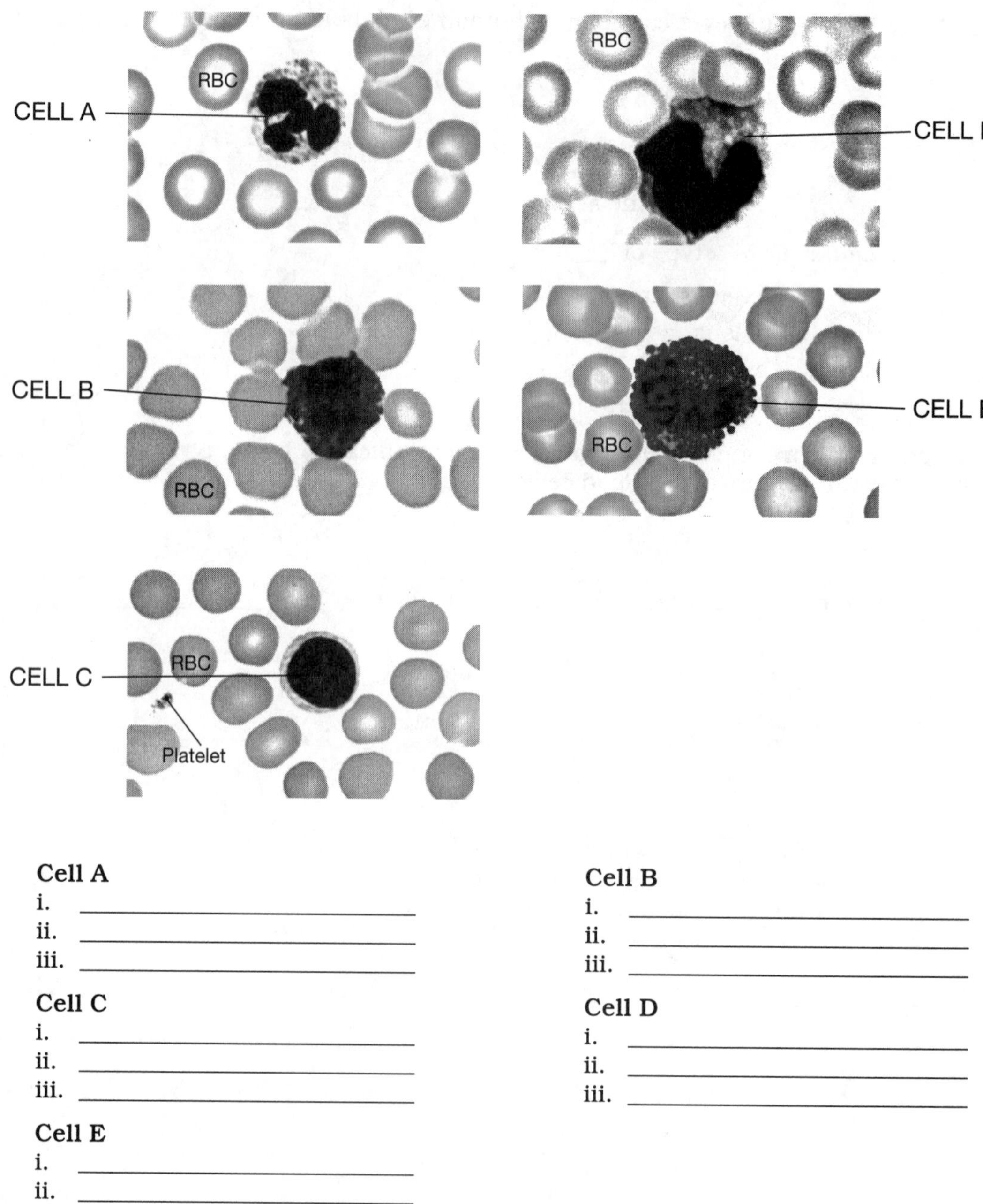

Cell A
i. ______
ii. ______
iii. ______

Cell B
i. ______
ii. ______
iii. ______

Cell C
i. ______
ii. ______
iii. ______

Cell D
i. ______
ii. ______
iii. ______

Cell E
i. ______
ii. ______
iii. ______

OBJECTIVE 7: Describe the formation of the formed elements in blood (textbook pp. 321–322).

After studying pp. 321–322 in the text, you should be able to describe how the formed elements of blood are created.

_____ 1. *Poiesis* refers to _____.

a. the formation of blood in general
b. the formation of platelets
c. the formation of red blood cells
d. the formation of something

_____ 2. Which of the following are considered to be the "formed elements" in blood?

a. red blood cells
b. white blood cells
c. platelets
d. all the above

_____ 3. What is the name for the formation of blood in general?

a. erythropoiesis
b. hemopoiesis
c. leukopoiesis
d. poiesis

_____ 4. Platelets are derived from _____.

a. bone marrow
b. megakaryocytes
c. stem cells
d. thrombocytes

_____ 5. What initiates red blood cell formation?

a. A condition of hypoxia causes kidney cells to release EPO.
b. High oxygen detected by kidney cells causes the release of EPO.
c. Low oxygen detected by kidney cells causes the release of EPO.
d. Both a and c are correct.

6. What is the name for the formation of red blood cells? ____________________

7. What is the name for the formation of white blood cells? ____________________

8. What is the name for the formation of platelets? ____________________

9. In Figure 14–4, identify the white blood cell, the red blood cell, platelets, and the megakaryocyte.

Figure 14–4 Blood Cells Entering into Circulation

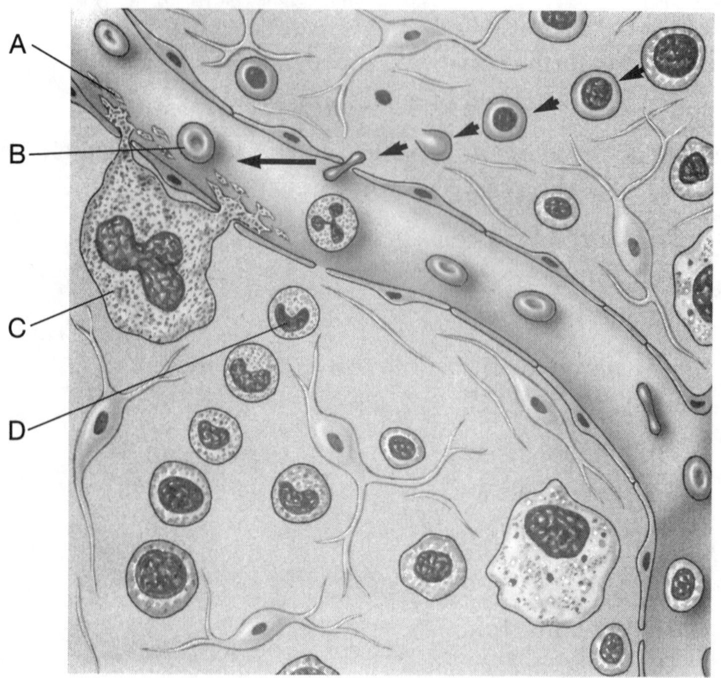

A ______________________ B ______________________

C ______________________ D ______________________

OBJECTIVE 8: Describe the mechanisms that control blood loss after an injury (textbook p. 323).

After studying p. 323 in the text, you should be able to explain blood clotting as a series of chemical reactions that are initiated by the platelets.

_____ 1. Structures that are involved in the clotting of blood are _____.

a. embolus
b. platelets
c. thrombocytes
d. both b and c

_____ 2. Which of the following events initiates the clotting of blood?

a. Fibrin converts to fibrinogen.
b. Platelets release a clotting factor.
c. Platelets stick to the wounded site.
d. Thrombin is produced.

_____ 3. Which of the following is a clotting protein found in the bloodstream and made by the liver?

a. fibrin
b. fibrinogen
c. prothrombin
d. thrombin

_____ 4. Platelets release _____ that convert _____ to _____.

a. clotting factors, prothrombin, thrombin
b. clotting factors, thrombin, prothrombin
c. prothrombin, thrombin, clotting factors
d. thrombin, prothrombin, clotting factors

_____ 5. Clotting factors plus _____ eventually form _____, which converts _____ to _____.

a. calcium ions, thrombin, fibrin, fibrinogen
b. calcium ions, thrombin, fibrinogen, fibrin
c. platelets, thrombin, fibrinogen, fibrin
d. thrombin, prothrombin, fibrinogen, fibrin

6. Look at Figure 14-3 in the text. Where do the clotting factors originate?

7. Look at Figure 14-3 of the text. How is thrombin produced?

8. Look at Figure 14-3 in the text. What forms fibrin, which is responsible for making the actual clot?

OBJECTIVE 9: Explain what determines blood type and why blood types are important (textbook pp. 324–328).

In the 1800s it was thought that anyone could donate blood to anyone. After all, it was thought, blood is blood. In the 1900s, it was determined that this was not the case. It is now known that there are different kinds of blood. After studying pp. 324–328 in the text, you should be able to explain the factors that determine blood type and the importance of assuring compatible blood types when transfusing blood.

_____ 1. On the surface of red blood cells are glycolipids called _____ that identify the cell.

a. antibodies
b. antigens
c. immunoglobulins
d. both a and c

_____ 2. If a type A person were to donate to a type B person, the B person's blood would "attack" the type A blood. What does the attacking?

a. Leukocytes attack the A blood.
b. Type B antigens attack the cells with type A antigens.
c. Type B plasma antibodies attack the cells with A antibodies.
d. Type B plasma antibodies attack the cells with type A antigens.

_____ 3. Which of the following blood types *does not* contain plasma antibodies?

a. A
b. B
c. AB
d. O

_____ 4. Which of the following statements best explains why a type O person can donate blood to a type A person?

a. A type O person does not have any antigens to be attacked by the type A blood.
b. A type O person does not have any antigens to attack the type A blood.
c. A type O person does not have any plasma antibodies to be attacked by the type A blood.
d. A type O person is a universal donor.

_____ 5. Which of the following statements best explains why a type B person can donate blood to a type AB person?

a. A type AB person does not have any plasma antibodies to attack the type B blood.
b. A type AB person does not have any antigens to attack the type B blood.
c. A type AB person is a universal recipient.
d. Type B cannot donate to AB.

6. Explain the difference between blood clumping and blood clotting.

7. A person with AB type blood has A antigens and B antigens, not AB antigens. In Figure 14–5, identify which letter represents the A antigen and which letter represents the B antigen of type AB blood.

Figure 14–5 Blood Types and Antigens

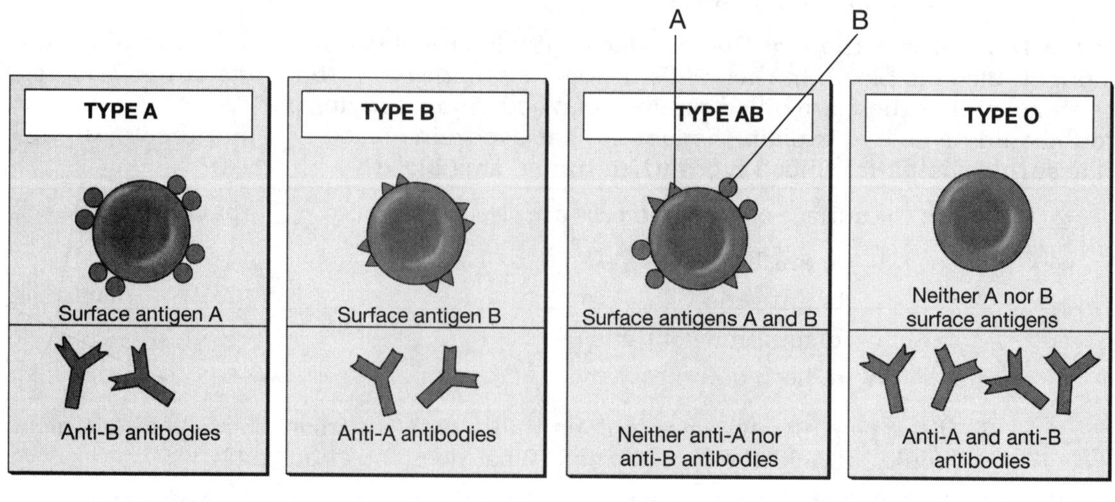

(A) ______________________ (B) ______________________

8. In Figure 14–6, identify which letters represent a safe blood donation and which letters represent an unsafe blood donation. In the corresponding blanks, write the words "safe" or "unsafe."

Figure 14–6 Cross-Reactions between Different Blood Types

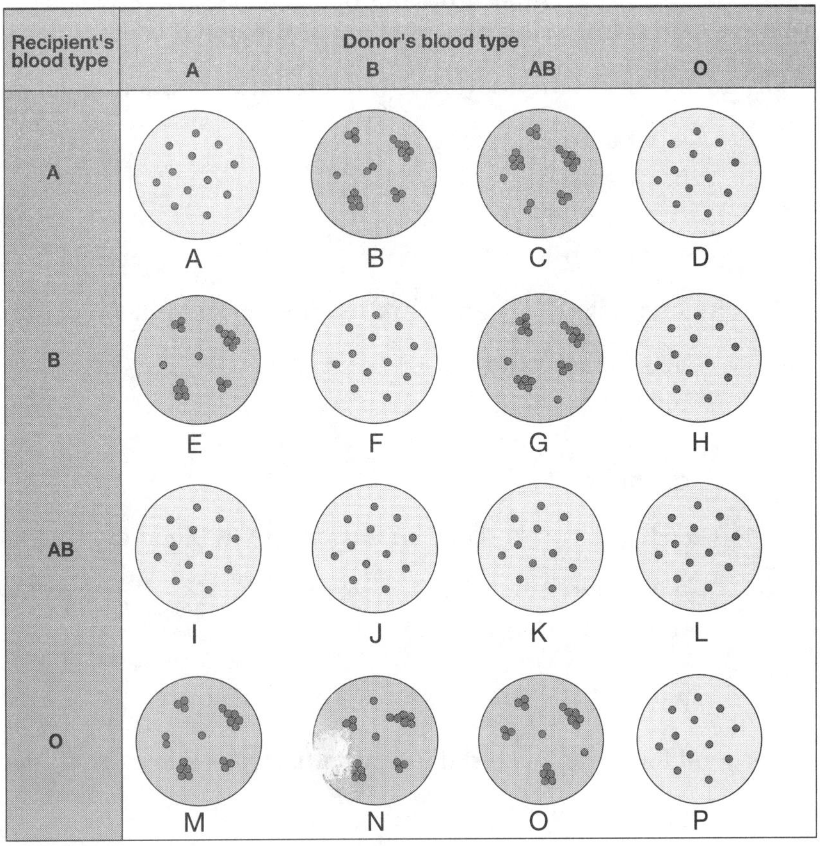

Recipient's blood type	Donor's blood type A	B	AB	O
A	A	B	C	D
B	E	F	G	H
AB	I	J	K	L
O	M	N	O	P

A ______________________ B ______________________

C ______________________ D ______________________

E ______________________ F ______________________

G ______________________ H ______________________

I ______________________ J ______________________

K ______________________ L ______________________

M ______________________ N ______________________

O ______________________ P ______________________

OBJECTIVE 10: Give an example of each of the primary blood disorders (textbook pp. 328–335).

The blood has numerous functions and is considered to be our second line of defense. Blood consists of platelets and blood clotting factors, and it transports oxygen and carbon dioxide. Perhaps a negative aspect of blood is that it also serves as a mode of transport for disease-causing organisms. After studying pp. 328–335 in your text, you should be able to describe the significant properties of blood and blood disorders.

_____ 1. Normal erythrocyte formation is hindered if the patient is lacking _____.

a. vitamin B_{12}
b. vitamin C
c. vitamin D
d. vitamin K

_____ 2. A malfunctioning _____ can result in a decrease in blood clotting factors.

a. adrenal gland
b. liver
c. pituitary gland
d. spleen

_____ 3. A decrease in _____ can result in a decrease in blood clotting factors.

a. vitamin A
b. vitamin C
c. vitamin D
d. vitamin K

_____ 4. Puerperal fever is a bacterial infection that spreads to and infects _____.

a. the blood
b. the brain
c. the child during childbirth
d. the uterus

_____ 5. The protozoan *Plasmodium* is transmitted to human blood by a mosquito and causes _____.

a. malaria
b. thalassemia
c. puerperal fever
d. pernicious anemia

OBJECTIVE 11: Distinguish among the different types of anemia (textbook pp. 332–335).

The numerous types of anemia all have the same end result – hypoxia at the tissues. The differences among each of the anemias is the causative factor. After studying in the text, pp. 332–335, you should be able to describe the different conditions leading to a specific type of anemia.

_____ 1. Which of the following may lead to hemolytic anemia?

a. hemophilia
b. malaria
c. sickle cell anemia
d. thalassemia

_____ 2. Which of the following statements about sickle cell anemia is true?

a. Sickle cells undergo phagocytosis, thus destroying normal cells.
b. The cells are sickle-shaped but become normal after adequate oxygen is added.
c. The cells cannot transport adequate amounts of oxygen due to their abnormal shape.
d. The cells transport adequate amounts of oxygen but become sickle shaped after giving up the oxygen.

_____ 3. Failure of the bone marrow to produce red blood cells results in a type of anemia called _____.

a. aplastic anemia
b. hemolytic anemia
c. sickle cell anemia
d. thalassemia

_____ 4. A deficiency in vitamin B_{12} results in _____.

a. aplastic anemia
b. pernicious anemia
c. sickle cell anemia
d. thalassemia

_____ 5. The inability to produce adequate amounts of protein subunits for hemoglobin is known as _____.

a. aplastic anemia
b. hemoglobinuria
c. pernicious anemia
d. thalassemia

PART II: CHAPTER-COMPREHENSIVE EXERCISES

Match the terms in column A with the descriptions or statements in column B.

MATCHING I

	(A)	(B)
_____	1. antigens	A. The WBCs that make up less than 1% of the WBC population and are responsible for inflammation
_____	2. basophil	B. The WBCs that make up about 3% of the WBC population and respond to allergens
_____	3. carbon monoxide	C. The hormone that is released when the kidneys detect a decrease in oxygen
_____	4. clump	D. There are about 5 million erythrocytes per _____.
_____	5. clot	E. The deadly molecule that binds to the hemoglobin molecule and prevents oxygen from doing so
_____	6. cubic millimeter	F. The glycolipids on the surface of red blood cells. We identify them as type A and B.
_____	7. eosinophil	G. The result of a series of reactions initiated by platelets
_____	8. erythropoietin	H. The result of a reaction between the antigens of an erythrocyte and the plasma antibodies
_____	9. hemophilia	I. The lack of blood clotting factor VIII
_____	10. septicemia	J. The presence of pathogens in blood

MATCHING II

	(A)	(B)
_____	1. hemoglobin	A. The WBCs that make up about 50–70% of the WBC population and respond to bacteria
_____	2. iron	B. The WBCs that make up about 8% of the WBC population and respond to fungi
_____	3. lymphocyte	C. The WBCs that make up about 20–30% of the WBC population and respond to viruses
_____	4. monocyte	D. The part of the hemoglobin molecule to which oxygen binds
_____	5. neutrophil	E. The cellular products that initiate blood clotting
_____	6. nucleus	F. The structure replaced by hemoglobin in a mature red blood cell
_____	7. pernicious anemia	G. The enzyme that converts fibrinogen to fibrin
_____	8. platelets	H. Oxygen and carbon dioxide bind to this molecule for transportation.
_____	9. thalassemia	I. Anemia die to vitamin B_{12} deficiency
_____	10. thrombin	J. Anemia due to abnormal formation of hemoglobin

CONCEPT MAP

This concept map summarizes and organizes the information in Chapter 14. Use the following terms to complete the map by filling in the boxes identified by the circled numbers, 1–8.

albumin	**eosinophil**	**oxygen**
solutes	**viruses**	**lymphocyte**
neutrophil	**leukocytes**	

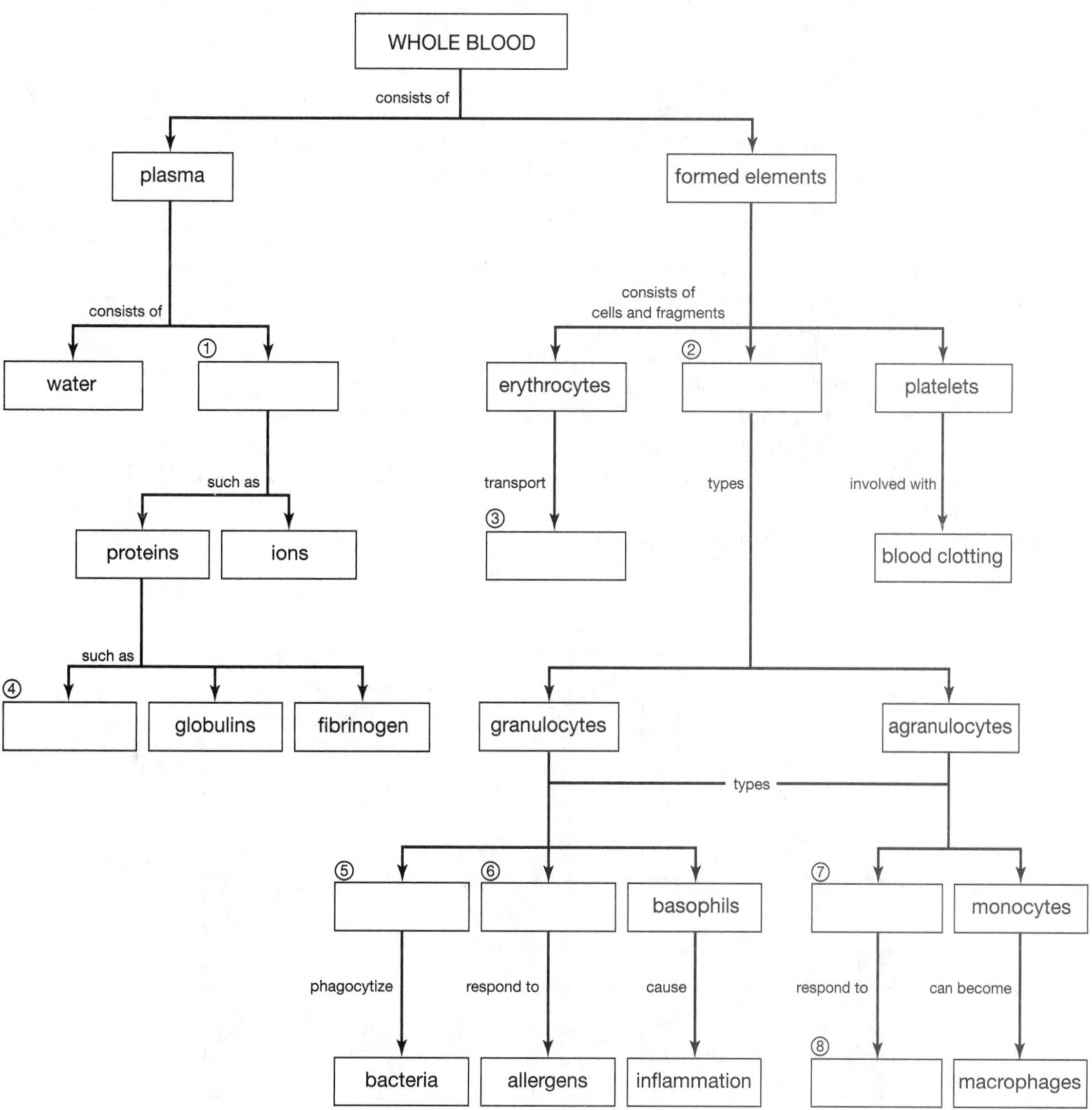

CROSSWORD PUZZLE

The following crossword reviews the material in Chapter 14. To complete the puzzle, you have to know the answers to the clues given, and must be able to spell the terms correctly.

ACROSS

3. The most common WBC in a healthy person
7. The type of anemia that occurs if the red blood cells do not receive enough vitamin B_{12}
9. Term for the formation of red blood cells
10. The type of anemia that occurs when hemoglobin is not manufactured properly by the red blood cells
12. A moving clot
15. The fluid portion of blood
16. Type AB blood has A and B _____.
17. A normal count of white blood cells is 6000–9000 per cubic _____.
18. After about 120 days, red blood cells begin to _____.

DOWN

1. Term for the formation of white blood cells
2. Term for the formation of blood
4. _____ cells contain only the antigens and not the antibodies.
5. Type B whole blood contains antigens and _____.
6. Type O packed cell blood is considered to be the universal _____.
8. The spread of pathogens and their toxins through the blood
11. The process of blood clotting
13. If type A blood were given to a person with type B blood, the type B blood would begin to _____.
14. The rarest WBC in healthy person

FILL-IN-THE-BLANK NARRATIVE

Use the following terms to fill in the blanks of this summary of Chapter 14.

carbon monoxide	**cubic millimeter**	**differential**
eosinophils	**erythroblasts**	**erythrocytes**
erythropoietin	**fungicides**	**homeostasis**
iron	**median cubital**	**antecubital**
monocytes	**neutrophils**	**lymphocytes**
20–30 percent	**39 percent**	**6000–9000/mm^3**

You go to the doctor because you don't feel well; in other words, your body is out of (1) __________. To determine what's causing you to feel ill, the doctor draws some blood from your (2) __________ vein. (This vein is located in the (3) __________ region of your arm). Results of lab tests on the blood show that you have a total white blood cell count of 25000/mm^3. The normal value is (4) __________. The lab technician then does a (5) __________ count to identify which white blood cell is in excess. The technician counts and identifies 100 leukocytes and calculates the percentage of each type of white blood cell. The technician reports that out of 100 leukocytes, 53 were neutrophils, 1 was a basophil, 1 was an eosinophil, 6 were monocytes, and 39 were (6) __________. The doctor is not going to prescribe allergy medicine because the (7) __________ were in the normal range of values. The doctor will not prescribe an antibiotic to kill bacteria because the (8) __________ were in the normal range. The doctor does not suspect any large foreign invaders like fungus because the (9) __________ were in the normal range. The doctor suspects that a virus is causing your illness because a normal lymphocyte percentage is (10) __________, and the technician reported (11) __________. Allergy medicines, antibiotics, and (12) __________ will *not* kill viruses. The doctor will probably send you home and tell you to drink fluids and get plenty of rest and watch to make sure your fever doesn't get too high.

The doctor notes that the technician also reported an erythrocyte count of 6 million per (13) __________ (5 million/mm^3 is considered normal). The doctor asks if you smoke. The doctor asks this because cigarettes produce (14) __________, which binds to the (15) __________ portion of hemoglobin, thereby taking the place of oxygen. Because the kidneys sense low oxygen concentrations, the kidney cells release (16) __________. This hormone targets the (17) __________ found in bone marrow. These stem cells begin the production of (18) __________. As red blood cells increase in number, more oxygen can be transported to make up the loss due to carbon monoxide.

The doctor tells you the best way to get back into homeostasis is to quit smoking and let the lymphocytes take care of your virus.

CLINICAL CONCEPT

The following clinical concept applies the information in Chapter 14. Following the application is a set of questions to help you understand the concept.

James, an African American, has been experiencing excessive fatigue. James eats right, exercises regularly, gets 8 hours of sleep every night, and appears to be in good health. James does not understand why he feels so tired, so he goes to the doctor to get some answers.

The doctor orders a blood test, and the results show normal values for leukocytes, erythrocytes, and hemoglobin. However, many erythrocytes were not spherical in shape but were shaped like a sickle. Even with normal erythrocyte and hemoglobin counts, having sickle-shaped erythrocytes will cause fatigue.

The doctor explains to James that he has sickle cell anemia. When the blood contains abundant oxygen, the erythrocytes appear normal. However, after the erythrocytes drop the oxygen off at the tissue, they lose their donut shape and become sickle-shaped. The sickle shapes have pointy edges that may get stuck in the capillary walls and eventually cause a blockage at the capillaries. The result is reduced blood flow and therefore reduced oxygen transport. Without oxygen, the mitochondria within the cells making up the various tissues cannot make ATP, and James feels fatigued.

The doctor schedules James for a blood transfusion to replace the sickle-shaped cells with normal cells. Over time, James will experience the same problem again because sickle cell disease is a genetic abnormality.

1. The clinical concept indicated that James is an African American. What is the significance of that information?

2. True or False: Having sickle cell anemia results in fatigue because the erythrocytes cannot pick up and transport sufficient amounts of oxygen to the tissues of the body.

3. True or False: Sickle cell anemia can be cured by blood transfusions.

4. What is the main problem that occurs with sickle cell anemia?

THIS CONCLUDES CHAPTER 14 EXERCISES

CHAPTER

15

The Heart and Circulatory System

There are several trillion cells in the human body. Each one of those cells needs a constant supply of oxygen and nutrients. The pumping action of the heart will push blood through a network of vessels. The nutrients and oxygen in the blood will pass through the vessel walls to the cells and tissues. Waste material from the cells will enter into the bloodstream and be carried away by the flow of blood.

The heart is an efficient and durable pump, beating 60–80 times per minute every single day. Chapter 15 reviews the structure of the heart and its role in pumping blood to all parts of the body through a network of blood vessels including arteries, capillaries, and veins.

The most common form of heart disease stems from problems with the blood vessels. Chapter 15 discusses numerous problems of the blood vessels ranging from inflammation to degenerative disorders to hypertension.

Disorders of the heart typically cause the heart to work harder thus resulting in an enlarged heart or enlarged chambers. Many disorders can lead to heart attacks. But if everything is in homeostasis, the heart will beat rhythmically for many years.

hapter Objectives:

- Describe the location and general features of the heart.
- Trace the flow of blood through the heart, identifying the major blood vessels, chambers, and heart valves.
- Identify the layers of the heart wall.
- Describe the events of a typical heartbeat, or cardiac cycle.
- Describe the components and functions of the conducting system of the heart.
- Describe the structure and function of arteries, capillaries, and veins.
- Describe how tissues and various organ systems interact to regulate blood flow and pressure in tissues.
- Distinguish among the types of blood vessels on the basis of their structure and function.
- Identify the major arteries and veins and the areas they serve.
- Describe the age-related changes that occur in the cardiovascular system.
- Give examples of the major disorders of the blood vessels.
- Describe the major disorders of the heart and distinguish among the various forms of shock.

PART I: OBJECTIVE-BASED QUESTIONS

OBJECTIVE 1: Describe the location and general features of the heart (textbook pp. 341–345).

After studying pp. 341–345 in the text, you should be able to identify the general features of the heart and be familiar with the terminology associated with the heart's internal and external structure.

_____ 1. The base of the heart is _____.

a. the pointy area of the heart
b. the superior area of the heart
c. the inferior area of the heart
d. Both a and c are correct.

_____ 2. Which pericardial tissue is nearest the heart tissue itself?

a. visceral pericardium
b. parietal pericardium
c. epicardium
d. both a and c
e. both b and c

_____ 3. The mitral valve is also known as the _____.

a. tricuspid valve
b. bicuspid valve
c. pulmonary semilunar valve
d. aortic semilunar valve

4. The AV valves are located between the superior and inferior chambers of the heart. AV stands for ____________________________.

5. The valve leading into the ascending aorta is the ____________________________.

6. In Figure 15–1, identify the base, apex, left side of the heart, right side of the heart, aortic arch, pulmonary trunk, and coronary vessels.

Figure 15–1 External View of the Heart

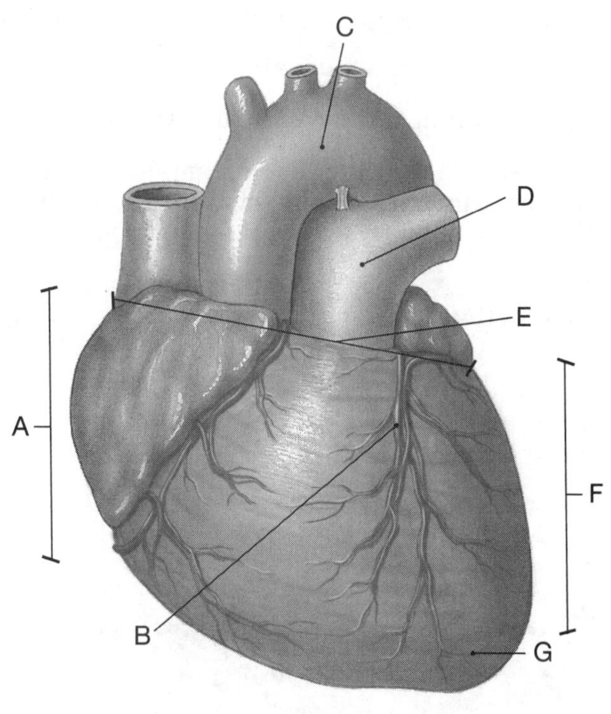

A ______________________ B ______________________

C ______________________ D ______________________

E ______________________ F ______________________

G ______________________

OBJECTIVE 2: Trace the flow of blood through the heart, identifying the major blood vessels, chambers, and heart valves (textbook pp. 342–346).

After studying pp. 342–346 in the text, you should be able to trace a drop of blood through the heart and identify the structures associated with the path of the circulating blood.

____ 1. Which chamber sends blood to the lungs?

a. right atrium
b. left atrium
c. right ventricle
d. left ventricle

____ 2. Blood going to the lungs travels through the ____.

a. aortic arch
b. pulmonary arteries
c. pulmonary veins
d. none of the above

____ 3. Which chamber receives blood from the systemic circuit?

a. left atrium
b. right atrium
c. left ventricle
d. right ventricle

4. The valve located between the right atrium and the right ventricle is called ____.

5. Blood entering into the ascending aorta came from what chamber of the heart? ____

6. In Figure 15–2, identify the right atrium, left atrium, left ventricle, right ventricle, bicuspid valve, tricuspid valve, interventricular septum, pulmonary semilunar valve, aortic arch, pulmonary trunk, pulmonary artery, pulmonary vein, superior vena cava, and inferior vena cava.

Figure 15–2 Internal View of the Heart

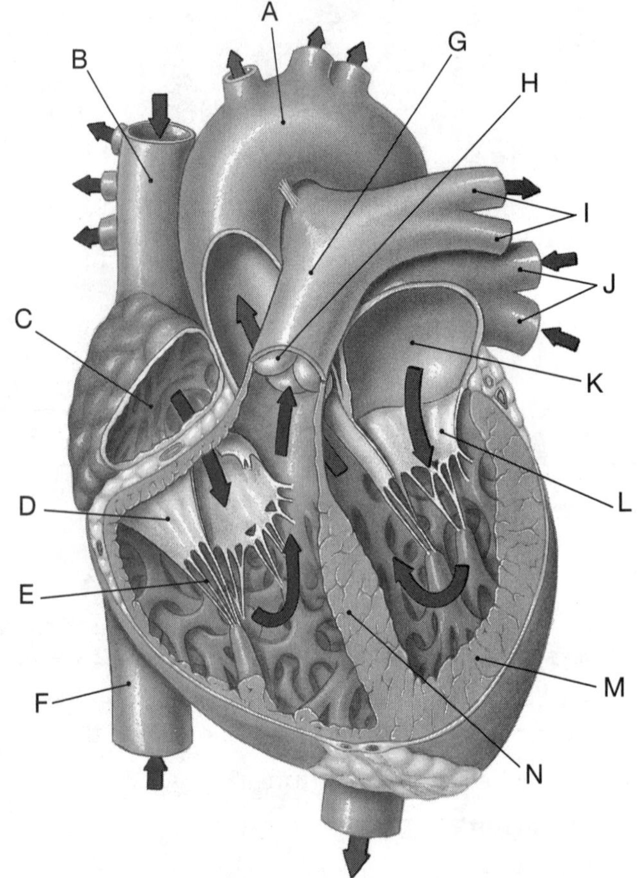

A ____________________ B ____________________

C ____________________ D ____________________

E ____________________ F ____________________

G ____________________ H ____________________

I ____________________ J ____________________

K ____________________ L ____________________

M ____________________ N ____________________

OBJECTIVE 3: Identify the layers of the heart wall (textbook p. 345).

After studying p. 345 in the text, you should be able to locate the three distinct layers of the heart: the endocardium, myocardium, and the epicardium.

_____ 1. The outer wall of the heart is called the _____.

a. cardiac layer
b. endocardium
c. epicardium
d. myocardium

_____ 2. The muscular portion of the heart wall is called the _____.

a. cardiac layer
b. endocardium
c. epicardium
d. myocardium

_____ 3. The cardiac cells can be found in which layer of the heart wall?

a. cardiac layer
b. endocardium
c. epicardium
d. myocardium

4. The myocardium is made of contractile tissue (muscle tissue). Of what kind of tissue (previously studied in Chapter 4) is the epicardium and the endocardium made? ____________________

5. In Figure 15–3, identify the endocardium, myocardium, epicardium, and intercalated disk.

Figure 15–3 Layers of the Heart Wall

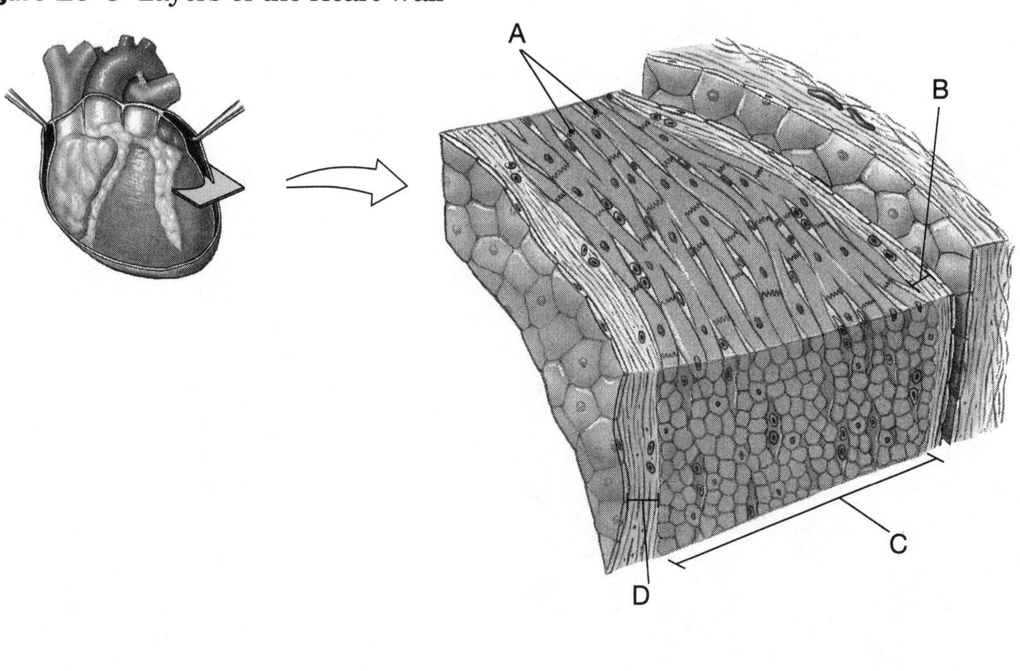

A ____________________ B ____________________

C ____________________ D ____________________

OBJECTIVE 4: Describe the events of a typical heartbeat, or cardiac cycle (textbook pp. 346–347).

After studying pp. 346–347 in the text, you should be able to explain what creates the "thump-thump" or "lubb-dubb" sounds of a typical heartbeat.

_____ 1. Systole and diastole refer to _____.

a. contraction and relaxation of the heart
b. relaxation and contraction of the heart
c. atrial contraction and ventricular contraction
d. ventricular contraction and atrial contraction

_____ 2. The atrioventricular valves close during _____.

a. atrial systole
b. atrial diastole
c. ventricular systole
d. ventricular diastole

_____ 3. The two sounds we hear as the heart contracts are due to _____.

a. the opening and closing of the AV valves and the opening and closing of the semilunar valves
b. the closing of the AV valves and the closing of the semilunar valves
c. blood rushing through the heart
d. the opening of the AV valves and the opening of the semilunar valves

4. We listen to the heartbeat with an instrument called a(an) ________________________.

5. The AV valves prevent the backflow of blood from the ventricles into the atria. What prevents the backflow of blood into the ventricles? ________________________

6. In Figure 15–4, identify the bicuspid valve, tricuspid valve, pulmonary semilunar valve, and aortic semilunar valve. For letter E, identify which valves (the atrioventricular or the semilunar valves) create the first sound we hear, and for letter F, identify the valves that create the second sound we hear when they close.

Figure 15–4 The Valves of the Heart

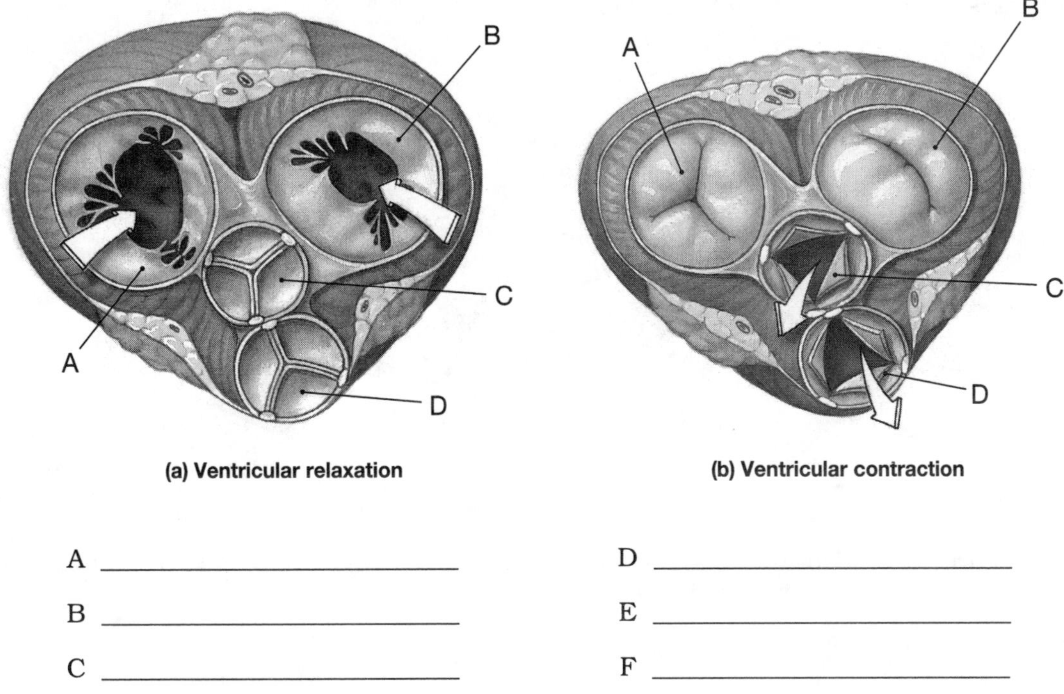

A ____________________ D ____________________

B ____________________ E ____________________

C ____________________ F ____________________

OBJECTIVE 5: Describe the components and functions of the conducting system of the heart (textbook pp. 347–350).

Cardiac muscle tissue contracts on its own without stimulation from hormones or nerves. After studying pp. 347–350 in the text, you should be able to describe the conducting system and its specialized cells that control and coordinate the activities of the heart's muscle cells. Refer to Figure 15–7 of the textbook while answering the following questions.

_____ 1. The ventricles contract when the impulse travels through the _____.

a. AV node
b. SA node
c. AV bundle
d. Purkinje fibers

_____ 2. The P wave on an ECG recording is associated with _____.

a. ventricular contraction
b. atrial contraction
c. atrial relaxation
d. none of the above

_____ 3. The QRS wave on an ECG recording is associated with _____.

a. ventricular contraction
b. atrial contraction
c. ventricular diastole
d. atrial systole

4. The sinoatrial node is located in which chamber of the heart?

____

5. Which nodal cell is considered the pacemaker? ______________________

6. In Figure 15–5, identify the sinoatrial node, atrioventricular node, bundle branches, and the Purkinje fibers.

Figure 15–5 The Conducting System of the Heart

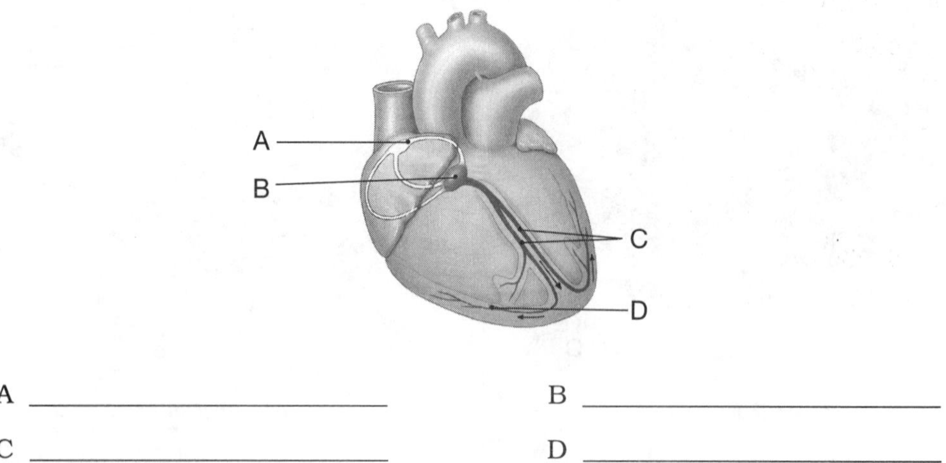

A ______________________ B ______________________

C ______________________ D ______________________

7. In Figure 15–6, identify which part of the ECG recording (1, 2, 3, etc.) correlates with heart A and which part correlates with heart B?

Figure 15–6 An Electrocardiogram

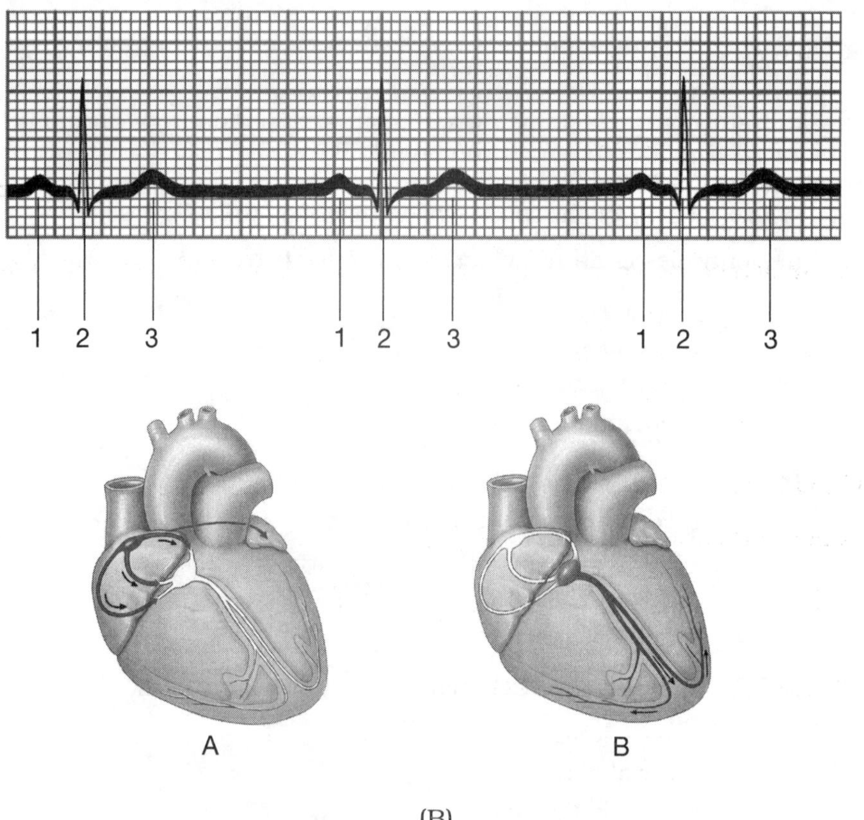

(A) ______________________ (B) ______________________

8. In Figure 15–6, number 1 represents the _____ wave of an ECG recording.

9. In Figure 15–6, number 2 represents the _____ wave of an ECG recording.

10. In Figure 15–6, number 3 represents the _____ wave of an ECG recording.

OBJECTIVE 6: Describe the structure and function of arteries, capillaries, and veins (textbook pp. 350–351).

After studying pp. 350–351 in the text, you should be able to use the terminology associated with the structure and function of blood vessels.

_____ 1. **The smallest** vessels are called _____.

a. arterioles
b. capillaries
c. venules
d. none of the above

_____ 2. **Gas exchange** occurs in the _____.

a. arterioles
b. capillaries
c. venules
d. heart

_____ 3. **When blood** leaves the capillaries, it flows into _____.

a. arteries
b. arterioles
c. venules
d. veins

_____ 4. **Which of the following** vessels can withstand the most pressure?

a. arteries
b. capillaries
c. veins
d. They can all withstand the same amount of pressure.

_____ 5. **Which of the following** contain a thicker layer of smooth muscle?

a. arteries
b. arterioles
c. capillaries
d. veins

6. Which blood vessels have valves? ____________________

7. Explain how varicose veins occur. ____________________

OBJECTIVE 7: Describe how tissues and various organ systems interact to regulate blood flow and pressure in tissues (textbook pp. 351–356).

After studying pp. 351–356 in the text, you should be able to explain how the nervous system and the endocrine system work together to regulate blood flow.

_____ 1. Which division of the autonomic nervous system slows down the heart rate?

a. sympathetic nervous system
b. parasympathetic nervous system
c. Both a and b will slow down the heart.
d. Neither a nor b will slow down the heart.

_____ 2. Stimulation of the vagus nerve will _____.

a. slow the heart rate
b. increase the heart rate
c. The vagus nerve does not affect the heart.
d. The vagus nerve will either speed up the heart or slow down the heart depending on environmental conditions.

_____ 3. Blood pressure is recorded as _____.

a. diastolic/systolic in units of millimeters of mercury
b. systolic/diastolic in units of millimeters of mercury
c. diastolic/systolic in units of cubic millimeters of mercury
d. systolic/diastolic in units of cubic millimeters of mercury

_____ 4. Which of the following is a normal blood pressure reading?

a. 130/82
b. 82/130
c. 120
d. Both a and b are typical readings.

5. Indicate whether each of the following hormones increases blood pressure or decreases blood pressure.

a. antidiuretic hormone ____________________
b. aldosterone ____________________
c. erythropoietin ____________________
d. atrial natriuretic peptide ____________________

OBJECTIVE 8: Distinguish among the types of blood vessels on the basis of their structure and function (textbook pp. 356–364).

After studying pp. 356–364 in the text, you should be able to identify the different types of blood vessels in the body according to their structure and their function. Probably the biggest misconception about blood vessels is that all arteries carry oxygenated blood and all veins carry deoxygenated blood. After studying this objective, you will find out and give the true definition of arteries and veins.

____ 1. Which vessels tend to have thicker walls?

a. arteries
b. veins
c. arterioles
d. venules

____ 2. Which of the following vessels supply nutrients and oxygen to the local tissues?

a. arteries
b. capillaries
c. veins
d. all the above

____ 3. Which vessels can withstand the highest pressure?

a. arteries
b. veins
c. arterioles
d. venules

____ 4. Vessels that carry blood away from the heart are called ____.

a. arteries
b. capillaries
c. veins
d. It depends on whether the blood is oxygenated or deoxygenated.

____ 5. Vessels that carry blood to the heart are called ____.

a. arteries
b. capillaries
c. veins
d. It depends on whether the blood is oxygenated or deoxygenated.

6. In Figure 15–7, identify the artery, arteriole, capillaries, venules, and vein.

Figure 15–7 Types of Blood Vessels

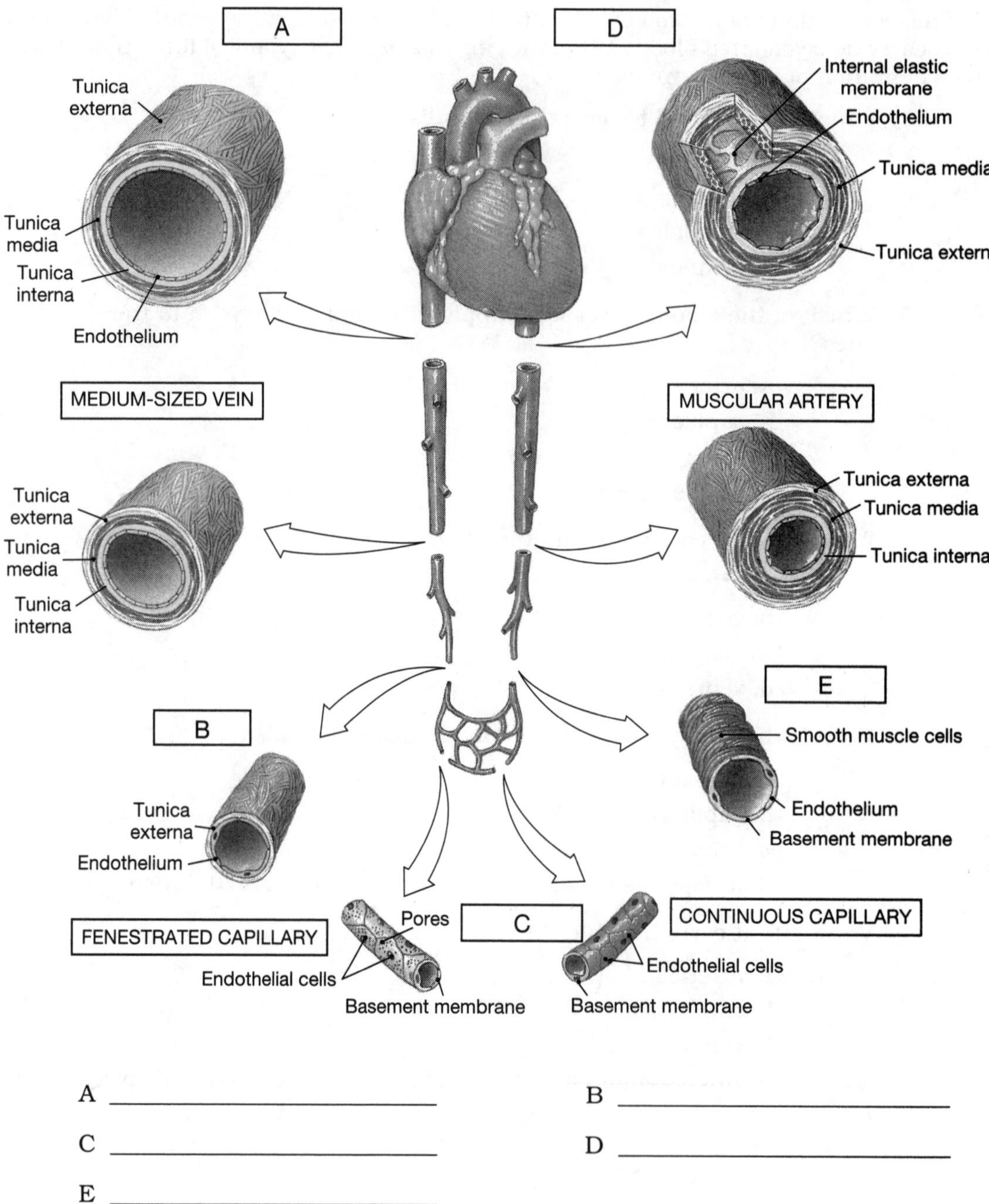

A ____________________ B ____________________

C ____________________ D ____________________

E ____________________

OBJECTIVE 9: Identify the major arteries and veins and the areas they serve (textbook pp. 356–366).

After studying pp. 356–366 in the text, you should be able to follow a drop of blood through the major blood vessels of the body keeping in mind that arteries carry blood away from the heart and veins carry blood to the heart.

____ 1. Blood flows from the popliteal vein to the ____.

a. anterior tibial vein
b. femoral artery
c. femoral vein
d. popliteal artery

____ 2. Blood flows from the brachiocephalic artery to the ____ and ____.

a. axillary, brachial artery
b. left subclavian, left carotid artery
c. right subclavian, axillary artery
d. right subclavian, right carotid artery

____ 3. Blood flows from the renal vein to the ____.

a. descending aorta
b. inferior vena cava
c. kidneys
d. renal artery

____ 4. Blood flows from the basilic vein to the ____.

a. axillary vein
b. brachial vein
c. cephalic vein
d. subclavian vein

____ 5. Blood flows from the jugular vein to the ____.

a. brachiocephalic vein
b. subclavian vein
c. superior vena cava
d. none of the above

____ 6. Blood flows from the brachial artery to the ____ and ____.

a. antecubital, antebrachial arteries
b. brachial vein, subclavian vein
c. cephalic, basilic veins
d. radial, ulnar arteries

____ 7. Blood flows from the great saphenous vein to the ____.

a. anterior or posterior tibial veins
b. femoral vein
c. iliac vein
d. popliteal vein

_____ 8. Blood flows from the femoral vein to the _____.

a. anterior or posterior tibial veins
b. great saphenous vein
c. iliac vein
d. popliteal vein

_____ 9. Blood flows from the femoral artery to the _____.

a. anterior or posterior tibial arteries
b. femoral vein
c. iliac artery
d. popliteal artery

_____ 10. Blood flows from the radial vein to the _____.

a. brachial vein
b. cephalic vein
c. median cubital vein
d. ulnar vein

_____ 11. Blood flows from the subclavian artery to the _____.

a. axillary artery
b. brachial artery
c. brachiocephalic artery
d. carotid artery

_____ 12. Blood flows from the cephalic vein to the _____.

a. axillary vein
b. basilic vein
c. brachial vein
d. subclavian vein

_____ 13. Blood flows from the iliac vein to the _____.

a. femoral vein
b. inferior vena cava
c. popliteal vein
d. none of the above

_____ 14. The subclavian arteries serve which of the following body areas?

a. the lower limbs and abdominal area
b. the upper limbs and chest area
c. the upper limbs and lower limbs
d. the thoracic area only

_____ 15. The mesenteric arteries serve which of the following body areas?

a. the abdominal organs
b. the brain area
c. the thoracic area
d. none of the above

____ 16. The carotid arteries serve which of the following body areas?

a. the abdominal region
b. the head region
c. the pelvic region
d. none of the above

17. Identify and label the blood vessels in Figure 15–8. Be sure to identify them as arteries or as veins where appropriate.

Figure 15–8 Pulmonary Circulation

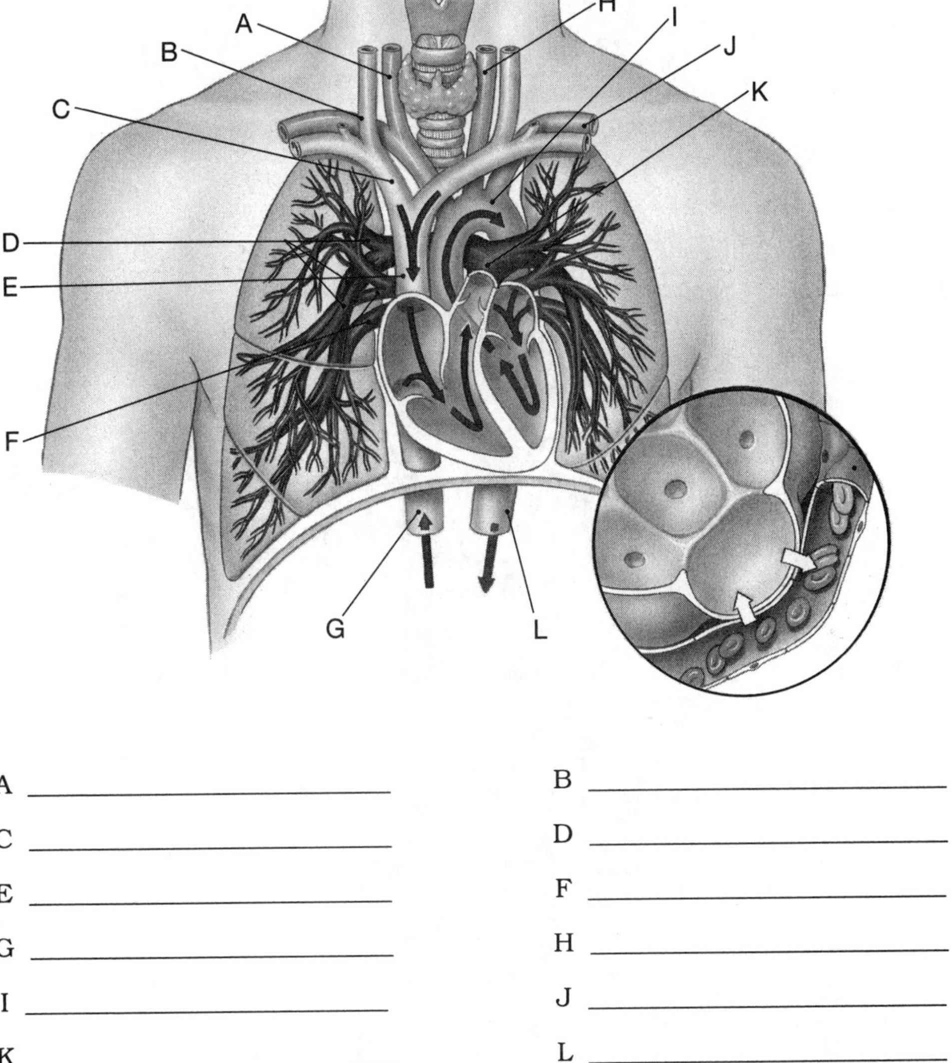

A ______________________ B ______________________

C ______________________ D ______________________

E ______________________ F ______________________

G ______________________ H ______________________

I ______________________ J ______________________

K ______________________ L ______________________

18. Identify and label the arteries in Figure 15–9.

Figure 15–9 Major Arteries of the Body

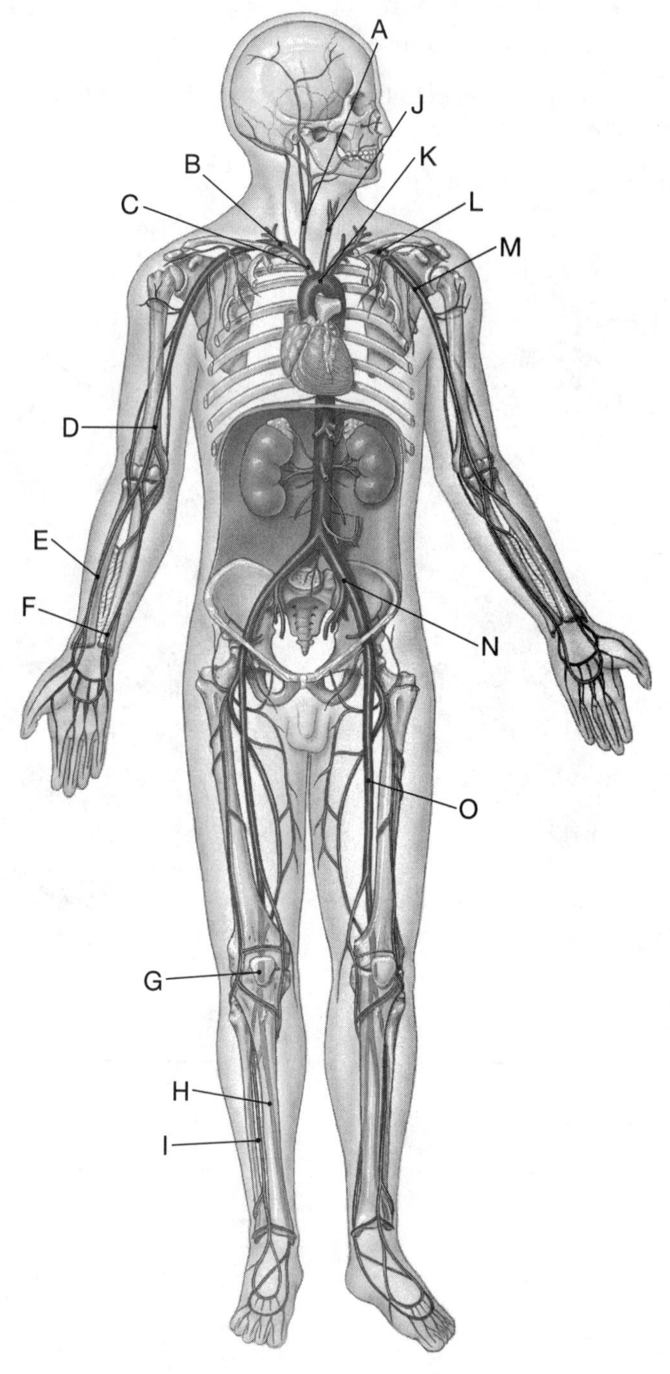

A ______________________ B ______________________ C ______________________

D ______________________ E ______________________ F ______________________

G ______________________ H ______________________ I ______________________

J ______________________ K ______________________ L ______________________

M ______________________ N ______________________ O ______________________

19. Identify and label the veins in Figure 15–10.

Figure 15–10 Major Veins of the Body

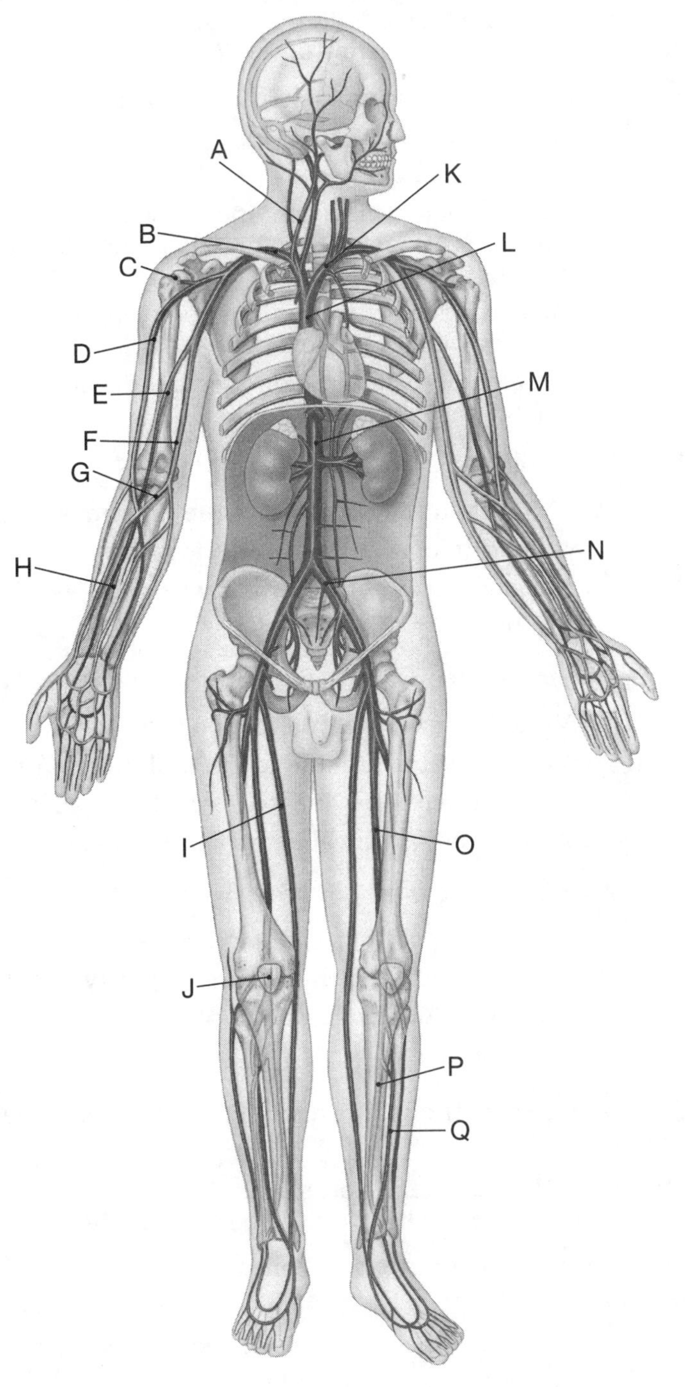

A ________________ B ________________ C ________________

D ________________ E ________________ F ________________

G ________________ H ________________ I ________________

J ________________ K ________________ L ________________

M ________________ N ________________ O ________________

P ________________ Q ________________

OBJECTIVE 10: Describe the age-related changes that occur in the cardiovascular system (textbook p. 366).

After studying p. 366 in the text, you should be able to list the age-related changes in the cardiovascular system, which include a buildup of plaque in the walls of the blood vessels and a consequent decrease in the efficiency of the heart.

_____ 1. The difference between arteriosclerosis and atherosclerosis is that _____.

a. Arteriosclerosis is a type of atherosclerosis that causes blood vessel constriction.

b. Arteriosclerosis is a condition in which the vessels begin to thicken. Atherosclerosis is one type of arteriosclerosis, which causes blood vessel constriction.

c. Artherosclerosis is an aneurysm, whereas arteriosclerosis is just a thickening of the arteries.

d. Atherosclerosis is plaque formation, and arteriosclerosis is an aneurysm.

_____ 2. Which of the following occur(s) with age?

a. changes in the activities of the SA nodes and AV nodes

b. a buildup of scar tissue on the heart

c. a constriction of the coronary vessels

d. all the above

_____ 3. With age, there is a tendency to develop varicose veins. Why does this occur?

a. Varicose veins are due to blood clot formation.

b. Varicose veins are due to the buildup of fatty tissue in our legs.

c. Varicose veins are due to the weakening of the valves in the veins.

d. All the above can cause varicose veins.

_____ 4. Which of the following does not occur with age?

a. a constriction of arteries and veins

b. a decrease in oxygen-carrying capacity

c. thrombus formation due to plaque buildup

d. all the above

OBJECTIVE 11: Give examples of the major disorders of the blood vessels (textbook pp. 368–371).

After studying pp. 368–371 in the text, you should be able to describe the major blood vessel disorders and the effects they have on the body as a whole.

_____ 1. An inflammation of an artery is called _____, and an inflammation of a vein is called _____.

a. arteriosclerosis, atherosclerosis

b. arteriosclerosis, phlebitis

c. arteritis, phlebitis

d. phlebitis, thrombophlebitis

_____ 2. Which of the following statements is true?

a. Atherosclerosis is a type of arteriosclerosis.

b. Arteriosclerosis is a type of atherosclerosis.

c. Focal calcification is a type of atherosclerosis.

d. Atherosclerosis is a type of focal calcification.

_____ 3. Hemorrhoids are a type of _____.

a. aneurysm
b. arteriosclerosis
c. atherosclerosis
d. varicose vein

_____ 4. Which of the following is an indication of high blood pressure?

a. a diastolic value of 80
b. a diastolic value of 90 or more
c. a systolic value of 80
d. a systolic value of 90 or more

_____ 5. Streptococcus bacteria cause rheumatic fever, which has an effect on _____.

a. the epicardium
b. the heart valves
c. the skin of the forehead region
d. the throat area

_____ 6. Which of the following therapies is being described that reduces the heart rate by reducing the sympathetic activity on the heart?

a. beta-blockers
b. calcium channel blockers
c. diuretics
d. vasodilators

OBJECTIVE 12: Describe the major disorders of the heart and distinguish among the various forms of shock (textbook pp. 371–374).

Many disorders of the heart will appear as symptoms of shock. After studying pp. 371–374 in the text, you should be able to describe three types of shock associated with heart problems.

_____ 1. Which of the following describes the effects of myocarditis?

a. The cells of the heart become easier to depolarize.
b. The cells of the heart become easier to repolarize.
c. The cells of the heart become more difficult to depolarize.
d. The cells of the heart become more difficult to repolarize.

_____ 2. The _____ will typically enlarge due to working harder because the foramen ovale is still slightly open some time after birth.

a. left atrium
b. left ventricle
c. right atrium
d. right ventricle

_____ 3. The _____ will typically enlarge if the patient has tetralogy of Fallot.

a. left atrium
b. left ventricle
c. right atrium
d. right ventricle

_____ 4. Vasodilators allow the heart to work easier. Which of the following is a vasodilator?

a. atrial natriuretic peptide
b. beta-blockers
c. metoprolol
d. nitroglycerin

_____ 5. Massive peripheral dilation of vessels due to the response of medication, for example, will cause the patient to become very pale. This condition is known as _____.

a. anaphylactic shock
b. cardiogenic shock
c. circulatory shock
d. myxoma

PART II: CHAPTER-COMPREHENSIVE EXERCISES

Match the terms in column A with the definitions in column B.

MATCHING I

	(A)	(B)
_____	1. arrhythmia	A. Carry blood away from the heart
_____	2. arteries	B. Small vessels that allow for gas and nutrient exchange
_____	3. base	C. The superior portion of the heart
_____	4. bradycardia	D. The membrane that is *not* adjacent to the heart's outer surface
_____	5. capillaries	E. The left atrioventricular valve
_____	6. coronary arteries	F. A condition in which the AV valves do not close properly
_____	7. diastole	G. The main muscle portion of the heart
_____	8. heart murmur	H. Vessels that supply oxygen to the heart
_____	9. mitral valve	I. Heart relaxation
_____	10. myocardium	J. Abnormally slow heart rate
_____	11. parietal pericardium	K. Abnormal cardiac rhythms
_____	12. phlebitus	L. Inflammation of a vein

MATCHING II

	(A)	(B)
_____	1. neurogenic	A. Carries blood to the heart
_____	2. pulmonary circuit	B. The circulatory system involving the lungs
_____	3. Purkinje fibers	C. The circulatory system involving the body's organs and tissues
_____	4. rheumatic	D. The membrane adjacent to the heart's outer surface
_____	5. sinoatrial node	E. Heart contraction
_____	6. systemic circuit	F. The pacemaker of the heart
_____	7. systole	G. Once initiated, these cause ventricular contraction
_____	8. tachycardia	H. Abnormally fast heart rate
_____	9. umbilical arteries	I. Carries blood to the fetal heart
_____	10. umbilical veins	J. Carries blood to the placenta
_____	11. veins	K. _____ shock may result from a vagus nerve stimulation.
_____	12. visceral pericardium	L. A disorder in which the valves of the heart are infected by a bacterium

CONCEPT MAP

This following concept map summarize and organizes the material in Chapter 15. Use the following terms to complete the map by filling in the boxes identified by the circled numbers 1–8.

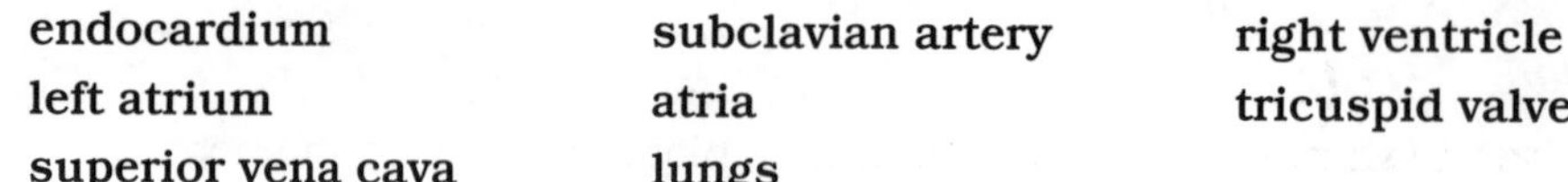

endocardium	**subclavian artery**	**right ventricle**
left atrium	**atria**	**tricuspid valve**
superior vena cava	**lungs**	

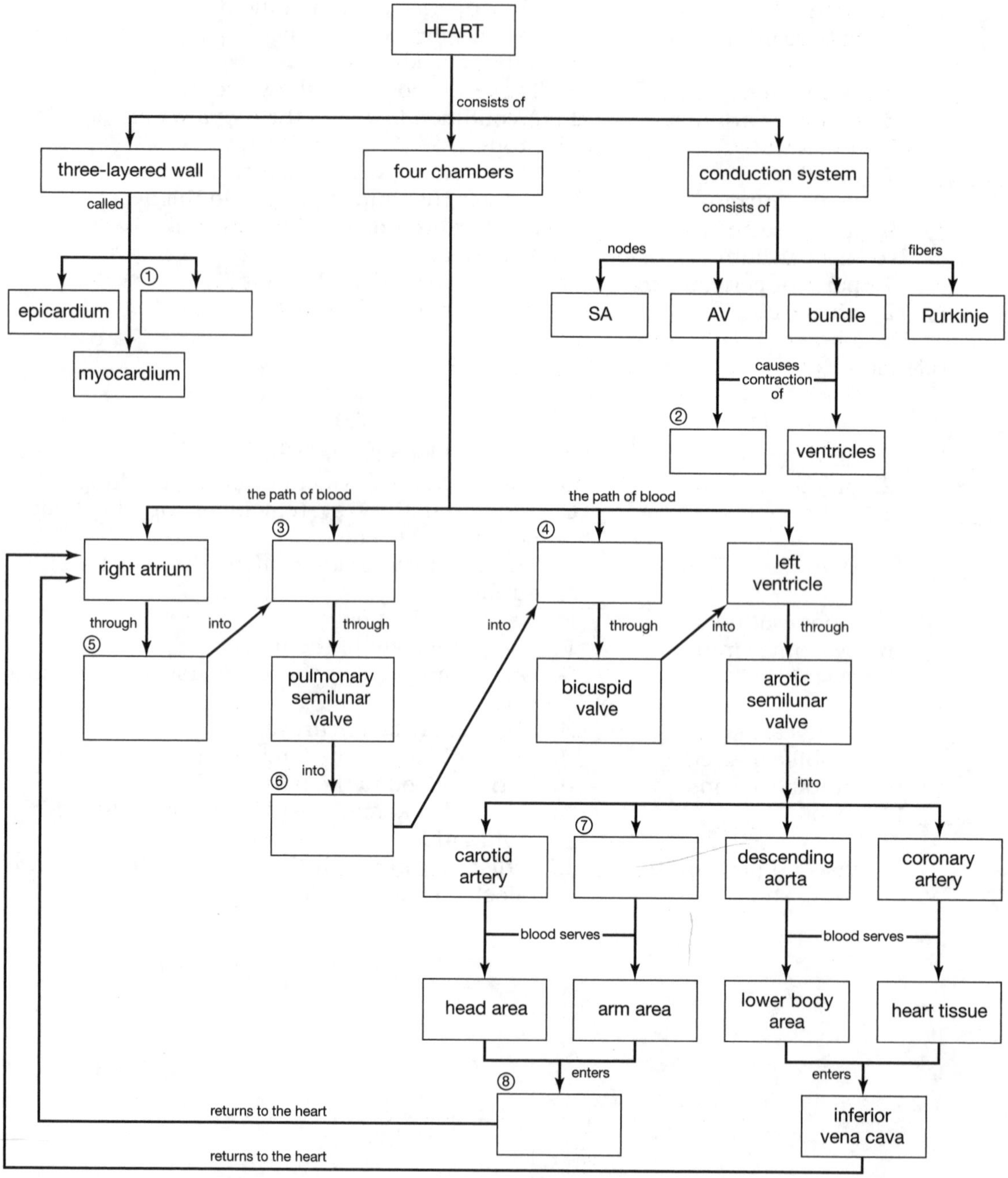

HEART
consists of
three-layered wall
four chambers
conduction system
called
epicardium
①
myocardium
consists of
nodes
fibers
SA
AV
bundle
Purkinje
causes contraction of
②
ventricles
the path of blood
the path of blood
③
④
right atrium
left ventricle
through
⑤
into
through
into
through
into
through
pulmonary semilunar valve
bicuspid valve
arotic semilunar valve
into
⑥
into
⑦
carotid artery
descending aorta
coronary artery
blood serves
blood serves
head area
arm area
lower body area
heart tissue
enters
⑧
enters
returns to the heart
inferior vena cava
returns to the heart

CROSSWORD PUZZLE

The following crossword puzzle reviews the material in Chapter 15. To complete the puzzle, you have to know the answers to the clues given, and must be able to spell the terms correctly.

ACROSS

2. Blood in the popliteal vein enters the _____ vein.
4. Arterioles flow into _____, then into venules.
8. Nitroglycerin is a _____.
10. Plaque buildup in blood vessels
13. Vessels that carry blood away from the heart
16. Blood in the _____ artery enters the brachial artery.
17. The aortic arch arches to the _____ as it emerges from the heart.
18. An opening between the two atria of a fetal heart is called the foramen _____.
20. When these fibers are activated, the ventricles will contract.

DOWN

1. Heart _____ are the sensations that indicate the heart has skipped a beat.
3. The main muscular portion of the heart
5. Blood entering the inferior vena cava could have come from the _____ vein.
6. The valve located between the right atrium and the right ventricle
7. The sinoatrial node is the heart's _____.
9. Blood leaving the femoral artery enters the _____ artery.
12. Another name for the bicuspid valve
14. Contraction of the heart
15. Is the aortic arch closer to the base or to the apex of the heart?
19. The hormone that helps lower blood pressure

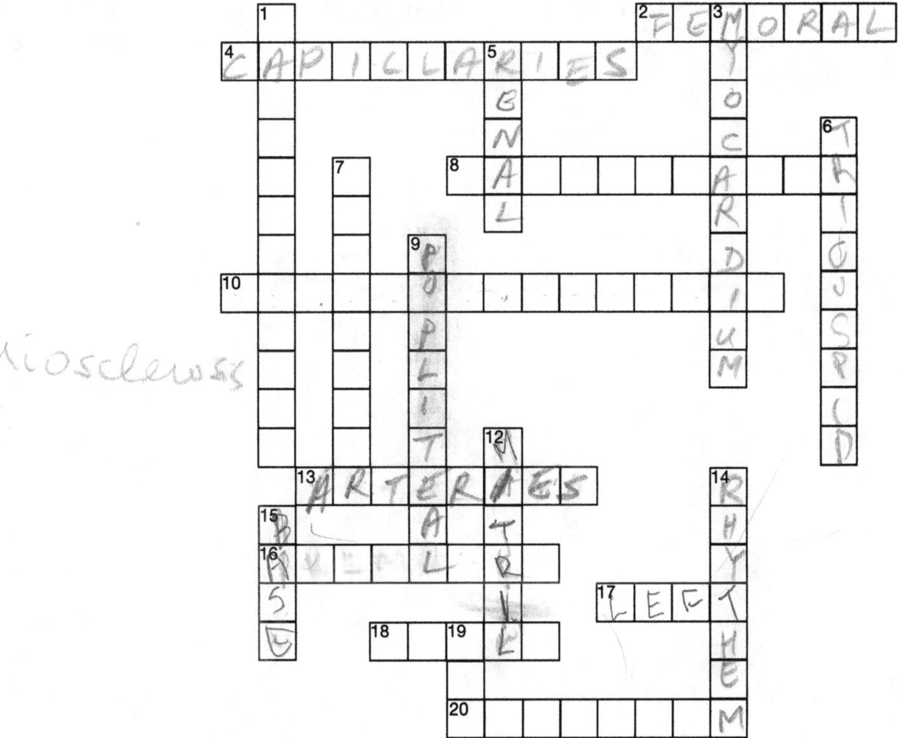

FILL-IN-THE-BLANK NARRATIVE

Use the following terms to fill in the blanks of this summary of Chapter 15.

superior vena cava	**right atrium**	**deoxygenated**
inferior vena cava	**left ventricle**	**Purkinje**
pulmonary veins	**sinoatrial**	**ventricles**
pulmonary arteries	**atrioventricular**	**oxygen**
carbon dioxide	**capillaries**	**apex**
aortic arch	**murmur**	**left**
atrium	**oxygenated**	**enlarged**

In order for the heart to contract, an impulse must be initiated by the (1) __________ node. As the impulse travels to the (2) __________ node, the two atria contract, which forces blood into the two ventricles. The impulse travels down the bundle fibers and then upward from the (3) __________ of the heart through the (4) __________ fibers, resulting in the contraction of the two (5) __________. Blood in the right ventricle enters the (6) __________ on its way to the lungs to exchange carbon dioxide for (7) __________. This oxygenated blood returns from the lungs and goes back to the heart (left atrium) via the (8) __________. Oxygenated blood from the (9) __________ enters the (10) __________ and is delivered to all tissues of the body. Once at the tissues, the oxygenated blood travels through (11) __________ that are small enough to allow oxygen to leave the blood and enter the tissues while allowing (12) __________ to leave the tissues and enter the blood. This blood is now (13) __________. It returns to the heart by entering the (14) __________ from the lower extremities, and the (15) __________ from the upper extremities. The deoxygenated blood enters into the (16) __________ of the heart, and the entire cycle is repeated.

A patient may have developed or perhaps was born with a stenosis of the bicuspid valve. This defect results in a heart (17) __________. This is a condition in which blood in the (18) __________ ventricle passes back through the valve and enters the left (19) __________. As a result of this "backflow" less (20) __________ blood goes out to the body tissues. This may cause the left ventricle to work harder to compensate for the oxygen deficiency, thus resulting in an (21) __________ left ventricle. If the condition does not improve or worsens, heart failure may occur.

CLINICAL CONCEPT

The following clinical concept applies the information in Chapter 15. Following the application is a set of questions to help you understand the concept.

Adam noticed that he felt dizzy from time to time when he sat up or stood up. He went to the doctor to have his blood pressure checked, which was 110/60. The doctor suggested that sometimes a low diastolic value will cause dizziness if you stand up too fast, but Adam's diastolic value was in the acceptable range of 60–80. Adam was reassured and didn't worry too much, since he felt dizzy only once in awhile.

Soon, Adam began to experience blackouts, so he returned to the doctor's office. His pulse was 70, and his blood pressure was 110/60. Because these values appeared to be normal, the doctor ordered a blood test and found a normal leukocyte count, a normal erythrocyte count, and a normal hemoglobin count. The doctor then ordered an X ray. The results showed that Adam had a rib growing on cervical vertebra number 7.

The doctor explained to Adam that he was experiencing blackouts because the cervical rib was putting pressure on a major artery emerging from his heart. As the rib grew, more pressure was put on the aorta, resulting in a constriction and therefore a decreased flow of blood to the brain. The remedy was surgery to remove the cervical rib.

1. How many pairs of cervical ribs, thoracic ribs, and lumbar ribs does the body normally have?

2. What might be a major artery that the cervical rib is constricting?

3. Which number in the reading 110/60 represents the diastolic value?

4. The doctor reported that the leukocytes and erythrocytes were within normal ranges. What are the normal ranges?

THIS CONCLUDES CHAPTER 15 EXERCISES

CHAPTER

16

The Lymphatic System and Immunity

The world around us contains an assortment of viruses, bacteria, fungi, and parasites capable of not only surviving but thriving inside our bodies and potentially causing us great harm in the process. Many organs and systems work together in an effort to keep us alive and healthy. In this ongoing struggle, the **lymphatic system** plays a central role. The lymphatic system includes lymphoid organs and tissues, such as lymph nodes, spleen, thymus, appendix, and tonsils. These organs and tissues contain large numbers of specialized cells called lymphocytes. These organs, tissues, and the venous system are linked together by a network of lymphatic vessels that contain the fluid lymph. The three primary functions of the lymphatic system are

- protection against disease causing organisms primarily by the activities of lymphocytes;
- the return of fluid and solutes from peripheral tissues to the blood;
- the distribution of hormones, nutrients, and waste products from their tissues of origin to venous circulation.

The exercises in Chapter 16 focus on topics that include the organization of the lymphatic system, the interaction of the lymphatic system with cells and tissues of other systems to defend the body against infection and disease, and patterns of the immune response.

After concentrating on the major functions of the lymphatic system, you will then learn about the ramifications of a lymphatic system that is out of homeostasis.

Chapter Objectives:

1 Identify the major components of the lymphatic system and explain their functions.

2 Discuss the importance of lymphocytes and describe where they are found in the body.

3 List the body's nonspecific defenses and explain how each functions.

4 Describe the different categories of immunity.

5 Distinguish between cell-mediated immunity and antibody-mediated immunity.

6 Discuss the different types of T cells and the role played by each in the immune response.

7 Describe the primary and secondary immune responses to antigen exposure.

8 Describe the changes in the immune system that occur with aging.

9 Describe examples of disroders of the lymphatic vessels, lymphoid tissue, lymphoid organs, and lymphocytes.

10 Distinguish among the three types of abnormal immune responses.

PART I: OBJECTIVE-BASED QUESTIONS

OBJECTIVE 1: Identify the major components of the lymphatic system and explain their functions (textbook pp. 381–385).

The lymphatic system helps return fluid from body tissues to the blood and also serves as a defense for the body. After studying pp. 381–385 in the text, you should be able to identify the organs and tissues of the lymphatic system and understand how these organs and tissues function to help maintain homeostasis in the body.

_____ 1. Which of the following are functions of the lymphatic system?

a. It helps transport lipids from the digestive tract to the bloodstream.
b. It helps maintain normal blood volume.
c. It helps protect the body against pathogens.
d. All the above are functions of the lymphatic system.

_____ 2. Which of the following are organs or tissues of the lymphatic system?

a. spleen
b. thymus gland
c. tonsils
d. all the above

_____ 3. Lymphocytes are produced in the _____, and some mature in the _____ to become T cells.

a. bone marrow, lymph gland
b. bone marrow, thymus gland
c. spleen, thymus gland
d. thymus gland, bone marrow

_____ 4. The lymphatic system returns fluid from body tissues to the circulatory system. The lymph from the lower regions returns to the bloodstream by entering the _____, and lymph from the upper regions returns to the bloodstream by entering the _____.

a. inferior vena cava, superior vena cava
b. left subclavian artery, right subclavian artery
c. left subclavian vein, right subclavian vein
d. thoracic duct, right lymphatic duct

_____ 5. The hormone produced by the thymus gland that causes some lymphocytes to become T cells is called ____________________.

6. In Figure 16–1, identify the cervical nodes, thymus gland, spleen, inguinal nodes, thoracic duct, and axillary nodes.

Figure 16–1 The Lymphatic System

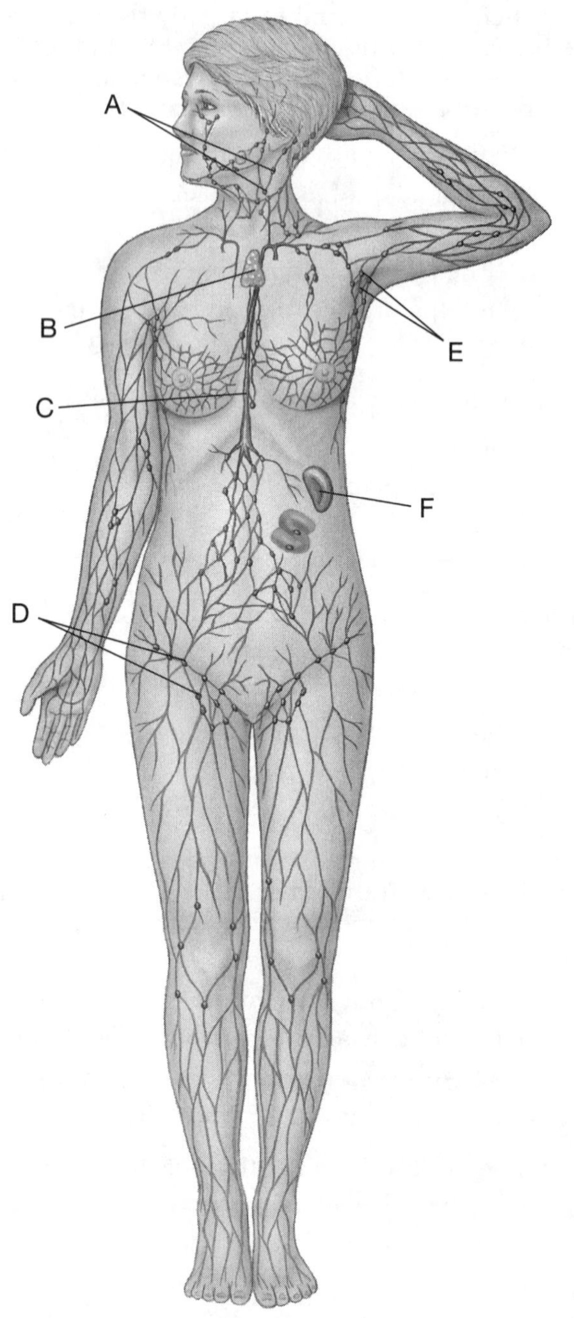

A ______________________ B ______________________

C ______________________ D ______________________

E ______________________ F ______________________

OBJECTIVE 2: Discuss the importance of lymphocytes and describe where they are found in the body (textbook pp. 385–386).

In Chapter 14, the lymphocytes were discussed in relationship to the cardiovascular system. After studying pp. 385–386 in the text, you should be able to discuss the lymphocytes in more detail in reference to their specific functions in the defense system.

_____ 1. T and B cells are both types of _____. The _____ cells mature in the thymus gland, and the _____ cells mature in the bone marrow.

a. immunoglobulins, B, T
b. lymphocytes, B, T
c. lymphocytes, T, B
d. natural killer cells, T, B

_____ 2. Which of the following types of lymphocytes produce antibodies?

a. B cells
b. NK cells
c. T cells
d. all the above

_____ 3. Which of the following statements about lymphocyte activity are true?

a. B cells can leave the circulatory system to enter the body tissues to "fight" invaders.
b. NK cells can leave the circulatory system to enter body tissues to "fight" invaders.
c. T cells can leave the circulatory system to enter body tissues to "fight" invaders.
d. All the above are true.

4. NK cells are produced in the ______________________________. The letters NK stand for ______________________________.

OBJECTIVE 3: List the body's nonspecific defenses and explain how each functions (textbook pp. 387–389).

The body has two main defense strategies that act together to defend and protect the human body. One is nonspecific and the other is specific. After studying pp. 387–389 in the text, you should be able to describe the nonspecific strategies that are present at birth and which do not discriminate between one pathogen and another but generally target all pathogens.

_____ 1. Which of the following is (are) not a part of the nonspecific immunity strategy?

a. B cells
b. fever
c. interferon
d. macrophages

_____ 2. Which of the following are *not* macrophages?

a. Kupffer cells
b. microglial cells
c. monocytes
d. All the above are macrophages.

____ 3. What is the name of the protein that interferes with the replication of viruses?

a. complement proteins
b. heparin
c. interferon
d. pyrogens

____ 4. Which of the following best explains why a fever develops?

a. A fever develops in an effort to kill bacteria.
b. A fever occurs because some pathogens act as pyrogens and reset the body's thermostat.
c. A fever develops as the result of increased immune activity.
d. A fever develops when bacteria enter the body.

5. How is the inside lining of the digestive tract protected against foreign substances?

____

OBJECTIVE 4: Describe the different categories of immunity (textbook pp. 389–391).

After studying pp. 389–391 in the text, you should be able to differentiate between innate immunity and the various forms of acquired immunity.

____ 1. Which of the following types of immunity is present at birth?

a. artificially acquired immunity
b. innate immunity
c. naturally acquired immunity
d. passive immunity

____ 2. Which type of immunity is involved with a vaccination?

a. artificially acquired immunity
b. innate immunity
c. naturally acquired immunity
d. passive immunity

____ 3. You receive a letter from the preschool your child attends that one of the students has chickenpox. Two weeks later, your child develops chickenpox. After your child gets over the chickenpox, your child will be immune to chickenpox. Which type of immunity is being described in this scenario?

a. artificially acquired immunity
b. innate immunity
c. naturally acquired immunity
d. passive immunity

____ 4. In order to develop an acquired immunity against a specific pathogen, you must first be exposed to _____.

a. cytokines of the virus or bacterium
b. the antibody of the virus or bacterium
c. the antigen of the virus or bacterium
d. the pyrogens of the virus or bacterium

_____ 5. Which of the following types of immunity is considered to be nonspecific?

a. acquired immunity
b. innate immunity
c. passive immunity
d. All the above are considered to be types of specific immunity.

OBJECTIVE 5: Distinguish between cell-mediated immunity and antibody-mediated immunity (textbook p. 391).

After studying p. 391 in the text, you should be able to explain the difference between cell-mediated immunity and antibody-mediated immunity, and the differences in modes of action between T cells and B cells.

_____ 1. T cells are involved with _____ and _____ attack pathogens.

a. cell-mediated responses, directly
b. cell-mediated responses, indirectly
c. humoral responses, directly
d. humoral responses, indirectly

_____ 2. B cells are involved with _____ and create a chemical attack on _____.

a. cell mediated responses, antigens
b. cell mediated responses, pathogens
c. humoral responses, antibodies
d. humoral responses, antigens

3. Why is antibody-mediated immunity also called humoral immunity?

OBJECTIVE 6: Discuss the different types of T cells and the role played by each in the immune response (textbook p. 391).

After studying p. 391 in the text, you should be able to describe the process by which some lymphocytes differentiate and mature into T cells when they are exposed to thymosin, and the differentiation of T cells into a variety of specialized T cells.

_____ 1. Which cells stimulate the production of antibodies?

a. B cells
b. cytotoxic T cells
c. helper T cells
d. memory T cells

_____ 2. Macrophages phagocytize pathogens and present the _____ to the helper T cells. This presentation activates the helper T cells.

a. antibodies
b. antigens
c. cytokines
d. interferons

_____ 3. Helper T cells activate _____.

a. B cells
b. cytotoxic T cells
c. memory T cells
d. plasma cells

_____ 4. Which cells kill pathogens by producing poisonous compounds?

a. B cells
b. cytotoxic T cells
c. suppressor T cells
d. None of the above produce poisonous compounds.

OBJECTIVE 7: Describe the primary and secondary immune responses to antigen exposure (textbook p. 391).

After studying p. 391 in the text, you should be able to explain the processes involved when you are exposed to an antigen the first time and the processes involved when you are exposed to the same antigen a second time.

_____ 1. _____ will differentiate into plasma cells if they are exposed to the same antigen a second time.

a. Helper T cells
b. Memory B cells
c. Memory T cells
d. NK cells

_____ 2. Which of the following conditions produces the highest number of circulating antibodies?

a. during the primary response
b. the first exposure to an antigen
c. the second exposure to an antigen
d. Both a and b are correct.

_____ 3. Which of the following statements about the immune response mechanisms is true?

a. The primary response produces more antibodies than the secondary response.
b. The secondary response is slower than the primary response.
c. The secondary response produces antibodies immediately.
d. All the above are true.

OBJECTIVE 8: Describe the changes in the immune system that occur with aging (textbook pp. 397–399).

After studying pp. 397–399 in the text, you should be able to describe the changes that take place in the immune system with age and their consequences for the elderly.

_____ 1. With age, _____ cells become less active due to a reduced number of _____.

a. B, antibodies
b. B, cytotoxic T cells
c. B, helper T cells
d. T, B cells

_____ 2. Which of the following cells have reduced activity with age?

a. B cells
b. helper T cells
c. cytoxic T cells
d. all the above

3. Why is it quite common for an elderly person to become afflicted with cancer?

____

OBJECTIVE 9: Describe examples of disorders of the lymphatic vessels, lymphoid tissue, lymphoid organs, and lymphocytes (textbook pp. 393–397).

The major role of the entire lymphatic system is that of helping to maintain homeostasis. After studying pp. 393–397 in the textbook, you should be able to describe some of the challenges that the lymphatic system faces.

_____ 1. Sometimes an infection of a fingernail may result in a red streak going up the arm. This red streak is due to _____.

a. blood poisoning
b. erysipelas
c. pathogens traveling through superficial blood vessels
d. pathogens traveling through the lymph vessels

_____ 2. Blockage of the lymph vessels resulting in tremendous swelling of the appendages for example, is _____.

a. cellulitis
b. filariasis
c. Lyme disease
d. lymphangitis

_____ 3. Lyme disease is caused by _____.

a. bacteria
b. mice
c. a tick
d. a virus

_____ 4. One function of the spleen is to phagocytize old red blood cells. _____ is a condition in which the spleen enlarges and may result in _____. This in turn causes _____.

a. Hypersplenism, splenomegaly, anemia
b. Hyposplenism, splenomegaly, anemia
c. Splenomegaly, hypersplenism, anemia
d. Splenomegaly, hyposplenism, anemia

_____ 5. Infectious mononucleosis is caused by a _____ and may cause _____.

a. bacterium, lymphadenopathy
b. bacterium, splenomegaly
c. parasite, enlarged spleen
d. virus, splenomegaly

OBJECTIVE 10: Distinguish among the three types of abnormal immune responses (textbook pp. 397–399).

Most disorders of the lymphatic system can be placed into one of three categories. After studying pp. 397–399 in the text, you should be able to describe the three categories and their characteristics.

_____ 1. The virus that results in AIDS affects _____. This disorder is known as an _____.

a. B cells, autoimmune response
b. helper T cells, autoimmune response
c. helper T cells, immunodeficiency disease
d. suppressor T cells, immunodeficiency disease

_____ 2. Excessive radiation may lead to_____.

a. autoimmune response
b. excessive immune response
c. hypersensitivity
d. immunodeficiency

_____ 3. Allergens trigger inflammatory response by activating _____.

a. antibodies
b. B cells
c. basophils
d. prostaglandins

_____ 4. Cells of the body contain glycolipids (antigens). In _____ B cells treat those normal antigens as foreign material.

a. an acquired immune response
b. an autoimmune disorder
c. an excessive immune response
d. an induced disorder

PART II: CHAPTER-COMPREHENSIVE EXERCISES

Match the terms in column A with the descriptions or statements in column B.

MATCHING I

	(A)	(B)
_____	1. acquired immunity	A. Molecules located on the surface of pathogens that ultimately cause the production of antibodies
_____	2. antibodies	B. The protein molecules produced by B cells
_____	3. antigen	C. Chemical messengers released by cells that defend the body
_____	4. cytokines	D. A type of cytokine that interferes with viral replication
_____	5. elephantiasis	E. Immunity present at birth
_____	6. innate immunity	F. Immunity that develops due to exposure to an antigen from the environment
_____	7. interferon	G. A ruptured appendix could result in _____.
_____	8. lymphoma	H. A condition resulting from blockage of the lymph vessels by a roundworm
_____	9. peritonitis	I. Malignant tumor of the lymph nodes

MATCHING II

	(A)	(B)
_____	1. auquired	A. Lymphocytes involved in humoral immunity
_____	2. autoimmune	B. Lymphocytes involved in cellular immunity
_____	3. B cells	C. Macrophages that process pathogenic antigens for T cell activation
_____	4. diapedesis	D. Cells that produce a poison that disrupts the metabolism of pathogens
_____	5. Kupffer cells	E. Macrophage cells of the liver
_____	6. microglial cells	F. Macrophage cells of the nervous system
_____	7. monocytes	G. The movement of macrophages in and out of circulation
_____	8. NK cells	H. AIDS is an _____ disorder of the immune system.
_____	9. T cells	I. Lupus is an _____ disorder of the immune system.

CONCEPT MAP

This concept map summarizes and organizes the material in Chapter 16. Use the following terms to complete the map by filling in the boxes identified by the circled numbers, 1–11.

antibodies	**helper T**	**interferon**
skin	**nonspecific**	**NK cell activity**
plasma B	**inflammation**	**innate**
AIDS	**autoantibodies**	

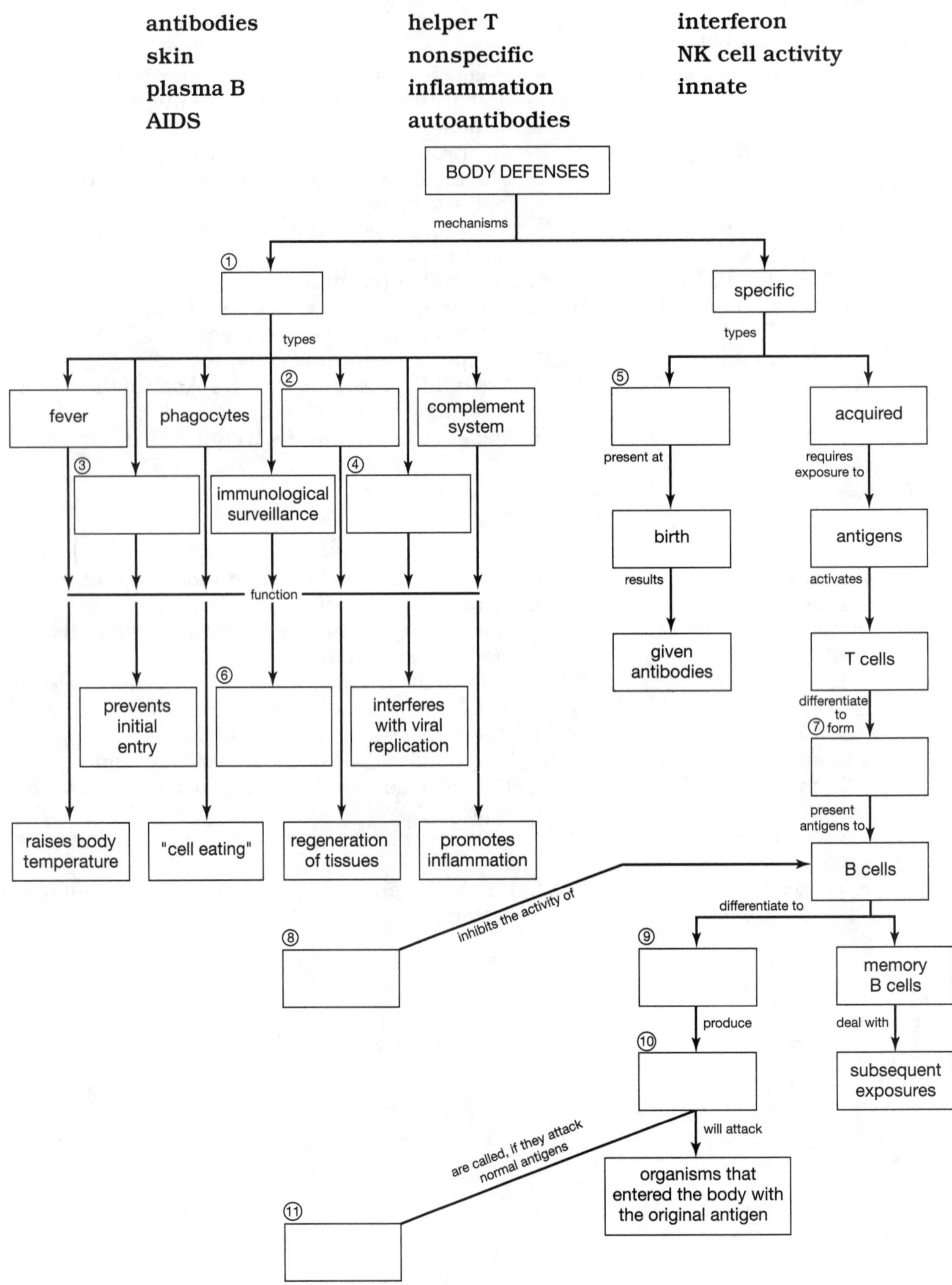

CROSSWORD PUZZLE

The following crossword puzzle reviews the material in Chapter 16. To complete the puzzle, you have to know the answers to the clues given, and must be able to spell the terms correctly.

ACROSS

1. These cells produce antibodies.
3. SCIDS is the acronym for _____ combined immunodeficiency syndrome.
8. This virus attacks helper T cells
9. AIDS is the acronym for _____ immune deficiency syndrome.
10. The type of white blood cell that can differentiate into T or B cells
12. Inborn immunity is called _____ immunity.
13. Filarasis is caused by the blockage of lymph vessels. This blockage is initiated by a _____.
14. The _____ gland converts lymphocytes to T cells.

DOWN

2. Vaccination is a type of _____ acquired immunity.
4. Injecting an antigen from a weakened or dead pathogen into the body is a process called _____.
5. When activated by an allergen, basophils will release _____.
6. Lyme disease is caused by a _____.
7. The normal number of leukocytes is 5,000 to 10,000 per cubic _____.
9. Antibodies target and "attack" specific _____.
11. A compound that promotes a fever

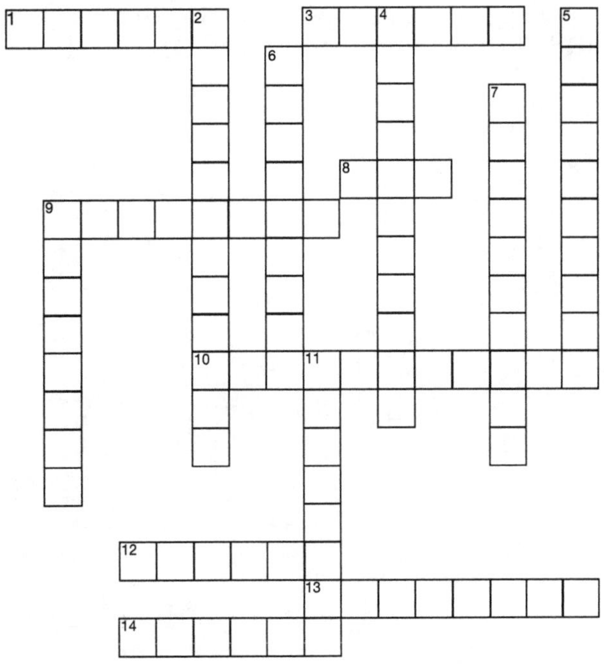

FILL-IN-THE BLANK NARRATIVE

Use the following terms of this summary of Chapter 16 to fill in the blanks.

helper T cells	**B cells**	**immunity**
innate	**plasma cells**	**macrophages**
antigen	**pathogens**	**secondary response**
antibodies	**specific**	**AIDS**
HIV	**secondary**	

There are two major types of immunity against (1) ___________. We may be born with some immunity, known as (2) ___________ immunity, or we may acquire the immunity by being exposed to the (3) ___________. For example, when individuals are exposed to the chickenpox virus, he or she is actually exposed to the antigen of the virus. (4) ___________ will phagocytize the virus and present the antigen to (5) ___________, which bind to (6) ___________, thereby activating them. The B cells differentiate into (7) ___________ and memory B cells. The plasma cells begin manufacturing (8) ___________ that are (9) ___________ for the chickenpox virus. The memory B cells will remember the viral antigen of chickenpox. It usually takes about 2 weeks for this process to occur, so, for about 2 weeks, these persons are sick with the virus. They soon get better and they will have now developed a(n) (10) ___________ against the chickenpox viral antigen.

A few years later, some of these persons become exposed to chickenpox again. As soon as the virus enters their body, the memory B cells are activated and immediately begin to produce antibodies specific for the chickenpox virus. This process occurs rapidly, so they won't feel any ill symptoms. This reaction is known as the (11) ___________.

As mentioned above, helper T cells are actively involved in maintaining homeostasis. The helper T cells of a patient with (12) ___________ are infected by a virus called (13) ___________. Without the activity of the helper T cells, the antigens of chickenpox or any other pathogen cannot be presented to the B cells. Generally, a patient infected with HIV does not die of AIDS directly but rather from a (14) ___________ disease.

CLINICAL CONCEPT

The following clinical concept applies the information in Chapter 16. Following the application is a set of questions to help you understand the concept.

Sam, a local businessman, doesn't want to get the flu, so he chooses to get the flu vaccine. Two months later, Sam gets the flu. Sam is upset because he was vaccinated so he would be immune to the flu virus. The doctor explains to Sam that he is indeed immune, but to only one specific type of flu. Each "flu bug" consists of different antigens. The flu shot he received caused him to develop antibodies against only those specific antigens. However, there are many viruses that are responsible for causing the flu, each with specific antigens. Therefore, Sam became sick due to his exposure to viruses with antigens different from those against which he was vaccinated.

1. What identifies each individual flu virus?

2. Sam got a vaccination for a specific type of flu. How does a vaccination cause a person to become immune to a specific flu virus?

3. Why did Sam get the flu even though he was vaccinated against the flu?

4. Sam became immune to the flu against which he was vaccinated. This type of immunity is called (natural acquired immunity or artificially acquired immunity)?

5. Sam became immune to the "flu bug" he contracted 2 months later. This type of immunity is called (natural acquired immunity or artificially acquired immunity)?

THIS CONCLUDES CHAPTER 16 EXERCISES

CHAPTER 17

The Respiratory System

Chapter Objectives:

1 Describe the primary functions of the respiratory system.
2 Explain how the delicate respiratory exchange surfaces are protected from pathogens, debris, and other hazards.
3 Relate respiratory functions to the structural specializations of the tissues and organs in the system.
4 Describe the process of breathing.
5 Describe the actions of respiratory muscles on respiratory movements.
6 Describe how oxygen and carbon dioxide are transported in the blood.
7 Describe the major factors that influence the rate of respiration.
8 Describe the changes that occur in the respiratory system with aging.
9 Describe the characteristic symptoms and signs of respiratory system disorders.
10 Give examples of different disorders of the respiratory system and their causes.

The human body can survive without food for several weeks and without water for several days, but death becomes imminent if the oxygen supply is cut off for more than a few minutes. All cells and tissues in the body require a continuous supply of oxygen and must get rid of carbon dioxide, a gaseous waste product. The respiratory system consists of the passageways that deliver oxygen from air to the blood and removes carbon dioxide. It also plays an important role in regulating the pH of the body fluids. Many realize the importance of inhaling properly but few realize that exhaling properly is just as important. If exhalation does not occur properly, the blood pH changes, the blood pressure may begin to rise, and the heart rate may increase, just to cite a few possible problems.

The respiratory system is divided into an **upper** and **lower** respiratory tract. The upper part includes the nose, nasal cavity, paranasal sinuses, and pharynx. The lower part includes the larynx, trachea, bronchi, bronchioles, and alveoli of the lungs. The terminal bronchioles and alveoli of the lower respiratory system are involved with the gaseous exchange of oxygen and carbon dioxide, whereas all the other structures of the respiratory tracts serve as passageways for incoming and outgoing air and protection of the lungs, specifically the alveoli.

Proper exchange of oxygen and carbon dioxide can be hindered if any part of the respiratory system is affected by pathogens. Within this chapter, numerous respiratory disorders and their effects are discussed.

The exercises in this chapter reinforce concepts involving the mechanics of breathing and the transport and exchange of oxygen and carbon dioxide among the air, blood, and tissues, as well as respiratory disorders.

PART I: OBJECTIVE-BASED QUESTIONS

OBJECTIVE 1: Describe the primary functions of the respiratory system (textbook p. 405).

After studying p. 405 in the text, you should be able to explain the primary functions of the respiratory system: (1) pulmonary ventilation, (2) gas exchange, and (3) gas pickup and transport. The following questions address these three functions.

____ 1. The breathing process is called ____.

a. cellular respiration
b. pulmonary ventilation
c. respiration
d. systemic ventilation

____ 2. The two gases typically involved in respiration are ____.

a. carbon dioxide and nitrogen
b. oxygen and carbon dioxide
c. oxygen and nitrogen
d. all the above

____ 3. The movement of the gases into the blood from the lungs is a process of ____.

a. cellular respiration
b. diffusion
c. metabolism
d. osmosis

4. External respiration involves the diffusion of gases between the ______________________ and the blood.

5. Internal respiration involves the diffusion of gases between the ______________________ and the blood.

6. In Figure 17–1, identify the pulmonary circuit, systemic circuit, bloodstream, an alveolus, oxygen going into the bloodstream, oxygen going into a muscle cell, carbon dioxide going into an alveolus, and carbon dioxide leaving a muscle cell.

Figure 17–1 External and Internal Respiration

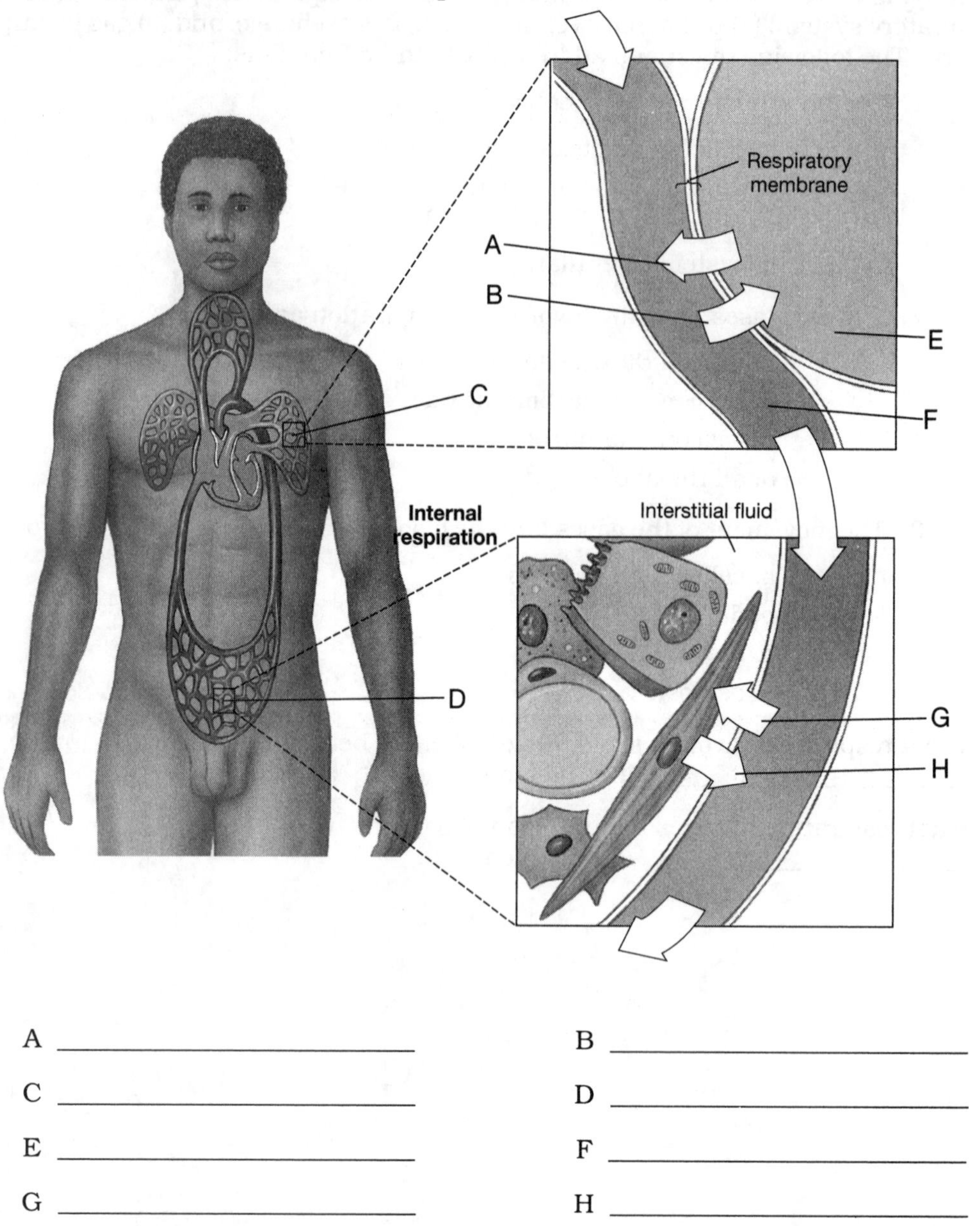

A ____________________ B ____________________

C ____________________ D ____________________

E ____________________ F ____________________

G ____________________ H ____________________

OBJECTIVE 2: Explain how the delicate respiratory exchange surfaces are protected from pathogens, debris, and other hazards (textbook pp. 405–411).

After studying pp. 405–411 in the text, you should be able to describe how the body protects the delicate respiratory structures called alveoli in the lungs. There are about 300 million alveoli, and they are only one squamous cell layer thick. Because they consist of only a single layer of cells, they are very vulnerable to damage and destruction.

_____ 1. What is the name of the structures responsible for warming the air before it gets to the lungs?

a. goblet cells
b. nasal conchae
c. nasopharynx
d. paranasal sinuses

____ 2. What is the name of the structures responsible for sweeping foreign material away from the lungs in order to protect them?

a. cilia
b. lobules
c. mucus
d. villi

____ 3. What is the name of the structure that folds over the opening of the trachea to prevent choking while swallowing food?

a. epiglottis
b. glottis
c. larynx
d. vocal folds

____ 4. What protects the alveolar sacs from debris that has made it that far into the respiratory system?

a. macrophages
b. mucus
c. surfactant
d. Nothing protects the alveolar sacs. Debris will be stopped before it gets that far.

5. Name at least two things the nose does for the respiratory system.
(1) ____________________ (2) ____________________

6. In Figure 17–2, identify the nasal conchae, pharynx, tracheal cartilage, and larynx.

Figure 17–2 Components of the Upper Respiratory System

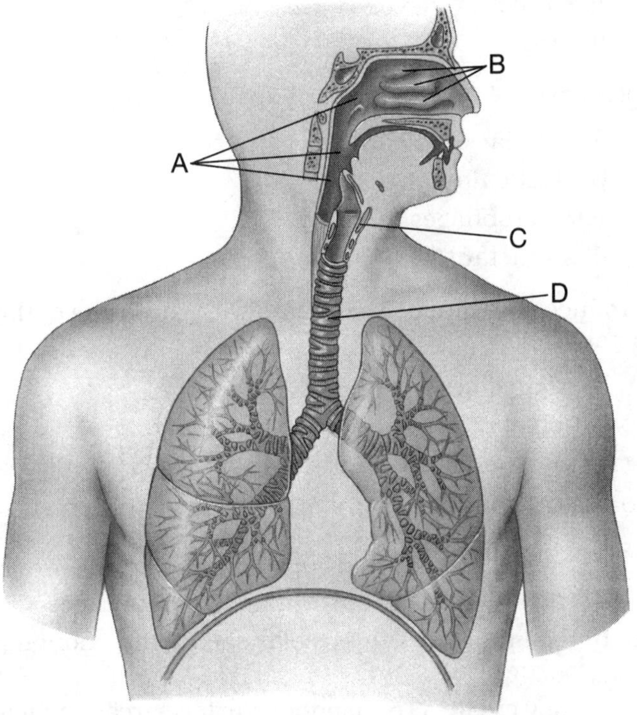

A ____________________ B ____________________

C ____________________ D ____________________

7. In Figure 17–3, identify the mucus layer, goblet cell, and ciliated columnar cells.

Figure 17–3 Ciliated Respiratory Epithelium that Protects the Respiratory Structures

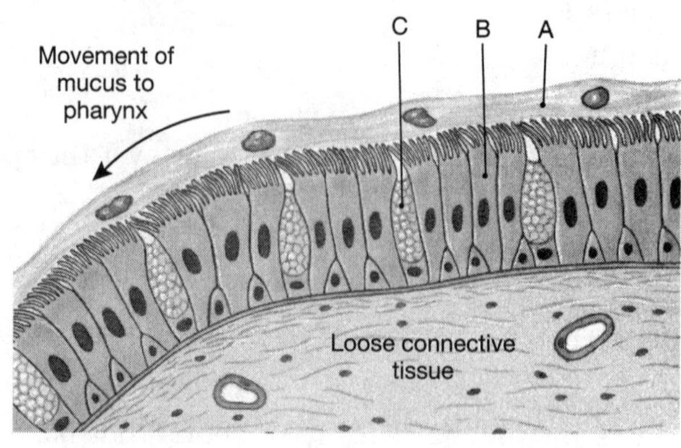

A ______________________ B ______________________

C ______________________

OBJECTIVE 3: Relate respiratory functions to the structural specializations of the tissues and organs in the system (textbook pp. 405–411).

After studying pp. 405–411 in the text, you should be able to explain the specific role of each tissue and/or organ of the respiratory system.

_____ 1. Gas exchange in the respiratory system occurs in the _____.

a. alveoli
b. bronchi
c. lobules
d. lungs

_____ 2. What prevents the alveoli from collapsing?

a. alveolar epithelium
b. elastic fibers
c. macrophages
d. surfactant

_____ 3. What produces the mucus associated with the lining of the trachea that traps foreign particles?

a. ciliated cells
b. goblet cells
c. macrophages
d. surfactant cells

_____ 4. Alveoli are very efficient at gas exchange because _____.

a. each alveolus is surrounded by capillaries
b. the lining of each alveolar sac is thin enough for gas exchange to occur
c. they create a tremendous surface area at which gas exchange can occur
d. all the above

____ 5. Which of the following has (have) ciliated columnar cells that are involved in protecting the alveoli?

a. trachea
b. bronchi
c. bronchioles
d. all the above

6. Once oxygen reaches the alveoli, it diffuses into the bloodstream and carbon dioxide diffuses out of the bloodstream into the alveoli. In Figure 17–4, identify the highest concentration of oxygen, the highest concentration of carbon dioxide, the movement of oxygen into the blood, and the movement of carbon dioxide into the alveolus.

Figure 17–4 Gas Exchange Between an Alveolus and the Bloodstream

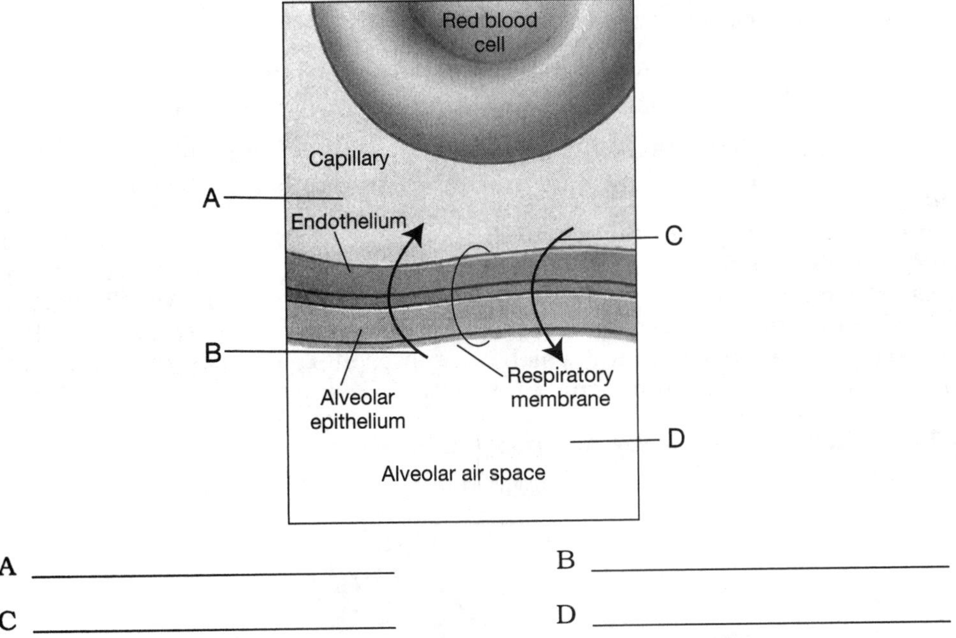

A ____________________ B ____________________

C ____________________ D ____________________

OBJECTIVE 4: Describe the process of breathing (textbook pp. 411–413).

After studying pp. 411–413 in the text, you should be able to describe the breathing process in terms of volume/pressure relationships between the atmosphere and the lungs. Inhaling and exhaling is not just a simple matter of sucking air in and forcing air out.

____ 1. Which of the following statements best describes vital capacity?

a. the maximum amount of air we can exhale, which is about 3000 to 4500 mL of air
b. the maximum amount of air we can exhale, which is between 1900 and 3300 mL of air
c. the total amount of air we can exhale, which is about 4000 to 5000 mL of air
d. the total amount of air we can exhale, which is about 1900 to 3300 mL of air

____ 2. Which of the following terms refers to normal passive breathing?

a. expiratory reserve
b. residual volume
c. tidal volume
d. vital capacity

_____ 3. In order to inhale, we _____ the size of our thoracic cavity, which _____ internal pressure.

a. decrease, decreases
b. decrease, increases
c. increase, decreases
d. increase, increases

_____ 4. In order to exhale, we _____ the size of our thoracic cavity, which _____ internal pressure.

a. decrease, decreases
b. decrease, increases
c. increase, decreases
d. increase, increases

_____ 5. To increase the size of our thoracic cavity, _____.

a. our diaphragm muscle does down and our ribs go up
b. our diaphragm muscle goes down and our ribs also go down
c. our diaphragm muscle goes up and our ribs also go up
d. our diaphragm muscle goes up and our ribs go down

6. Assume that the boxes in Figure 17–5 represent the various sizes of the thoracic cavity. The dots inside the boxes represent air molecules. The arrows inside the boxes represent the movement of air molecules. As air molecules move, they create pressure. Identify which box represents an increased thoracic cavity size and therefore a decreased internal pressure, and which box represents a decreased thoracic cavity size and therefore an increased internal pressure.

Figure 17–5 Thoracic Cavity Size and Internal Pressure

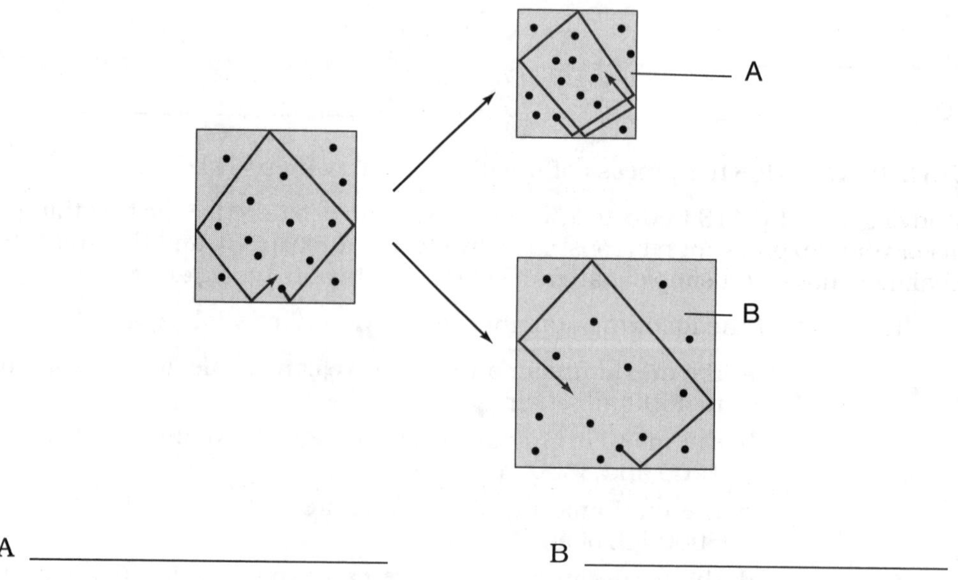

A ______________________ B ______________________

7. In Figure 17–5, identify which letter represents inhalation and which letter represents exhalation.

A ______________________ B ______________________

8. In summary, the containers in Figure 17–5 represent our thoracic cavity. To inhale, we must increase the size of our thoracic cavity, which (A) __________ the internal pressure, and air enters our lungs. To exhale, we must decrease the size of our thoracic cavity, which (B) __________ the internal pressure, thus forcing air out of our lungs.

OBJECTIVE 5: Describe the actions of respiratory muscles on respiratory movements (textbook pp. 411–413).

After studying pp. 411–413 in the text, you should be able to explain how respiratory muscles work to change the size of the thoracic cavity to accomplish the task of breathing.

____ 1. The diaphragm muscle used for breathing is located ____.

a. inferior to the lungs but superior to the stomach
b. superior to the lungs but inferior to the heart
c. in the thoracic cavity
d. in the abdominal cavity

____ 2. When we inhale, ____.

a. our diaphragm muscle moves downward to increase our thoracic cavity size
b. our diaphragm muscle moves upward to increase our thoracic cavity size
c. our diaphragm muscle moves downward to lower our thoracic cavity size
d. our diaphragm muscle moves upward to lower our thoracic cavity size

____ 3. Which of the following statements describes exhalation?

a. The diaphragm muscle moves upward and ribs move downward.
b. The diaphragm muscle moves downward and ribs move upward.
c. The diaphragm muscle moves downward and ribs move downward.
d. The diaphragm muscle moves upward and ribs move upward.

____ 4. Which of the following is the amount of air moved into or out of the lungs in a single passive respiratory cycle?

a. expiratory reserve volume
b. residual volume
c. tidal volume
d. vital capacity

____ 5. The amount of air we can exhale with one forceful breath is called ____.

a. expiratory reserve volume
b. residual volume
c. tidal volume
d. vital capacity

OBJECTIVE 6: Describe how oxygen and carbon dioxide are transported in the blood (textbook pp. 414–416).

After studying pp. 414–416 in the text, you should be able to explain how oxygen and carbon dioxide are transported throughout the body via the bloodstream, and discuss the role of hemoglobin in the process. Oxygen and carbon dioxide are transported by binding to hemoglobin molecules. Oxygen transport in the bloodstream is relatively simple, but the transport of carbon dioxide presents possible problems such as the production of hydrogen ions.

_____ 1. What percentage of the carbon dioxide we generate binds to hemoglobin in the red blood cells?

a. 7%
b. 23%
c. 70%
d. 93%

_____ 2. What percentage of the carbon dioxide we generate is converted to bicarbonate ions in the red blood cells?

a. 7%
b. 23%
c. 70%
d. 93%

_____ 3. What percentage of the carbon dioxide we generate is transported to the alveoli and then exhaled?

a. 7%
b. 23%
c. 70%
d. 93%

_____ 4. An increase in CO_2 will _____ the concentration of H^+ in the red blood cells, which will _____ the pH of the blood.

a. increase, decrease
b. increase, increase
c. decrease, decrease
d. decrease, increase

5. Finish this sentence to make a complete thought: Twenty-three percent of the carbon dioxide we generate will bind to hemoglobin, and then it will be

_________________________.

6. How does an increased level of carbon dioxide affect blood pH?

____

7. In Figure 17–6, identify the letters that represent 7% of the carbon dioxide staying in the plasma, 23% of the carbon dioxide being transported by hemoglobin, and 70% of the carbon dioxide converting to the bicarbonate ion, which ionizes to form H^+.

Figure 17–6 Carbon Dioxide Transport in the Blood

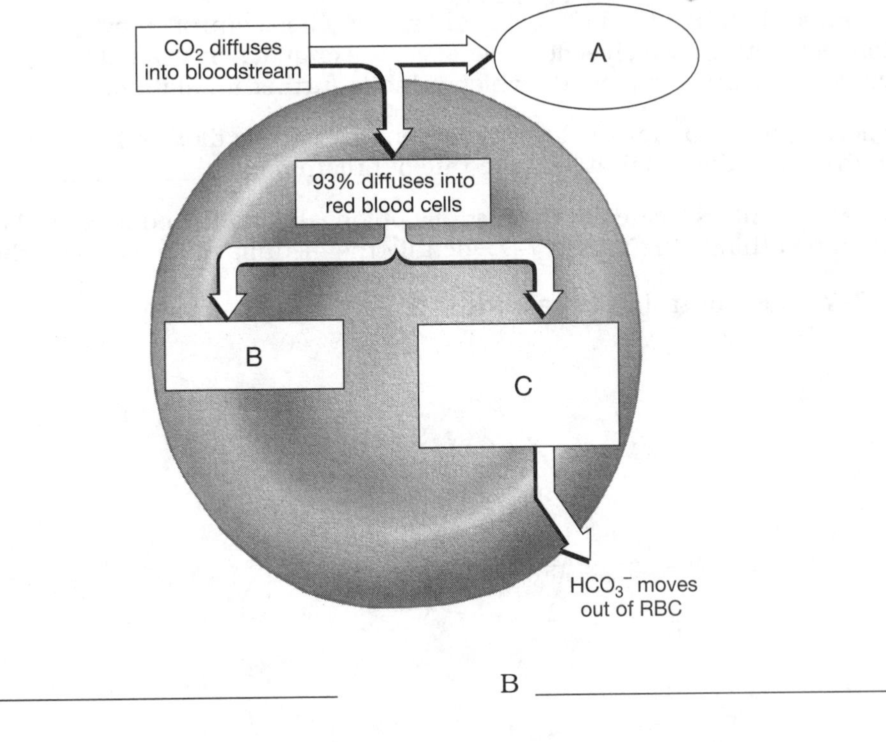

A ______________________ B ______________________

C ______________________

OBJECTIVE 7: Describe the major factors that influence the rate of respiration (textbook pp. 416–418).

After studying pp. 416–418 in the text, you should be able to explain how breathing is controlled by respiratory centers located in the medulla oblongata and the pons, which constantly receive messages from mechanical receptors (pressure receptors and stretch receptors) and chemical sensory receptors in the body.

____ 1. The respiratory centers that regulate breathing are located in the ____.

a. brainstem and cerebrum
b. cerebrum and cerebellum
c. lungs
d. pons and medulla oblongata

____ 2. When you hold your breath, you can hold it only for a certain amount of time. Your body "makes" you breathe. This is because your body ____.

a. needs oxygen
b. needs to exhale carbon dioxide
c. recognizes that your lungs have stretched to maximum size
d. recognizes that blood pH is rising

____ 3. Which receptors are located in the carotid arteries?

a. chemoreceptors
b. pressure receptors
c. stretch receptors
d. both a and b

all the above

4. The volume of air breathed in during an inhalation is controlled by __________ receptors of the lungs, which send a signal via the cranial nerve called the __________ nerve to the medulla oblongata, which inhibits further inhalation.

5. Chemoreceptors monitor the level of oxygen and carbon dioxide in the blood. However, of these two the gas that sets the rate of breathing is __________.

6. Respiratory centers located in the medulla oblongata and the pons control the rate and depth of breathing. In Figure 17–7, identify the medulla oblongata and the pons.

Figure 17–7 The Control of Breathing

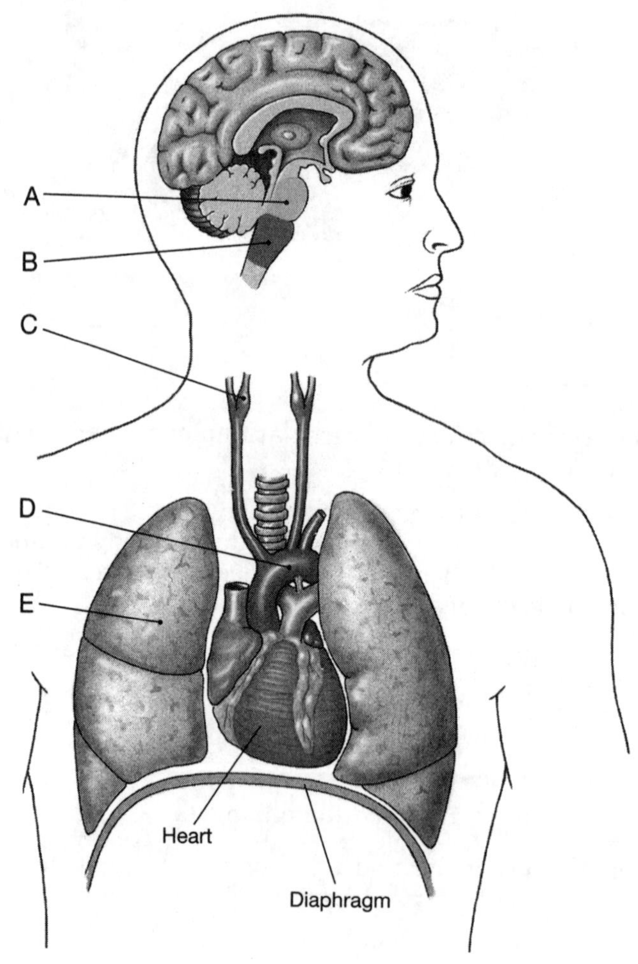

A ________________ B ________________

7. Receptors located in the lungs, aorta, carotid arteries, and the medulla oblongata are responsible for detecting the amount of inflation, pH, and blood pressure. Which letter in Figure 17–7 represents the location(s) of stretch receptors? _____

8. Which letter or letters in Figure 17–7, represent the location(s) of chemoreceptors? _____

9. Which letter or letters in Figure 17–7, represent the location(s) of blood pressure receptors? _____

OBJECTIVE 8: Describe the changes that occur in the respiratory system with aging (textbook p. 418).

After studying p. 418 in the text, you should be able to explain the many factors that may reduce the efficiency of the respiratory system in elderly individuals, including blood pressure, narrowing of the arteries, decreased elasticity of the lungs, and some emphysema like characteristics.

_____ 1. The destruction of alveoli due to age or by respiratory irritants is called _____.

a. asthma
b. emphysema
c. pneumonia
d. pulmonary embolism

_____ 2. The elasticity of the lungs diminishes with age. This loss of elasticity decreases the _____.

a. residual volume
b. tidal volume
c. vital capacity
d. all the above

_____ 3. The most probable cause of restricted movements of the chest cavity in the elderly is _____.

a. excessive smoking
b. increased incidence of asthma
c. arthritic changes in the rib joints
d. the presence of pulmonary emboli

OBJECTIVE 9: Describe characteristic symptoms and signs of respiratory system disorders (textbook pp. 418–420).

The analysis and interpretation of the various signs and symptoms associated with breathing difficulties helps a physician pinpoint the type of respiratory problem the patient is experiencing so that corrective measures can be taken. After studying pp. 418–420 in the text, you should be able to describe the characteristic signs and symptoms of the common respiratory system disorders.

_____ 1. A breathing rate of 22 breaths per minute is classified as _____.

a. bradypnea
b. dyspnea
c. stridor
d. tachypnea

_____ 2. Which of the following is a symptom of possible respiratory problems?

a. chest pains
b. cough
c. tachypnea
d. wheezing

_____ 3. Which of the following is a lower respiratory disorder?

a. obstructive disorder
b. restrictive disorder
c. both a and b
d. neither a nor b

_____ 4. Pulmonary fibrosis is a _____ of a (an) respiratory disorder.

a. sign, lower
b. sign, upper
c. symptom, lower
d. symptom upper

_____ 5. Which of the following are signs of respiratory disorders?

a. chest pain and wheezing
b. coughing and dyspnea
c. cyanosis and clubbing of the fingers
d. dyspnea and chest pains

OBJECTIVE 10: Give examples of different disorders of the respiratory system and their causes (textbook pp. 420–429).

After studying pp. 420–429 in the text, you should be able to describe common respiratory disorders and their causes and the effects they have on homeostasis of the body.

_____ 1. Strep throat is also called _____ and is caused by the _____ bacterium.

a. bronchitis, staphylococcus
b. laryngitis, streptococcus
c. pharyngitis, mycoplasma
d. pharyngitis, streptococcus

_____ 2. Pneumonia is characterized by _____.

a. collapse of the alveoli
b. fluid entering the alveoli
c. mucus buildup in the bronchi
d. the development of scar tissue on the lungs

_____ 3. Which of the following diseases is characterized by the development of scar tissue on the lungs during the immune response?

a. diphtheria
b. pneumonia
c. pneumothorax
d. tuberculosis

_____ 4. A collapsed lung is a condition in which _____.

a. air enters the pleural cavity from the atmosphere
b. air enters the pleural cavity from within the lungs
c. air leaves the lungs to the atmosphere
d. a blood clot forms and prevents circulation of blood through the lung tissue

_____ 5. Which of the following best describes asthma?

a. An irritation causes the columnar cells of the respiratory tubes to lose their cilia so that they are no longer able to move mucus.
b. An irritation of the lungs causes the lungs to lose their elasticity.
c. An irritation of the respiratory tubes causes the tubes to accumulate mucus, thereby causing constriction.
d. An irritation of the respiratory tubes causes the tubes to dilate.

_____ 6. Which of the following identifies the condition being described: A defective gene causes a buildup of mucus that plugs the respiratory tubes because the columnar cells cannot move the mucus?

a. asthma
b. cystic fibrosis
c. respiratory distress syndrome
d. chronic obstructive pulmonary disease

_____ 7. Respiratory distress syndrome is a condition in which there is less _____ production than normal, thus resulting in the collapse of_____.

a. alveoli, the lungs
b. mucus, respiratory tubes
c. surfactant, alveoli
d. surfactant, the lungs

_____ 8. The disorder in which alveoli are destroyed by irritants such as cigarette smoke is _____.

a. bronchitis
b. cystic fibrosis
c. emphysema
d. PEEP

PART II: CHAPTER-COMPREHENSIVE EXERCISES

Match the terms in column A with the description in column B.

MATCHING I

	(A)	(B)
_____	1. alveoli	A. The place where gas exchange occurs
_____	2. bronchi	B. Structures that warm inhaled air
_____	3. bronchiole	C. The opening to the trachea
_____	4. diaphragm	D. The cartilage tissue that closes off the trachea to prevent food from entering
_____	5. diptheria	E. The windpipe
_____	6. epiglottis	F. The two tracheal branches that enter each lung
_____	7. glottis	G. The tracheal tube that terminates with a lobule
_____	8. intercostal	H. Rib muscles
_____	9. nasal conchae	I. A breathing muscle that separates the thoracic cavity from the abdominal cavity
_____	10. trachea	J. Bacteria form a membrane-like material over the surface of the pharynx, thus causing suffocation in the disease called _____.
_____	11. tuberculosis	K. Macrophages try to barricade the causative agent, thus creating scar tissue, in the disease called _____.

MATCHING II

	(A)	(B)
_____	1. cellular respiration	A. The use of oxygen in the cells
_____	2. goblet	B. Special cells that produce mucus
_____	3. hypoxia	C. Material that prevents alveolar collapse
_____	4. pertussis	D. A lack of oxygen
_____	5. stridor	E. The amount of air that can be exhaled passively
_____	6. surfactant	F. The maximum amount of air that can be exhaled
_____	7. 21%	G. The amount of air left in the lungs after maximum exhalation
_____	8. 23%	H. The amount of atmospheric oxygen
_____	9. approximately 500 mL	I. The amount of CO_2 exhaled
_____	10. approximately 1200 mL	J. The DTP shot includes a vaccine for dyptheria, tetanus, and _____.
_____	11. approximately 3800 mL	K. An auscultation typical of patients with croup

CONCEPT MAP

This concept map summarizes and organizes the material in Chapter 17. Use the following terms to complete the map by filling in the boxes identified by the circled numbers, 1–9.

decreased internal air pressure	**bronchi**	**diaphragm**
increased internal air pressure	**bronchioles**	**glottis**
cystic fibrosis	**emphysema**	**pneumonia**

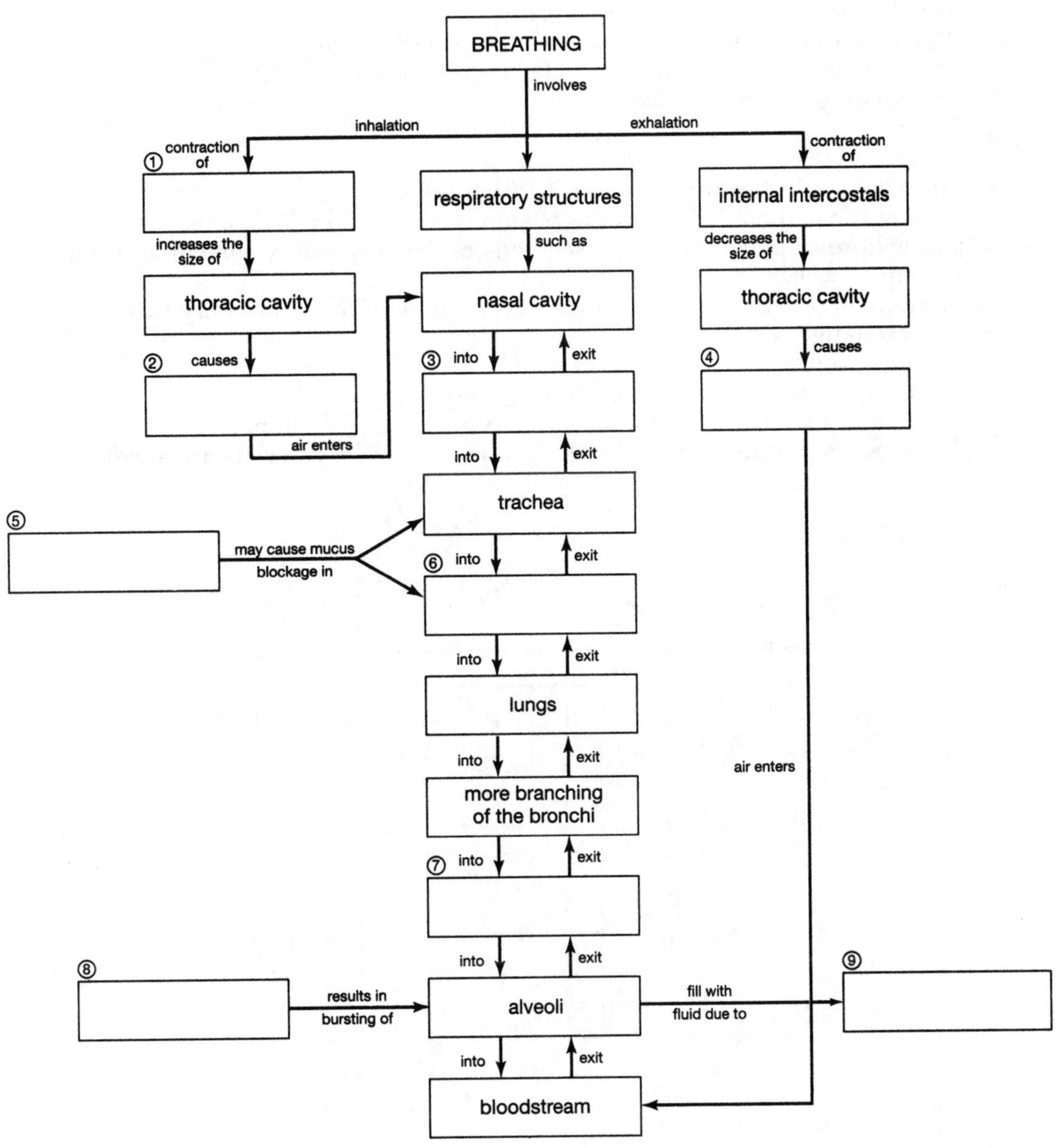

CROSSWORD PUZZLE

The following crossword puzzle reviews the material in Chapter 17. To complete the puzzle, you have to know the answers to the clues given, and you must be able to spell the terms correctly.

ACROSS

2. The tracheal tubes that enter each lung
5. The amount of air we can forcibly exhale
8. This volume of air in the lungs is equivalent to about 1200 mL.
9. About 21% of atmospheric air consists of _____.
13. When the diaphragm muscle is moving upward, we are _____.
14. The condition in which the alveoli begin to fill up with fluid
15. The opening into the trachea

DOWN

1. The genus of bacterium that causes tuberculosis
2. Most of the carbon dioxide in the blood is converted to _____ ions.
3. Cystic fibrosis affects the mucus glands of the respiratory system and the _____ system as well.
4. These cells produce mucus to protect the lining of the respiratory tubes.
6. Gas exchange occurs here.
7. Violent cough
10. Chemoreceptors are located in the _____ arteries.
11. The amount of air we exhale passively is called _____ volume.
12. The process that keeps the alveoli open until surfactant levels are adequate

FILL-IN-THE BLANK NARRATIVE

Use the following terms to fill in the blanks of this summary of Chapter 17.

pulmonary arteries	**bronchi**	**bronchioles**
diaphragm	**pulmonary veins**	**lobules**
epiglottis	**alveoli**	**nasal conchae**
cyctic fibrosis	**lumen**	**destruction**
pneumonia	**homeostasis**	**cilia**

During inhalation, the (1) __________ muscle moves downward, thus increasing the size of the thoracic cavity and decreasing the internal air pressure. Air enters the nasal area and becomes turbulent due to the (2) __________. As the air becomes turbulent, it warms up.

Air will enter the trachea as long as the (3) __________ is open. Air then enters the (4) __________, which branch numerous times. Eventually air enters the (5) __________, which terminate with (6) __________ that consist of numerous (7) __________. Each alveolus is surrounded by capillaries. Oxygen diffuses from the alveoli and enters the capillaries. The blood carries oxygen to the heart via the (8) __________. The heart eventually sends this oxygen to all body parts.

At the tissues, the blood gives up oxygen and picks up carbon dioxide and transports it to the heart. The heart sends this blood to the lungs via the (9) __________. Once the carbon dioxide arrives at the alveoli of the lungs, the exhalation process begins. During exhalation, the diaphragm muscle moves upward, decreasing the size of the thoracic cavity, and increasing the internal pressure, which forces the air out. The cycle of breathing is now complete.

Breathing can be hindered by any number of respiratory disorders. (10) __________ is the result of a genetic defect that causes excess mucus accumulation in the respiratory tubes. Columnar cell (11) __________ cannot move mucus out of the respiratory tubes. Therefore, the (12) __________ of the respiratory tubes is constricted.

People who smoke will invariably experience emphysema. The cigarette smoke irritants will cause (13) __________ of alveoli, which decreases gas exchange. If the smoker also develops (14) __________, the remaining "good" alveoli will fill with fluid, which also decreases gas exchange. This patient then experiences heavier breathing to compensate for the loss of oxygen intake and carbon dioxide exhalation. The patient will be out of (15) __________.

CLINICAL CONCEPT

The following clinical concept applies the information presented in Chapter 17. Following the application is a set of questions to help you understand the concept.

Jason is at a family gathering and is sitting at the dinner table with 14 other family members. Jason puts food in his mouth and then, seeing an opportunity to get a word in, begins to talk. Suddenly, he begins to choke. One family member quickly reaches over and gives him a few hearty slaps on the back. You politely have this family member move to the side because you have been trained how to administer the Heimlich maneuver. You apply the maneuver, and the lodged food comes flying out of Jason's mouth. Jason's mother speaks up and reminds Jason that is why she always said, "Don't talk with food in your mouth." Fortunately, Jason suffers no ill effects.

In order for Jason to be able to talk, his epiglottis has to be open so air can pass through his vocal cords and eventually exit his mouth. Because Jason had food in his mouth when he tried to talk, the food could enter one of two tubes: the esophagus, which leads to the stomach, or the trachea, which leads to the lungs. If the food goes down the trachea, it may get lodged in the trachea or in the bronchi. If the blockage is sufficiently large, air cannot enter or exit. The family member who reached over and slapped Jason on the back could actually have forced the food even deeper into the tracheal tubes. The only way to get that food out would be to turn Jason upside down and then slap him on the back (this is the technique used for infants). This was not possible, since Jason weighs 150 pounds. With the Heimlich maneuver, Jason's diaphragm muscle was forced upward, thereby dislodging the trapped food.

1. How does the Heimlich maneuver increase internal thoracic pressure, thereby forcing the lodged food out of the tracheal tubes?

2. Why is it best not to slap somebody on the back when they are choking in a vertical position?

3. Why does talking at the same time you are trying to swallow food create a potential choking situation?

4. What does the epiglottis do when you prepare to swallow?

5. The Heimlich maneuver is a technique that is applied to what respiratory muscle?

THIS CONCLUDES CHAPTER 17 EXERCISES

CHAPTER

18

The Digestive System

By the time you are 65 years old, you will have consumed more than 50,000 meals. Most of this food will have been converted into chemical forms that cells can utilize for their metabolic functions. The nutrients in food are not ready for use by the cells. The digestive system is the "food processor" in the human body. The digestive system chemically breaks food down to smaller particles called nutrients. The nutrients are absorbed into the bloodstream and then transported to the cells and tissues of the body.

The digestive system consists of a muscular tube, **the digestive tract,** which includes the mouth, pharynx, esophagus, stomach, and the small and large intestines. **Accessory digestive organs,** such as the teeth, tongue, salivary glands, liver, gallbladder, and pancreas, assist in the digestive process as food moves through the digestive tract. Many of these steps can be hindered if any part of the gastrointestinal tract is affected by a pathogen.

The exercises in this chapter review the basic structure of the digestive system and the series of steps that transform food for use by cells, as well as numerous digestive disorders and their effects.

hapter Objectives:

- List the functions of the digestive system.
- Identify the organs of the digestive tract and the accessory organs of digestion.
- Explain how materials move along the digestive tract.
- Describe how food is processed in the mouth and how it is swallowed.
- Describe the structure of the stomach and its role in digestion and absorption.
- Describe digestion and absorption in the small intestine.
- Describe the structure and functions of the pancreas, liver, and gallbladder.
- Describe the structure and functions of the large intestine.
- Describe the digestion and absorption of the substances in food.
- Describe the changes in the digestive system that occur with aging.
- Describe the common symptoms and signs of digestive disorders.
- Describe examples of inflammation and infection, and other major disorders of the digestive system.

PART I: OBJECTIVE-BASED QUESTIONS

OBJECTIVE 1: List the functions of the digestive system (textbook p. 435).

After studying p. 435 in the text, you should be able to describe the following functions of the digestive system:

1. ingestion of food
2. mechanical processing of food
3. secretion of enzymes and hormones
4. digestion of food
5. absorption of food
6. excretion of wastes

_____ 1. The manipulation of food by the tongue and teeth is which of the following functions of the digestive system?

a. absorption
b. digestion
c. ingestion
d. mechanical processing

_____ 2. Which of the following functions of the digestive system must occur first before digestion begins?

a. absorption
b. defecation
c. ingestion
d. all the above

_____ 3. The digested food becomes nutrients, which leave the small intestine and enter the bloodstream in the process of _____.

a. absorption
b. digestion
c. excretion
d. secretion

_____ 4. The process by which the pancreas releases digestive enzymes into the small intestine is called _____.

a. absorption
b. ingestion
c. secretion
d. none of the above

5. What is the definition of digestion? ____________________________

OBJECTIVE 2: Identify the organs of the digestive tract and the accessory organs of digestion (textbook pp. 435–446).

After studying pp. 435–446 in the text, you should be able to name the organs of the digestive tract and the accessory organs that assist in the processing of the food as it passes through the gastrointestinal (GI) tract.

____ 1. Which of the following is an accessory structure of the digestive system?

a. mouth
b. pancreas
c. small intestine
d. stomach

____ 2. The stomach is located ____.

a. above the diaphragm muscle
b. in the middle of the abdominal area
c. to the left of the liver
d. to the right of the liver

____ 3. The small intestine consists of three regions. Which of the following sequences is correct regarding the passage of food through it?

a. duodenum, ileum, jejunum
b. duodenum, jejunum, ileum
c. ileum, jejunum, duodenum
d. jejunum, duodenum, ileum

____ 4. Which of the following is an accessory structure and produces a digestive enzyme?

a. gallbladder
b. mouth
c. pancreas
d. stomach

____ 5. Which of the following is not a part of the digestive system?

a. gallbladder
b. large intestine
c. pancreas
d. spleen

6. Identify the digestive organs in Figure 18–1.

Figure 18–1 The Digestive Tract

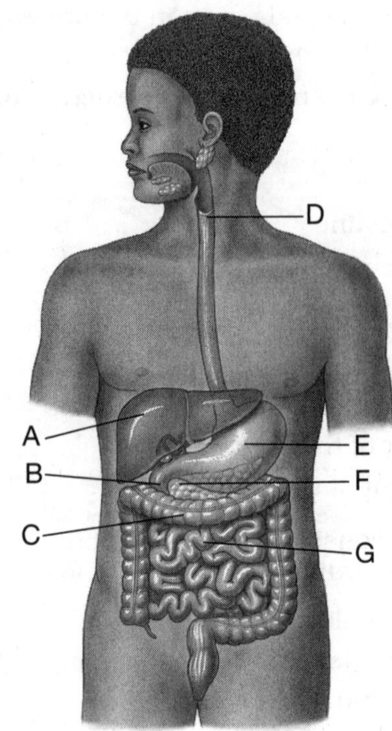

A ______________________ B ______________________

C ______________________ D ______________________

E ______________________ F ______________________

G ______________________

OBJECTIVE 3: Explain how materials move along the digestive tract (textbook pp. 435–446).

After studying pp. 435–446 in the text, you should be able to describe how food is passed along the digestive tract by the peristaltic activity of the smooth muscles.

_____ 1. Which of the following organs exhibits peristalsis?

a. esophagus
b. large intestine
c. small intestine
d. all the above

_____ 2. Peristalsis is the result of the action of _____.

a. involuntary muscles
b. skeletal muscles
c. smooth muscles
d. both a and c

_____ 3. Which layer of the digestive tract is made of the smooth muscles involved in peristalsis?

a. mucosa
b. muscularis externa
c. serosa
d. submucosa

4. The muscular wavelike action that forces food down the esophagus and through the entire digestive tract is called ______________________.

OBJECTIVE 4: Describe how food is processed in the mouth and how it is swallowed (textbook pp. 438–440).

After studying pp. 438–440 in the text, you should be able to explain how the mouth initiates the digestive process and prepares the food for swallowing by mastication (mechanical processing) and lubrication with mucus and salivary secretions (chemical processing).

_____ 1. The food is initially ground and torn into smaller pieces by the teeth. Adults typically have _____ incisors, _____ cuspids, _____ bicuspids, and _____ molars.

a. 4,2,8,6
b. 8,4,8,6
c. 8,4,8,12
d. 8,8,4,12

_____ 2. The _____ gland is the largest of the salivary glands, and it lies in the area of the _____ muscle.

a. parotid, masseter
b. parotid, platysma
c. sublingual, platysma
d. submandibular, platysma

_____ 3. Salivary amylase is an enzyme produced and released by the salivary glands that partially digests _____.

a. carbohydrates
b. fats
c. proteins
d. all the above

_____ 4. The tongue, controlled by the _____ nerve, pushes food toward the pharynx. (*Hint:* Refer to Chapter 12.)

a. abducens
b. facial
c. glossopharyngeal
d. vagus

_____ 5. When the _____ cells of the tongue are activated, they send signals to the brain via the _____ nerve to interpret the flavor of the food. (*Hint:* Refer to Chapter 12.)

a. gustatory, facial
b. papillary, trigeminal
c. tastebud, facial
d. none of the above

_____ 6. The food passing down the esophagus is called _____.

a. a bolus
b. chyme
c. hiatal material
d. nutrients

_____ 7. As we swallow, the _____ closes off the _____.

a. epiglottis, esophagus
b. epiglottis, trachea
c. glottis, esophagus
d. glottis, trachea

8. Identify the teeth in Figure 18–2.

Figure 18–2 The Teeth

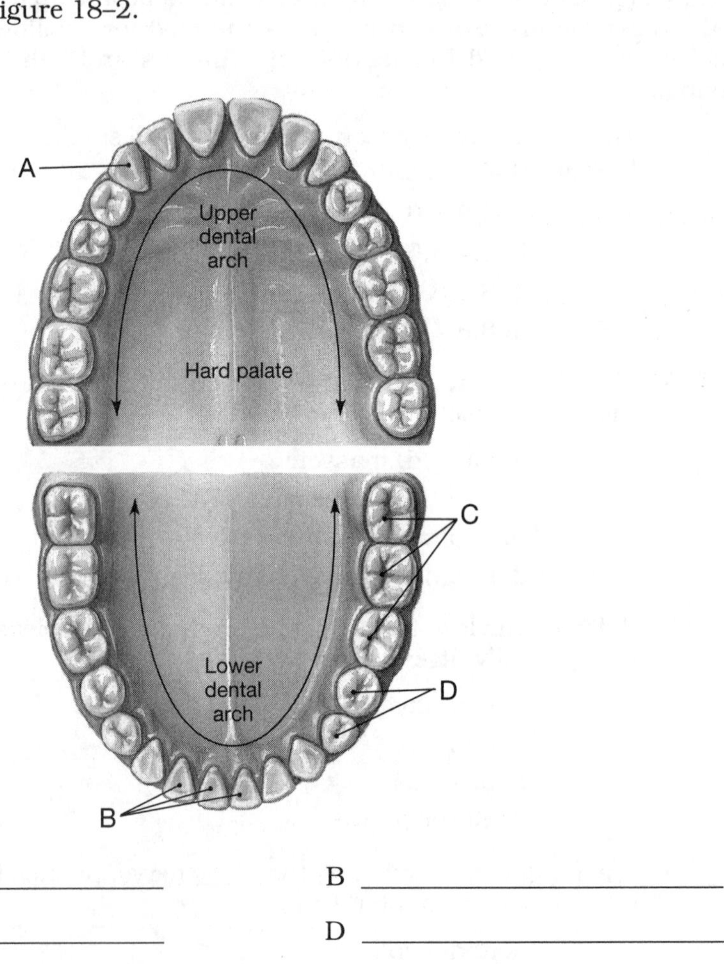

A _____________________ B _____________________

C _____________________ D _____________________

OBJECTIVE 5: Describe the structure of the stomach and its role in digestion and absorption (textbook pp. 440–441).

After studying pp. 440–441 in the text, you should be able to explain how the stomach prepares food for further digestion in the small intestine by secreting pepsin, an enzyme that digests protein, and hydrochloric acid, which assists in the formation of pepsin. (Hydrochloric acid will also kill microorganisms that may be present in the food.)

_____ 1. As food passes through the stomach it travels from the _____ to the _____.

a. cardia region, pyloric region
b. greater omentum, lesser omentum
c. lesser curvature, greater curvature
d. pyloric region, cardia region

____ 2. The stomach produces pepsin, which digests ____.

a. carbohydrates
b. fats
c. protein
d. all the above

____ 3. The hydrochloric acid in the stomach ____.

a. causes the release of gastrin
b. lowers the pH to kill microorganisms
c. raises the pH to kill microorganisms
d. both a and b are correct

____ 4. Gastrin is a(n) ____ that causes ____.

a. enzyme, the stomach to release HCl and other enzymes
b. hormone, the pyloric sphincter to open
c. hormone, the stomach to release HCl and enzymes
d. hormone, the stomach to stretch

5. After the food is mixed in the stomach with digestive juices and HCl, it is called ____________________.

6. In Figure 18–3, identify the duodenum, the body of the stomach, esophageal sphincter area, esophagus, fundus, rugae, and pyloric sphincter.

Figure 18–3 The Stomach

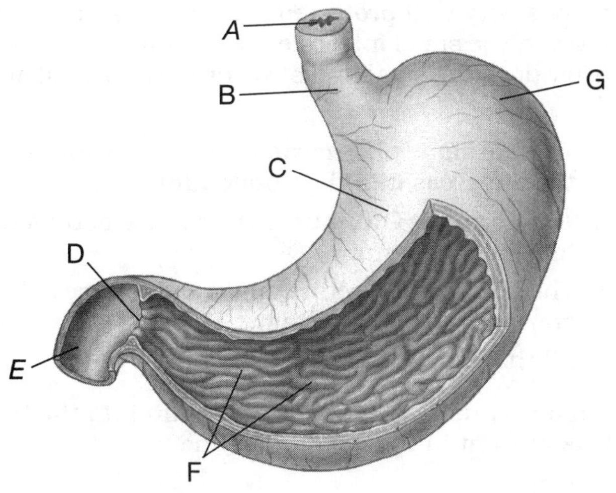

A ____________________ B ____________________

C ____________________ D ____________________

E ____________________ F ____________________

G ____________________

OBJECTIVE 6: Describe digestion and absorption in the small intestine (textbook pp. 442–443).

The small intestine serves as the main organ for digestion and absorption of nutrients. After studying pp. 442–443 in the text, you should be able to explain how nutrients are absorbed into the bloodstream and then transported to all the cells of the body.

_____ 1. The majority of the digestion in the small intestine occurs in the _____.

a. cecum
b. duodenum
c. ileum
d. jejunum

_____ 2. The majority of the absorption of the nutrients into the bloodstream occurs in the _____.

a. cecum
b. duodenum
c. ileum
d. jejunum

_____ 3. In order for the nutrients to leave the small intestine and enter the bloodstream, the nutrients must be absorbed into the _____.

a. gastric pits
b. lacteals
c. mucosal glands
d. villi

_____ 4. The stomach is very well protected, by a mucus lining, from the HCl it produces. However, the small intestine does not have the mucus protection from the acid. How does the small intestine protect itself against the HCl from the stomach?

a. The hormone cholecystokinin causes the release of buffers from the pancreas into the duodenum.
b. The hormone GIP causes the release of buffers from the pancreas into the duodenum.
c. The hormone secretin causes the release of buffers from the pancreas into the duodenum.
d. All the above are correct.

_____ 5. Which hormone causes the release of insulin into the bloodstream whenever glucose is present in the small intestine?

a. CCK
b. gastrin
c. GIP
d. secretin

_____ 6. The hormones cholecystokinin and secretin are released by the _____.

a. gallbladder
b. small intestine
c. pancreas
d. liver

OBJECTIVE 7: Describe the structure and functions of the pancreas, liver, and gallbladder (textbook pp. 443–445).

After studying pp. 443–445 in the text, you should be able to explain the roles that the accessory organs of the digestive system play in the digestive processes.

_____ 1. Which hormone causes the pancreas to release its digestive enzymes into the small intestine? _____

a. cholecystokinin
b. proteinase
c. secretin
d. none of the above

_____ 2. Which hormone causes the gallbladder to release bile?

a. cholecystokinin
b. proteinase
c. secretin
d. none of the above

_____ 3. More bile will be released into the small intestine if the small intestine contains large amount of _____.

a. protein
b. fat
c. carbohydrates
d. cholesterol

_____ 4. Bile consists of _____.

a. bile salts (lipids)
b. bilirubin (pigment from hemoglobin)
c. cholesterol
d. all the above

_____ 5. Enzymes that digest food are found in or come from the _____.

a. liver
b. pancreas
c. small intestine
d. both b and c

_____ 6. The pancreatic duct enters the duodenum along with the _____.

a. common bile duct
b. cystic duct
c. hepatic duct
d. none of the above

7. In Figure 18–4, identify the structures of the accessory glands.

Figure 18–4 The Accessory Glands of Digestion

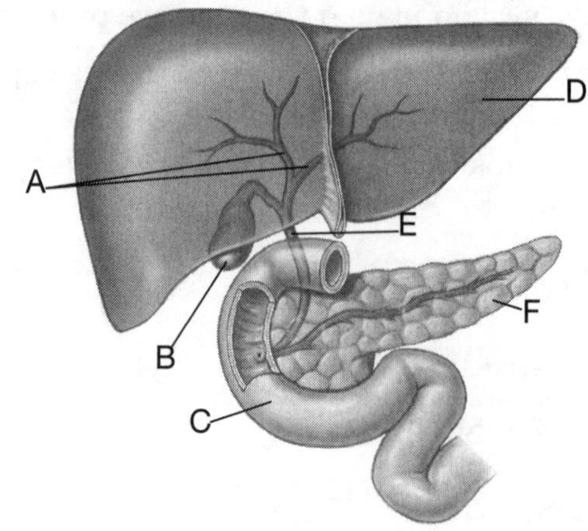

A ______________________ B ______________________

C ______________________ D ______________________

E ______________________ F ______________________

OBJECTIVE 8: Describe the structure and functions of the large intestine (textbook pp. 445–446).

After studying pp. 445–446 in the text, you should be able to explain the functions of the large intestine which are not limited to the removal of waste.

_____ 1. The large intestine _____.

a. releases nutrients into the bloodstream
b. absorbs water from the bloodstream
c. releases water into the bloodstream
d. serves only as an organ of excretion

_____ 2. The _____ of the small intestine joins with the _____ of the large intestine.

a. cecum, ileum
b. ileum, ascending colon
c. ileum, cecum
d. jejunum, cecum

_____ 3. Approximately what percentage of the water that enters the large intestine daily is recycled into the bloodstream?

a. 13%
b. 15%
c. 50%
d. 87%

_____ 4. Which of the following statements best describes one of the functions of the large intestine?

a. Bacteria in the large intestine manufacture some vitamins.
b. The large intestine makes some vitamins.
c. The large intestine removes some vitamins from the food we eat and puts the vitamins into the bloodstream.
d. The large intestine puts nutrients into the bloodstream.

____ 5. The action of peristalsis occurs in the ____.

a. esophagus
b. large intestine
c. small intestine
d. all the above

6. In Figure 18–5, identify the parts of the large intestine.

Figure 18–5 The Large Intestine

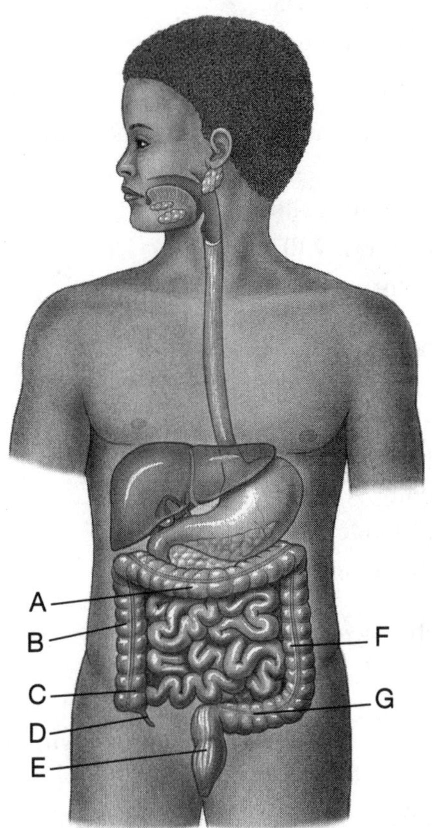

A ____________________ B ____________________

C ____________________ D ____________________

E ____________________ F ____________________

G ____________________

OBJECTIVE 9: Describe the digestion and absorption of the substances in food (textbook pp. 446–448).

After studying pp. 446–448 in the text, you should be able to identify the specific enzyme that each food that enters the digestive process requires in order to be digested to form nutrients and absorbed for use by cells in metabolism.

____ 1. Which of the following enzymes digests carbohydrates?

a. amylase
b. carboxypeptidase
c. lipase
d. pepsin

_____ 2. Which of the following enzymes digests fats?

a. amylase
b. carboxypeptidase
c. lipase
d. pepsin

_____ 3. Which of the following enzymes digests protein?

a. amylase
b. carboxypeptidase
c. hydrochloric acid
d. lipase

_____ 4. The process of _____ breaks down large molecules into small molecules so they can enter the _____.

a. absorption, cells
b. absorption, villi
c. digestion, cells
d. secretion, bloodstream

_____ 5. Once fats are digested they are absorbed into the _____ and then enter the _____.

a. bloodstream, lacteals
b. lacteals, villi
c. villi, bloodstream
d. villi, lacteals

OBJECTIVE 10: Describe the changes in the digestive system that occur with aging (textbook pp. 448–449).

After studying pp. 448–449 in the text, you should be able to list the many changes that occur in the digestive system with age, making it increasingly difficult to get proper nutrients to the body's cells.

_____ 1. With age, peristaltic actions slow down, resulting in _____.

a. more frequent constipation
b. more frequent diarrhea
c. more frequent heartburn
d. all the above

_____ 2. Dietary changes that affect the entire body in the elderly result from a _____.

a. decreased movement within the digestive tract
b. weakening of muscular sphincters
c. decline in sense of smell and taste
d. weaker peristaltic contractions

3. What is the relationship between tooth loss in the elderly and reduction in bone mass?

4. The types of cancers most common in elderly smokers are _____ and pharyngeal.

5. Decreased digestive tract movement and weak peristaltic contractions in the elderly may result in _____.

OBJECTIVE 11: Describe common symptoms and signs of digestive system disorders (textbook pp. 449–450).

The analysis and interpretation of the various signs and symptoms associated with the digestive system helps a physician determine the type of digestive problem the patient might have. Once the problem has been determined, corrective measures can be taken. After studying pp. 449–450 in the text, you should be able to describe the signs and symptoms of common digestive disorders.

_____ 1. Which of the following would be considered a sign of a digestive disorder?

a. indigestion
b. jaundice
c. nausea
d. vomiting

_____ 2. *Candida albicans* is a _____ that causes _____ in the mouth.

a. bacterium, gingivitis
b. bacterium, tooth decay
c. fungus, gingivitis
d. fungus, thrush

_____ 3. Having enlarged palatine tonsils may result in _____.

a. dyspepsia
b. dysphagia
c. emesis
d. jaundice

_____ 4. Pain in the upper right quadrant may indicate _____.

a. acute hepatitis
b. appendicitis
c. respiratory problems
d. a possible herniated esophagus

_____ 5. Which of the following physical examination techniques could be used to determine the presence of excess striae in the abdominal region?

a. auscultation
b. inspection
c. palpation
d. percussion

OBJECTIVE 12: Describe examples of inflammation and infection, and other major disorders of the digestive system (textbook pp. 451–461).

After studying in the textbook pp. 451–461, you should be able to describe the common digestive disorders and their causes, and the effects they have on homeostasis of the body.

_____ 1. Heartburn is a condition in which _____.

a. peptic ulcers develop in the stomach
b. stomach acid begins to "burn" the heart tissue
c. stomach acid refluxes into the esophagus
d. the heart tissue releases some acid material

_____ 2. Which of the following diseases characteristically spreads from the small intestine to the lymph nodes and gallbladder?

a. ascariasis
b. cholera
c. giardiasis
d. typhoid fever

_____ 3. Which of the following best describes giardiasis?

a. a condition in which protozoans cover the villi of the small intestine, thereby preventing absorption of nutrients
b. a bacterium that produces a toxin in the small intestine
c. a bacterium that spreads from the small intestine to the lymph and gallbladder
d. a worm that feeds on the contents of the small intestine

_____ 4. Which of the following hepatitis infections is not transmitted by body fluids?

a. hepatitis A
b. hepatitis B
c. hepatitis C
d. hepatitis D

_____ 5. Which of the following is a condition in which the sigmoid colon develops pockets in the mucosa?

a. cholecystitis
b. diverticulitis
c. irritable bowel syndrome
d. ulcerative colitis

_____ 6. A condition that causes the patient to experience projectile vomiting is called _____.

a. pyloric stenosis
b. cholitis
c. irritable bowel syndrome
d. diverticulitis

PART II: CHAPTER-COMPREHENSIVE EXERCISES

MATCHING

Match the terms in column A with the descriptions or statements in column B.

	(A)	(B)
_____	1. amylase	A. The part of the intestine where bile and pancreatic enzymes enter the small intestine
_____	2. bile	B. The part of the small intestine where most of the digestion and absorption occurs
_____	3. cecum	C. The part of the small intestine that joins the large intestine
_____	4. duodenum	D. Structures inside the small intestine that absorb the nutrients
_____	5. hepatitis A	E. An enzyme produced and released by the pancreas that digests fat
_____	6. hepatitis B	F. An enzyme produced and released by the pancreas that digests carbohydrates
_____	7. ileum	G. An enzyme produced and released by the pancreas that digests protein
_____	8. jejunum	H. A material that emulsifies fat
_____	9. lipase	I. The part of the large intestine to which the appendix is attached
_____	10. salmonella	J. A viral infection transmitted by body fluids that results in inflammation of the liver
_____	11. trypsin	K. A viral infection transmitted by fecal-contaminated water or food that results in inflammation of the liver
_____	12. vibrio	L. The bacterium responsible for typhoid fever
_____	13. villi	M. The bacterium responsible for cholera

CONCEPT MAP

This concept map summarizes and organizes the material in Chapter 18. Use the following terms to complete the map by filling in the boxes identified by the circled numbers, 1–10.

carbohydrates	**duodenum**	**pepsin**
protein	**lipase**	**salivary amylase**
CCK	**fat**	**pancreas**
large intestine		

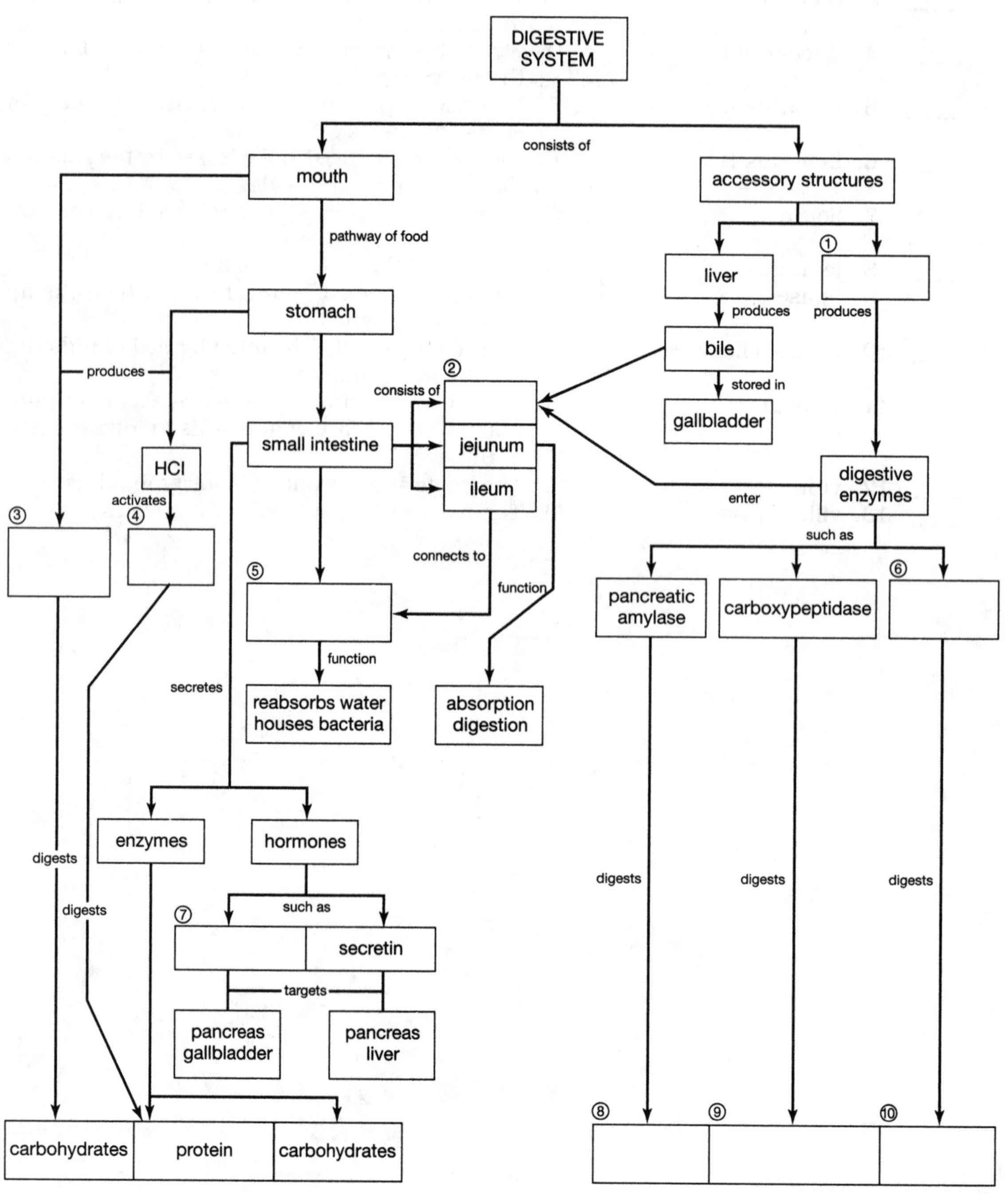

CROSSWORD PUZZLE

The following crossword puzzle reviews the material in Chapter 18. To complete the puzzle, you have to know the answers to the clues given, and you must be able to spell the terms correctly.

ACROSS

1. A condition that may occur when the liver malfunctions
4. The procedure that allows physicians to view internal body cavities
7. The process of moving nutrients from the digestive tract to the bloodstream
9. A type of hepatitis that is transmitted by body fluids
10. The _____ and the gallbladder secrete their digestive products into the duodenum of the small intestine.
12. Where most of the digestion in the digestive tract occurs
13. The muscular ridges found inside the stomach

DOWN

2. The pancreas and liver are considered to be _____ organs of the digestive system.
3. A condition in which bacterium produces a toxin that causes the intestines to lose chloride ions
5. The first portion of the small intestine
6. The fleshy protrusion hanging at the back of the throat
8. The gland that produces salivary amylase
9. Salivary amylase digests _____ to form glucose.
11. The part of the large intestine to which the appendix is attached

FILL-IN-THE-BLANK NARRATIVE

Use the following terms to fill in the blanks of this summary of Chapter 18.

carboxypeptidase	**cecum**	**epiglottis**
gastrin	**esophagus**	**cholecystokinin**
villi	**peristalsis**	**ileum**
pharynx	**proteinases**	**pepsin**
jejunum	**rugae**	**incisors**
esophageal	**pyloric**	**duodenum**
uvula	**diarrhea**	**dehydration**
constipation		

The foods we eat typically consist of a combination of carbohydrates, fats, and proteins. Let's follow a protein through the digestive system.

The protein is chewed and torn with the teeth (if it is a piece of meat). To tear the food we might use the (1) ___________ and cuspid teeth. We then push the food to the (2) ___________. Food brushes past the (3) ___________, which causes the (4) ___________ to close over the trachea. This ensures that food will go down the (5) ___________ instead of the trachea. The food continues down the esophagus by muscular action called (6) ___________. The protein passes through the (7) ___________ sphincter and enters the stomach. The (8) ___________ allows the stomach to stretch. This stretching causes some stomach cells to release the hormone (9) ___________, which ultimately causes other stomach cells to produce and release the digestive enzyme (10) ___________. This enzyme begins the digestion of protein. Continuing its journey, the food passes through the (11) ___________ sphincter and enters the (12) ___________. The presence of protein causes the small intestine to release the hormone (13) ___________, which targets the pancreas and causes it to release (14) ___________ such as trypsin and (15) ___________ into the duodenum. Protein is further digested in the (16) ___________. The digested particles can now be absorbed through the (17) ___________ and enter the bloodstream to be transported to the body's cells. The material that is not digested will leave the (18) ___________ and enter the (19) ___________ of the large intestine. As the waste products pass through the large intestine, water is reabsorbed into the bloodstream to prevent dehydration. Elimination of the waste products occurs through the rectum and the anus.

Waste products must pass through the large intestine at a normal pace to ensure the reabsorption of water. If peristalsis is rapid, (20) ___________ may occur. This results in (21) ___________. If the waste products pass through the large intestine too slowly, (22) ___________ may occur.

CLINICAL CONCEPT

The following clinical concept applies the information presented in Chapter 18. Following the application is a set of questions to help you understand the concept.

Steve really likes bacon. At breakfast, Steve can easily eat 10 slices of bacon. One day, he feels pain in the upper right quadrant. Upon examination, the doctor explains that Steve has gallstones and suggest using an ultrasound procedure to pulverize them so they can be easily passed. Steve undergoes the lithotripsy but several months later has another attack. Gallstones do recur, especially if the diet is high in cholesterol. The doctor suggests surgery to remove the gallbladder. Steve concurs. Now, at breakfast, Steve feels uncomfortable if he consumes 10 slices of bacon, but he feels fine if he eats 2 slices and no more. Steve decides to alter his diet to reduce his fat intake.

1. What does the gallbladder store?

2. How are gallstones produced?

3. How is the gallbladder indirectly related to fat digestion?

4. Why does Steve need to reduce his fat intake after gallbladder surgery?

5. After removal of the gallbladder, how does bile enter the small intestine? (To answer this question, refer to Figure 18-7 in the textbook.)

THIS CONCLUDES CHAPTER 18 EXERCISES

CHAPTER

19

Nutrition and Metabolism

Chapter Objectives:

1. Define metabolism and explain why cells need to synthesize new structures.
2. Define catabolism and anabolism.
3. Describe the catabolism and anabolism of carbohydrates, lipids, and proteins.
4. Describe some of the roles of minerals and vitamins in the body.
5. Explain what constitutes a balanced diet, and why it is important.
6. Contrast the energy content of carbohydrates, lipids, and proteins.
7. Define metabolic rate and discuss the factors involved in determining an individual's metabolic rate.
8. Discuss the homeostatic mechanisms that maintain a constant body temperature.
9. Describe specific examples of eating, metabolic, and thermoregulatory disorders.

All cells require energy for the anabolic and catabolic processes of metabolism. The nutrients and calories from food provide this energy and promote tissue growth and repair. Living cells are chemical factories that break down organic molecules to obtain energy; this energy is used to generate ATP. To carry out these **metabolic** reactions, the cells must have a constant supply of oxygen and nutrients, including water, vitamins, minerals, and organic substrates that initiate enzyme activity and promote chemical reactions involved in the work of metabolism.

Most students of anatomy and physiology are intrigued with the body's metabolic machinery because of the mystique associated with the fate of absorbed nutrients. The knowledge of what happens to the food we eat and how the nutrients from food are utilized by the body continues to capture the interest of those who have adopted the belief that "we are what we eat."

Much of the material you have learned in previous chapters will be useful to you as you begin to understand how the integrated processes of the digestive and cardiovascular systems exert a coordinated effort to initiate and advance the work of metabolism.

The process of metabolism can be altered by overeating, undereating, or malnutrition. Malnutrition can be due to undereating or to some metabolic disorder. Any disorder of metabolism will upset homeostasis.

The exercises in this chapter review some of the basic principles of nutrition and cellular metabolism, as well as nutritional and metabolic disorders.

PART I: OBJECTIVE-BASED QUESTIONS

OBJECTIVE 1: Define metabolism and explain why cells need to synthesize new structures (textbook pp. 467–468).

After studying pp. 467–468 in the text, you should be able to explain the processes collectively called metabolism, the sum of all the chemical reactions that occur in the body.

_____ 1. All the chemical reactions occurring in the body are collectively called _____.

a. anabolism
b. catabolism
c. cellular respiration
d. metabolism

_____ 2. Which of the following statements is true?

a. Anabolism and catabolism are a type of metabolism.
b. Catabolism is a type of anabolism, which is a type of metabolism.
c. Metabolism is a type of anabolism and catabolism.
d. Metabolism is the same thing as catabolism and anabolism.

_____ 3. Which of the following statements is true?

a. Metabolism involves the chemical reactions within the body.
b. Catabolism involves the chemical reactions within the body.
c. Anabolism involves the chemical reaction within the body.
d. All the above are true.

OBJECTIVE 2: Define catabolism and anabolism (textbook pp. 467–468).

After studying pp. 467–468 in the text, you should be able to define the two types of reactions of metabolism: catabolism, the series of reactions that breaks down organic molecules, and anabolism, the series of reactions that builds new organic molecules.

_____ 1. Which of the following terms refers to the breaking down of organic molecules?

a. anabolism
b. catabolism
c. metabolism
d. all the above

_____ 2. Which of the following terms refers to the building of new molecules?

a. anabolism
b. catabolism
c. metabolism
d. all the above

_____ 3. The production of adenosine triphosphate by the mitochondria is an example of _____.

a. anabolism
b. catabolism
c. metabolism
d. all the above

_____ 4. The bonding together of amino acids is an example of _____.

a. anabolism
b. catabolism
c. metabolism
d. all the above

_____ 5. The conversion of starch to glucose is an example of _____.

a. anabolism
b. catabolism
c. metabolism
d. all the above

OBJECTIVE 3: Describe the catabolism and anabolism of carbohydrates, lipids, and proteins (textbook pp. 468–470).

After studying pp. 468–470 in the text, you should be able to explain the main concept of catabolism, which is to break down the large organic molecules into smaller organic molecules that the cells can use, and the main concept of anabolism, which is to build new organic molecules that the cells can use.

_____ 1. Which of the following statements about the conversion of glucose to ATP is correct?

a. Glucose forms pyruvic acid under aerobic conditions. Pyruvic acid enters the mitochondria and eventually makes ATP. Oxygen is required.
b. Glucose forms pyruvic acid under aerobic conditions. Pyruvic acid enters the mitochondria and eventually makes ATP. Oxygen is not required.
c. Glucose forms pyruvic acid under anaerobic conditions. Pyruvic acid enters the mitochondria and eventually makes ATP. Oxygen is required.
d. Glucose forms pyruvic acid under anaerobic conditions. Pyruvic acid enters the mitochondria and eventually makes ATP. Oxygen is not required.

_____ 2. The anabolism of glucose by the body produces _____.

a. glycogen
b. monosaccharides
c. pyruvic acid
d. starch

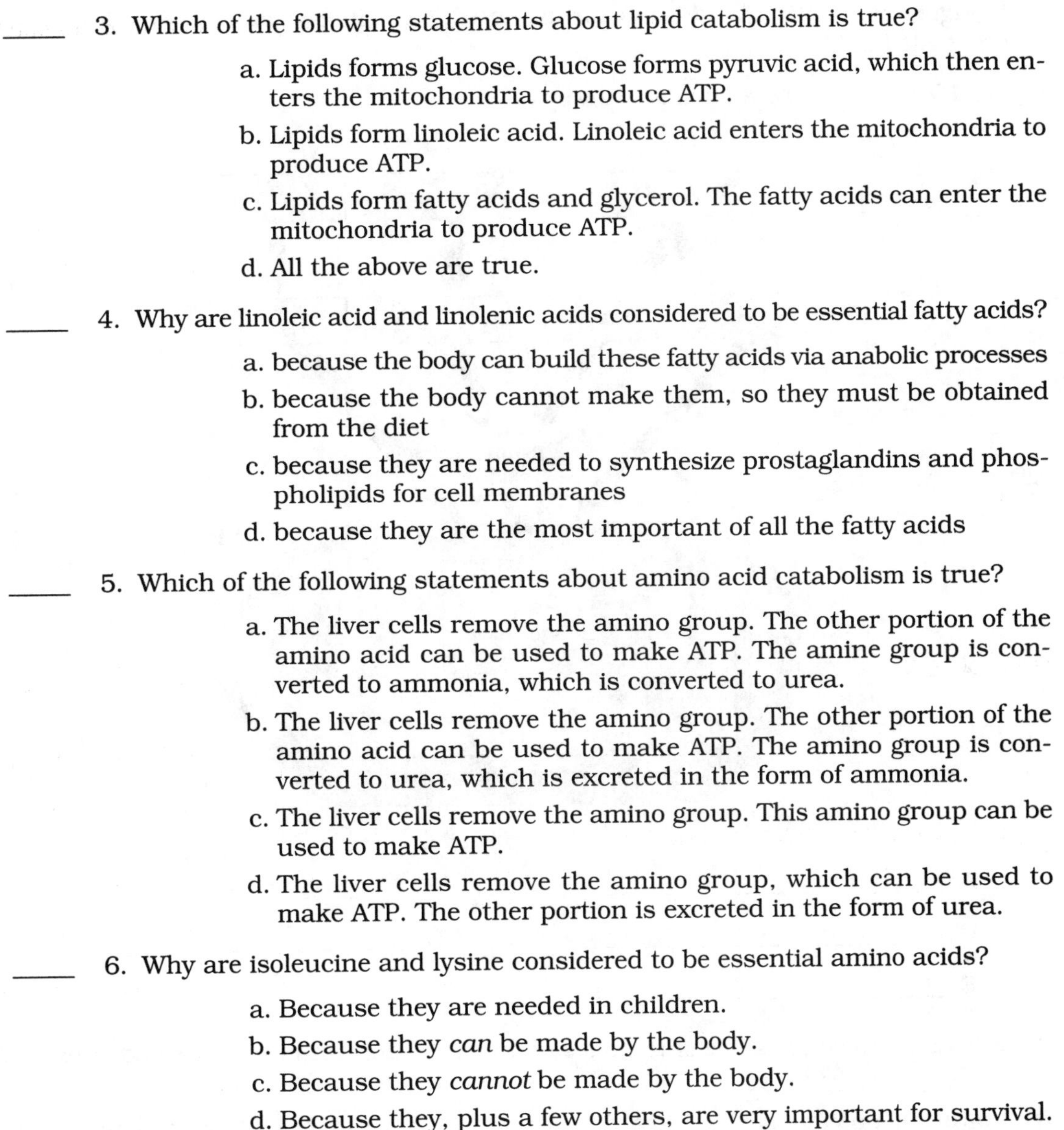

____ 3. Which of the following statements about lipid catabolism is true?

a. Lipids forms glucose. Glucose forms pyruvic acid, which then enters the mitochondria to produce ATP.

b. Lipids form linoleic acid. Linoleic acid enters the mitochondria to produce ATP.

c. Lipids form fatty acids and glycerol. The fatty acids can enter the mitochondria to produce ATP.

d. All the above are true.

____ 4. Why are linoleic acid and linolenic acids considered to be essential fatty acids?

a. because the body can build these fatty acids via anabolic processes

b. because the body cannot make them, so they must be obtained from the diet

c. because they are needed to synthesize prostaglandins and phospholipids for cell membranes

d. because they are the most important of all the fatty acids

____ 5. Which of the following statements about amino acid catabolism is true?

a. The liver cells remove the amino group. The other portion of the amino acid can be used to make ATP. The amine group is converted to ammonia, which is converted to urea.

b. The liver cells remove the amino group. The other portion of the amino acid can be used to make ATP. The amino group is converted to urea, which is excreted in the form of ammonia.

c. The liver cells remove the amino group. This amino group can be used to make ATP.

d. The liver cells remove the amino group, which can be used to make ATP. The other portion is excreted in the form of urea.

____ 6. Why are isoleucine and lysine considered to be essential amino acids?

a. Because they are needed in children.

b. Because they *can* be made by the body.

c. Because they *cannot* be made by the body.

d. Because they, plus a few others, are very important for survival.

7. In Figure 19–1, identify the amino acids, fatty acids, glucose, and glycerol. Identify cell organelle E.

Figure 19–1 Energy-Producing Pathways

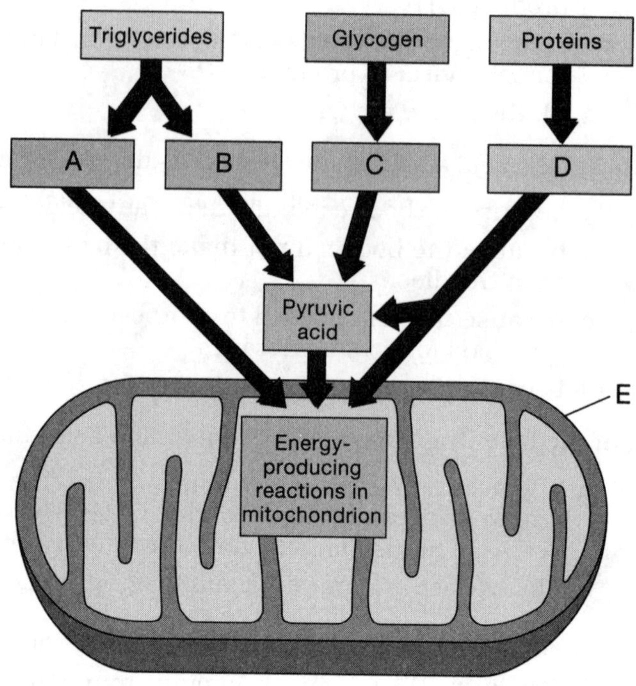

A ______________________ B ______________________

C ______________________ D ______________________

E ______________________

OBJECTIVE 4: Describe some of the roles of minerals and vitamins in the body (textbook p. 472).

While growing up you may have heard your parents say, "Eat a well-balanced meal so you can get all the vitamins and minerals you need to grow up big and strong." After studying p. 472 in the text, you should be able to explain the importance of vitamins and minerals in the diet.

_____ 1. The body cannot make _____.

a. minerals
b. vitamins
c. water
d. all the above

_____ 2. Enzymes are very important in the body because they assist numerous chemical reactions. Which of the following is a part of an enzyme?

a. ATP
b. minerals
c. vitamins
d. none of the above

_____ 3. Which of the following is (are) involved in transporting oxygen in the blood?

a. minerals
b. sodium chloride
c. vitamins
d. none of the above

_____ 4. Which of the following minerals is involved in transporting oxygen?

a. Ca^{2+}
b. Cl^{-}
c. Fe^{2+}
d. Na^{+}

OBJECTIVE 5: Explain what constitutes a balanced diet, and why it is important (textbook pp. 470–472).

Many people have a very hectic and rushed lifestyle because they are busy working or busy preparing for that final exam. Some people become so busy they fail to eat or take time to eat a balanced meal. They are so rushed that all they have time for is a can of pop and a candy bar. Nutritionists would not consider this to be a balanced meal. A balanced diet contains all the nutrients necessary to maintain homeostasis. After studying pp. 470–472 in the text, you should be able to describe a balanced diet, that is, one that contains all the nutrients necessary to maintain homeostasis.

_____ 1. Which of the following is *not* a part of the basic food groups?

a. fats
b. fruit
c. milk
d. They are all part of the basic food groups.

_____ 2. A complete protein is a _____.

a. protein that contains all the amino acids the body can make
b. protein that contains all the amino acids the body can use
c. protein that contains all the essential amino acids
d. all the above

_____ 3. An incomplete protein is a protein that _____.

a. lacks some of the amino acids the body can make
b. lacks some of the amino acids the body cannot make
c. lacks some of the essential amino acids
d. Both b and c are correct.

4. Why are fats and oils and sweets placed at the top of the food pyramid?

____

OBJECTIVE 6: Contrast the energy content of carbohydrates, lipids, and proteins (textbook p. 479).

After studying p. 479 in the text, you should be able to explain how the energy content of food is measured and the significance of this information.

_____ 1. Which of the following statement(s) about a Calorie is true?

a. It is the same as a kilocalorie.
b. It is the same as 1000 calories.
c. It is the amount of energy required to raise the temperature of 1 kg of water 1°C.
d. All the above are true.

_____ 2. If we eat a hamburger that weighs 1/4 lb (113 g), how many Calories will be released from that hamburger if it is pure protein?

a. about 109 C (113 - 4.32)
b. about 26 C (113/4.32)
c. about 4.32 C
d. about 488 C (113 X 4.32)

_____ 3. From which of the following do we obtain the most Calories on a per gram basis?

a. carbohydrates
b. fat
c. protein
d. sweets

_____ 4. Which of the following releases the most Calories when consumed?

a. a baked potato with a mass of 82 g
b. two slices of bread (15 g each) and a slice of meat (protein) at 30 g
c. 10 strips of bacon with 4 g of fat and 8 g of protein each
d. pop that contains 40 g of carbohydrates, 0 g of protein, and 0 g of fat

_____ 5. The unit used to measure the amount of energy released is _____.

a. a centigrade
b. a gram
c. ATP
d. the Calorie

OBJECTIVE 7: Define metabolic rate and discuss the factors involved in determining an individual's metabolic rate (textbook p. 479).

Sally says she has a higher metabolic rate than Jennifer does. What are they referring to? After studying p. 479 in the text, you should be able to explain metabolic rate and how it is measured.

_____ 1. Which of the following statements best defines metabolic rate?

a. It is the amount of energy a person gains per day.
b. It is the amount of energy a person loses in the form of heat.
c. It is the amount of energy a person uses during exercise.
d. It is the amount of energy a person uses per day.

____ 2. A person's BMR value represents _____.

a. a measurement of how much energy a resting person uses during a day
b. a measurement of how much energy is gained during a day by that person
c. a measurement of how much energy is lost during a day by that person
d. a measurement of how much energy is used during a day by that person

____ 3. Which of the following is a factor in calculating a person's BMR?

a. a person's age
b. a person's sex
c. a person's weight
d. all the above

OBJECTIVE 8: Discuss the homeostatic mechanisms that maintain a constant body temperature (textbook pp. 479–482).

After studying pp. 479–482 in the text, you should be able to explain the importance of keeping body temperature within acceptable ranges.

____ 1. The body loses heat to the environment. Which of the following mechanisms of heat transfer is involved?

a. conduction
b. convection
c. evaporation
d. radiation

____ 2. Somebody just got up from a chair. You sit on that same chair and you find it feels warm to you. Which of the following mechanisms of heat transfer is involved?

a. conduction
b. convection
c. evaporation
d. radiation

____ 3. When we sweat, water from our body goes to the surface of our skin, from which it leaves, taking the heat with it. As the heat leaves our body we feel cooler. Which of the following mechanisms of heat transfer is involved?

a. conduction
b. convection
c. evaporation
d. radiation

____ 4. Which of the following structures is involved in the regulation of body temperature?

a. hypothalamus
b. medulla oblongata
c. pituitary
d. skin

_____ 5. Which of the following will the body do on a cold day? _____

a. constrict the blood vessels

b. dilate the blood vessels

c. shiver

d. both a and c

OBJECTIVE 9: Describe specific examples of eating, metabolic, and thermoregulatory disorders (textbook pp. 475–479).

After studying pp. 475–479 in the text, you should be able to explain how eating disorders can result in metabolic and thermoregulatory disorders and thus affect homeostasis.

_____ 1. The excess production of ketones in the body is due to _____.

a. a decrease in pH

b. excess breakdown of fatty acids

c. excess breakdown of glucose

d. the accumulation of glucose

_____ 2. Kwashiorkor is a condition in which _____.

a. carbohydrate intake is excessive and protein intake is deficient

b. protein intake is excessive and carbohydrate intake is deficient

c. carbohydrate and protein intake are excessive, thus resulting in "large abdomens"

d. vitamin and mineral intake is deficient, resulting in abnormal swelling of the abdomen

_____ 3. Which of the following statements is true?

a. Phenylalanine needs to convert to tyrosine. Tyrosine is important in the formation of many proteins.

b. PKU is a metabolic enzyme.

c. PKU converts phenylalanine to tyrosine. Tyrosine is important in the formation of many proteins.

d. Without PKU, tyrosine will not convert to phenylalanine.

_____ 4. The introduction of sodium ions into the body can help restore homeostasis if the patient is experiencing _____.

a. dehydration

b. hypervitaminosis

c. low basal metabolic rate

d. overhydration

_____ 5. During a heat stroke, the brain may be affected because _____.

a. blood vessels are dilated in an effort to reduce the heat, thus resulting in a drop in blood pressure

b. blood vessels constrict, thus reducing the flow of blood to the brain

c. heat causes more blood to flow to the head region, resulting in the "flushed" look

d. heat causes the production of pyrogens

_____ 6. Hypothermia occurs faster in cold water than in cold air because _____.

a. water cools down the air temperature, so both water and air cause hypothermia at the same time
b. water transfers cold into the body faster than air does
c. water surrounds the body more evenly than air does
d. water causes heat to be removed from the body faster than air does

PART II: CHAPTER-COMPREHENSIVE EXERCISES

MATCHING

Match the terms in column A with the descriptions or statements in column B.

(A)	(B)
_____1. aerobic	A. A process in which large molecules are broken down into smaller molecules
_____2. anabolism	B. A process in which molecules are being built
_____3. anaerobic	C. Metabolism that *does not* require the use of oxygen
_____4. catabolism	D. Metabolism that *does* require the use of oxygen.
_____5. conduction	E. The catabolism of amino acids ultimately produces _____.
_____6. evaporation	F. Low body temperature
_____7. hypothermia	G. The loss or gain of heat by contact
_____8. ketoacidosis	H. The loss of heat via sweating
_____9. metabolism	I. Chemical reactions involving catabolism and anabolism
_____10. urea	J. Diabetes mellitus can lead to diabetic _____.
_____11. vasoconstrict	K. Vessels will generally _____ in an effort to retain heat within the body.
_____12. vasodilate	L. Vessels will generally _____ in an effort to lose heat from the body.

CONCEPT MAP

This concept map summarizes and organizes the information in Chapter 19. Use the following terms to complete the map by filling in the boxes identified by the circled numbers, 1–13.

amino acids	**anabolism**	**lipids**
carbohydrates	**glycerol**	**linoleic acid**
protein	**lysine**	**essential**
ketones	**tyrosine**	**phenylalanine**
PKU		

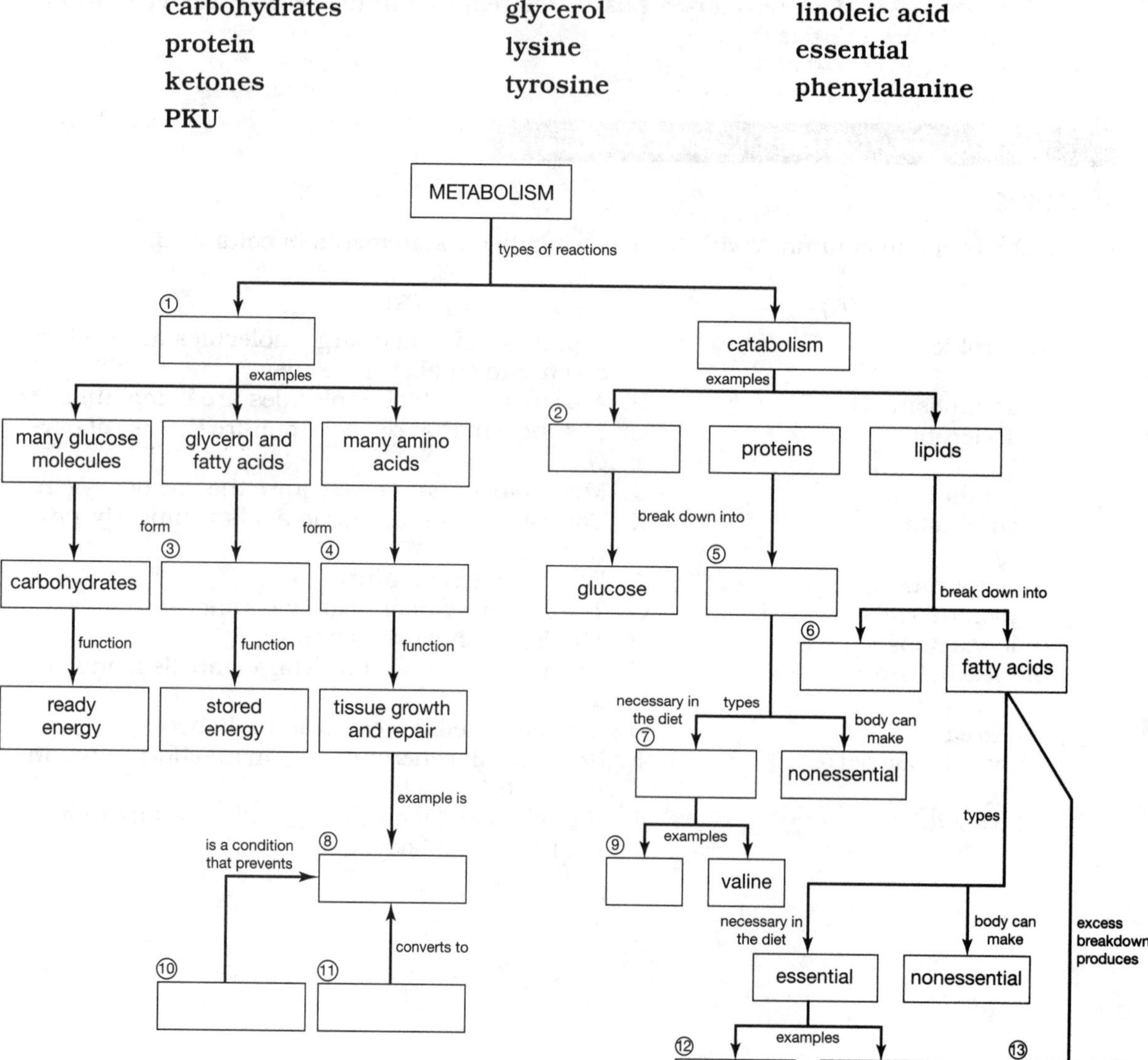

CROSSWORD PUZZLE

The following crossword puzzle is a review of the material in Chapter 19. To complete the puzzle, you have to know the answers to the clues given, and you must be able to spell the terms correctly.

ACROSS

2. Substances released by bacteria or viruses that cause a fever
4. One of the essential fatty acids
5. A molecule required by the body but which the body cannot make
7. Excess carbohydrate intake coupled with protein deficiency may result in a condition called _____.
8. The sum of all the chemical reactions in the body
10. One of the essential amino acids
11. A unit of measure of the amount of energy in a specific food

DOWN

1. The digestion of protein produces _____.
2. The condition in which lack of an enzyme results in a decreased conversion of phenylalanine to tyrosine is called _____.
3. Metabolic reactions that break down molecules
6. A food substance that has all the amino acids the body requires is known as a(n) _____ protein.
9. The digestion of starch produces _____.

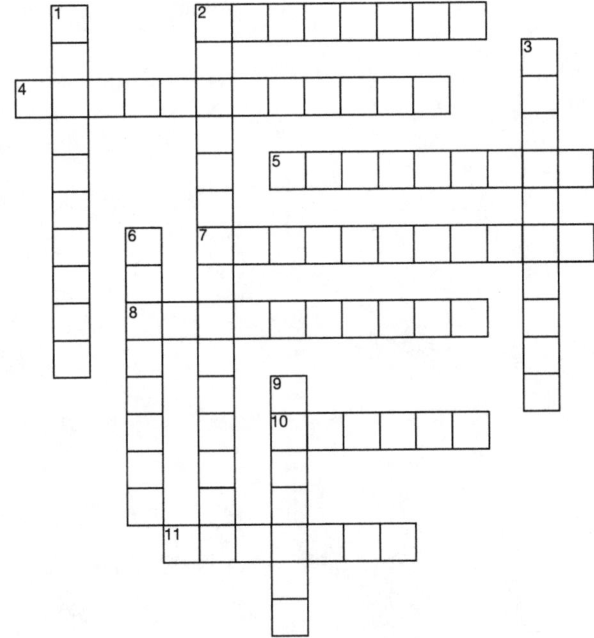

FILL-IN-THE-BLANK NARRATIVE

Use the following terms to fill in the blanks of this summary of Chapter 19.

aerobic	**anabolism**	**anaerobic**
catabolism	**metabolism**	**pH**
ketones	**enzymes**	

After food is ingested, the process of digestion breaks down large food particles into smaller ones. The process of (1) __________ breaks these molecules down to even smaller molecules that the cells of the body can use to manufacture necessary molecules by the process called (2) __________.

The (3) __________ of glucose proceeds initially by a(n) (4) __________ pathway but is completed via the (5) __________ pathway. Glucose enters the cell and eventually forms pyruvic acid. Pyruvic acid then enters the mitochondria, which, in a series of reactions, form carbon dioxide and adenosine triphosphate (ATP).

In order for these metabolic reactions to proceed properly, the pH has to be optimum. Some patients try to lose weight by participating in fad diets. These diets force the body to break down fatty acids. This causes the patients to lose weight; however, a byproduct of fatty acid breakdown is (6) __________. An excessive amount of ketones will lower the (7) __________. An alteration in pH will denature the (8) __________ that are necessary for metabolism. Thus, due to the diet, the patient is out of homeostasis.

CLINICAL CONCEPT

The following clinical concept applies the information presented in Chapter 19. Following the application is a set of questions to help you understand the concept.

Keith is a college student. He's extremely busy studying and working part-time. Keith sleeps in late and therefore does not take time for breakfast. He rushes off to class. At lunch, he has time for only french fries. When he gets home from school, he eats canned corn (only corn, nothing else) and washes it down with water. Then, he goes to work, only to start the same routine the next day.

When Keith is confronted about his poor diet, he responds, "The french fries are a good source of carbohydrates. Since they're fried in oil, I get my daily requirement of fat. At supper time, I eat corn and it has a lot of amino acids in it."

You, a medical student, suggest that Keith see a doctor. Keith complies and makes an appointment with a doctor. The doctor's report looks like this: Keith is pale and underweight, his urine and blood pH are low, and the ketones in the urine and blood are high. The Na^+ concentration in the blood is low. Keith seems to have possible nerve problems.

The doctor explains that Keith is pale due to anemic conditions brought on by the poor diet. With a poor protein diet, Keith's erythrocytes may not be making adequate hemoglobin. Because his diet is very low in carbohydrates, his body is metabolizing fat, which is causing the weight loss. Byproducts of fat metabolism are ketones, which are acidic. This accounts for the increase in ketones in the urine and the blood and a drop in the pH. A drop in blood pH will inhibit the small intestine from putting Na^+ into the bloodstream. Sodium ions are necessary for proper nerve function. In short, Keith needs to begin eating well-balanced meals.

1. Why is corn considered to be an incomplete protein?

2. How are ketones formed?

3. How are sodium ions involved in the nervous system?

4. How does this scenario relate to fad diets?

THIS CONCLUDES CHAPTER 19 EXERCISES

CHAPTER 20

The Urinary System and Body Fluids

Chapter Objectives:

1 Identify the components of the urinary system and their functions.

2 Describe the structural features of the kidneys.

3 Describe the structure of the nephron and the process involved in urine formation.

4 List and describe the factors that influence filtration pressure and the rate of filtrate formation.

5 Describe the changes that occur in the filtrate as it moves along the nephron and exits as urine.

6 Describe how the kidneys respond to conditions of changing blood pressure.

7 Describe the structures and functions of the ureters, urinary bladder, and urethra.

8 Discuss the process of urination and how it is controlled.

9 Explain the basic concepts involved in the control of fluid and electrolyte regulation.

10 Explain the buffering systems that balance the pH of the intracellular and extracellular fluids.

11 Describe the effects of aging on the urinary system.

12 Describe the general symptoms and signs of disorders of the urinary system.

13 Give examples of the various disorders of the urinary system.

14 Give examples of conditions resulting from abnormal renal function.

Many people might think that the only function of the kidneys is to excrete waste in the form of urine. The kidneys play many vital roles in the human body to help maintain homeostasis. The kidneys:

1. Excrete waste
2. Regulate blood pressure
3. Initiate red blood cell formation
4. Prevent dehydration
5. Filter the blood
6. Regulate blood ions
7. Regulate blood pH

The macroscopic anatomy of the urinary system consists of kidneys, ureters, urinary bladder, and urethra. It appears that the anatomy of the urinary system is simple. However, you will find that the microscopic anatomy of the urinary system consists of millions of tubules and specialized cells that are designed to carry out the functions listed above.

The kidneys play numerous roles in the effort to maintain homeostasis. Any alteration or disease that upsets the function of any part of the urinary system will affect other systems of the body.

The review exercises in Chapter 20 focus on the structured and functional organization of the urinary system, as well as abnormalities associated with the urinary system, and their effects on other systems of the body.

PART I: OBJECTIVE-BASED QUESTIONS

OBJECTIVE 1: Identify the components of the urinary system and their functions (textbook p. 487).

After studying p. 487 in the text, you should be able to name the organs of the urinary system and trace the flow of urine from the kidneys to the urethra.

_____ 1. The tubes leaving the kidneys are the _____.

a. hilus
b. nephrons
c. ureters
d. urethra

_____ 2. Urine enters the urinary bladder via the _____.

a. hilus
b. nephrons
c. ureters
d. urethra

_____ 3. Micturition occurs when urine leaves the _____.

a. kidneys
b. urethra
c. ureter
d. nephrons

4. Which kidney is slightly higher than the other? ____________________

5. Identify the organs of the urinary system in Figure 20–1 below.

Figure 20–1 The Urinary System

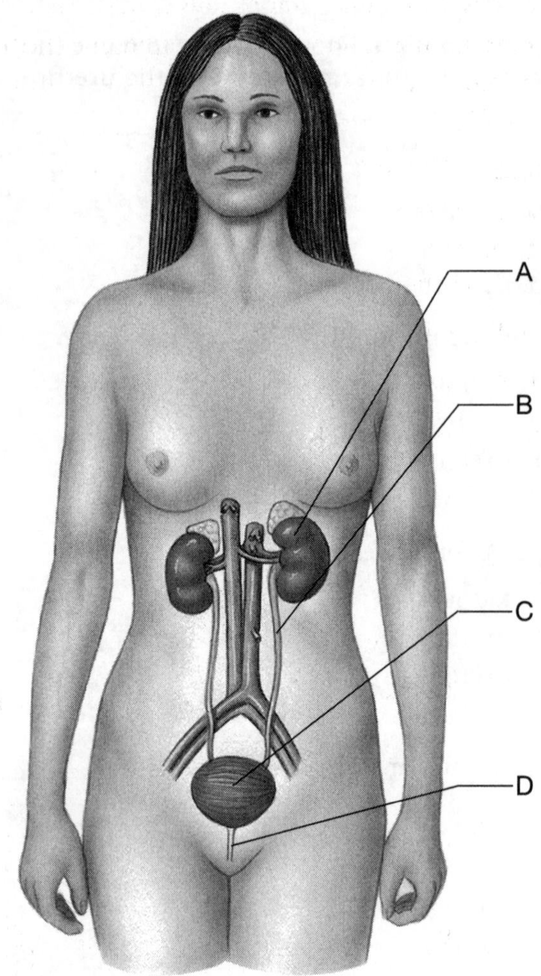

A ______________________ B ______________________

C ______________________ D ______________________

OBJECTIVE 2: Describe the structural features of the kidneys (textbook pp. 487–488).

After studying pp. 487–488 in the text, you should be able to describe the macroscopic internal anatomy of the kidney.

____ 1. The hilus is the point where ____.

a. a ureter exits the kidney

b. blood vessels enter and exit the kidney

c. the urethra exits the kidney

d. both a and b

_____ 2. Once urine is formed, in which of the following sequences does it pass through the kidneys?

a. renal pyramid, major calyx, minor calyx, renal pelvis, ureter
b. renal pyramid, minor calyx, major calyx, renal pelvis, ureter
c. renal pyramid, minor calyx, major calyx, renal pelvis, urethra
d. renal pyramid, renal cortex, minor calyx, major calyx, renal pelvis, ureter

_____ 3. The renal pyramids are located in the _____.

a. calyces
b. cortex
c. medulla
d. renal pelvis

4. Identify the labeled structures of the kidney in Figure 20–2.

Figure 20–2 Structure of the Kidney, Internal View

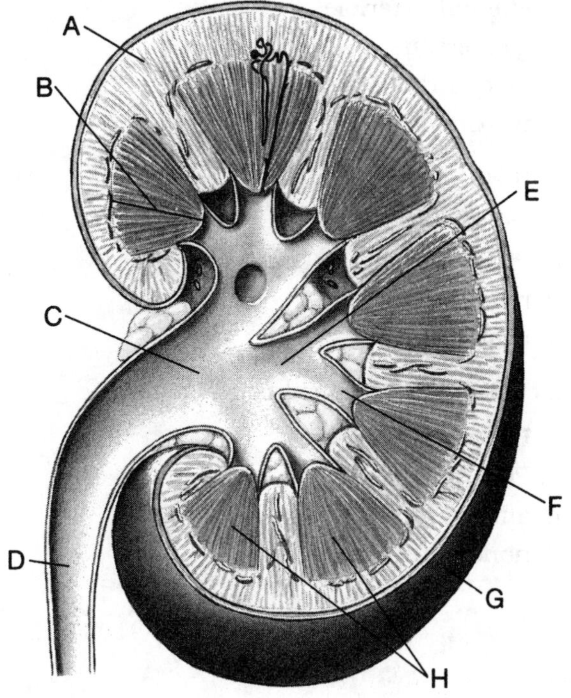

A _______________ B _______________

C _______________ D _______________

E _______________ F _______________

G _______________ H _______________

OBJECTIVE 3: Describe the structure of the nephron and the processes involved in urine formation (textbook pp. 488–491).

After studying pp. 488–491 in the text, you should be able to trace a drop of filtrate through a nephron and identify each segment involved in converting the filtrate into urine.

_____ 1. Fluid in the afferent arteriole will enter the _____.

a. Bowman's capsule
b. efferent arteriole
c. glomerulus
d. nephron

_____ 2. Fluid in the glomerulus will enter the _____.

a. Bowman's capsule
b. efferent arteriole
c. proximal convoluted tubule
d. both a and b

_____ 3. Fluid in the Bowman's capsule will flow into the _____.

a. efferent arteriole
b. glomerulus
c. loop of Henle
d. proximal convoluted tubule

_____ 4. The collecting duct collects urine products from several different _____.

a. DCTs
b. PCTs
c. renal pyramids
d. ureters

_____ 5. Red blood cells in the glomerulus will enter the _____.

a. Bowman's capsule
b. PCT
c. afferent arteriole
d. none of the above

6. In Figures 20–3 and 20–4, identify the afferent arteriole, collecting duct, distal convoluted tubule, efferent arteriole, glomerular capillaries, loop of Henle, proximal convoluted tubule, and renal corpuscle (glomerular capsule).

Figure 20–3 Diagrammatic View Figure of a Nephron

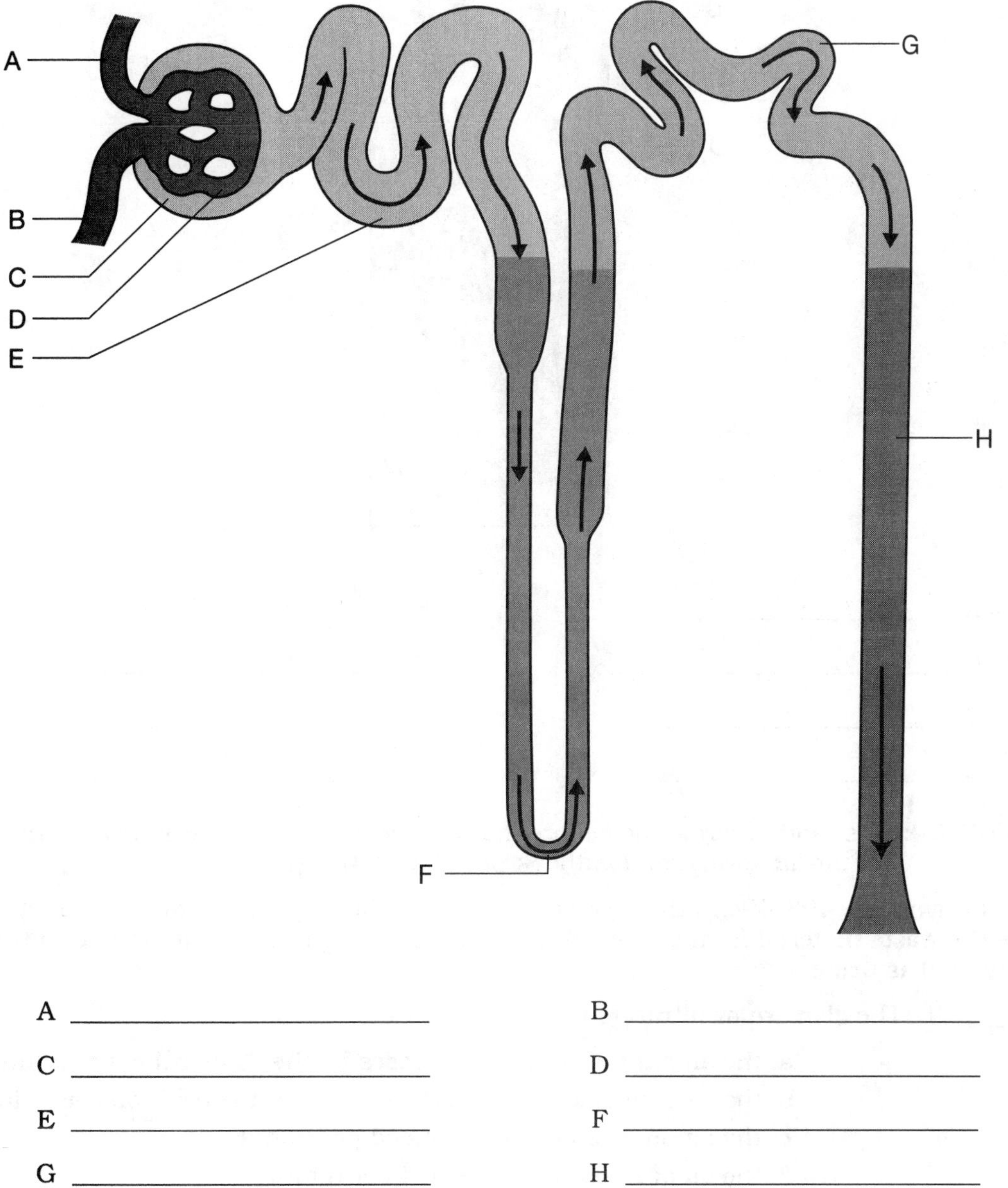

A ____________________ B ____________________

C ____________________ D ____________________

E ____________________ F ____________________

G ____________________ H ____________________

Figure 20–4 Structures of a Typical Nephron

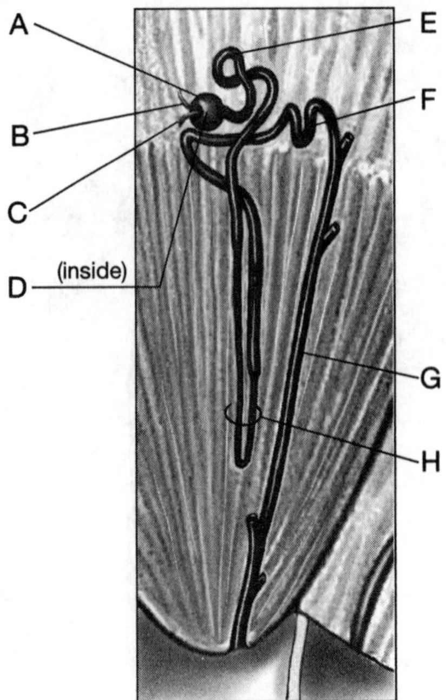

A ______________________ B ______________________

C ______________________ D ______________________

E ______________________ F ______________________

G ______________________ H ______________________

OBJECTIVE 4: List and describe the factors that influence filtration pressure and the rate of filtrate formation (textbook pp. 493–494).

After studying pp. 493–494 in the text, you should be able to explain how the kidneys remove the waste material from the blood and filter it through the nephrons to eventually be excreted as urine.

____ 1. The glomerular filtration rate is ____.

a. the amount of blood that passes by the glomerulus per minute
b. the amount of filtrate material that leaves the nephrons per minute
c. the amount of filtrate produced per minute
d. the amount of urine released per minute

____ 2. Which of the following best defines a filtrate in the body?

a. It is the material entering the ureters.
b. It is the material passing from the glomerulus to the Bowman's capsule.
c. It is the material passing from the glomerulus to the efferent arteriole.
d. It is the material passing from the nephron to the vasa recta.

_____ 3. The glomerular filtration rate (GFR) can be altered by _____.

a. a change in blood pressure
b. a change in hormone secretions
c. a change in the release of renin
d. all the above

_____ 4. How much water is lost by the body every 2 hours?

a. 0.075 L
b. 0.15 L
c. 15.0 L
d. 7.5 L

_____ 5. If blood pressure drops, the GFR will _____, and the release of renin will _____.

a. decrease, decrease
b. decrease, increase
c. increase, decrease
d. increase, increase

OBJECTIVE 5: Describe the changes that occur in the filtrate as it moves through the nephron and exits as urine (textbook pp. 494–496).

After studying pp. 494–496 in the text, you should be able to explain how the filtrate is coverted to urine as it passes from the Bowman's capsule to the PCT to the loop of Henle to the DCT and to the collecting duct.

_____ 1. Which of the following statements best describes the action of the PCT?

a. The PCT reabsorbs water and ions *into* the interstitial fluid area, thereby *increasing* the water and solute concentration in the PCT.
b. The PCT reabsorbs water and ions *into* the interstitial fluid area, thereby *decreasing* the water and solute concentration in the PCT.
c. The PCT reabsorbs water and ions *from* the interstitial fluid area, thereby *increasing* the water and solute concentration in the PCT.
d. The PCT reabsorbs water and ions *from* the interstitial fluid area, thereby *decreasing* the water and solute concentration in the PCT.

_____ 2. Which of the following hormones will not cause reabsorption of water from the nephrons?

a. ADH
b. aldosterone
c. ANP
d. EPO

_____ 3. Suppose 100 mL of water and 100 g of solute entered the PCT. How much of each will enter the DCT?

a. 1 mL of water, 99 g of solute
b. 20 mL of water, 15 g of solute
c. 80 mL of water, 85 g of solute
d. 85 mL of water, 80 g of solute

_____ 4. As the solute concentration increases in the interstitial fluid, _____.

a. the water concentration will decrease in the interstitial areas
b. water will leave the interstitial area and enter the nephron
c. water will leave the nephron and enter the interstitial region
d. both a and c

OBJECTIVE 6: Describe how the kidneys respond to conditions of changing blood pressure (textbook pp. 495–496).

The kidneys do more than excrete urine and prevent dehydration. After studying pp. 495–496 in the text, you should be able to explain how the kidneys are directly involved in helping to maintain blood pressure.

_____ 1.What part of the kidney responds to low blood pressure?

a. the cortex
b. the juxtaglomerular apparatus
c. the medulla
d. the nephron

_____ 2. Which of the following happens when renin enters the circulation?

a. Renin activates angiotensin II. Ultimately, angiotensin II stimulates aldosterone secretion.
b. Renin activates angiotensin II. Ultimately, angiotensin II stimulates ANP secretion.
c. Renin activates angiotensin II. Ultimately, angiotensin II stimulates ADH secretion.
d. Both a and c are correct.

_____ 3. Which of the following statements is true?

a. When blood pressure is high, the kidneys release ANP.
b. When blood pressure is low, the kidneys release renin.
c. When oxygen is low, the kidneys release erythropoietin.
d. All the above are correct.

OBJECTIVE 7: Describe the structures and functions of the ureters, urinary bladder, and urethra (textbook pages 496–497).

After studying pp. 496–497 in the text, you should be able to describe the path of urine as it exits the kidneys and is eliminated from the body.

_____ 1. As urine flows out of the kidneys it travels _____.

a. out the urinary bladder, then through the ureters, then through the urethra
b. through the ureters, then into the urinary bladder, then through the urethra
c. through the urethra, then into the urinary bladder, then through the ureters
d. through the urethra, then through the ureter, then out the urinary bladder

_____ 2. Where are the internal sphincters of the urinary system located?

a. at the distal end of the ureter
b. at the distal end of the urethra
c. inside the urinary bladder
d. none of the above

____ 3. The internal sphincter is located ____ and is under ____ control.

a. at the urethral entrance, involuntary
b. at the urethral entrance, voluntary
c. at the urethral exit, involuntary
d. at the urethral exit, voluntary

____ 4. The external sphincter is located ____ and is under ____ control.

a. at the urethral entrance, involuntary
b. at the urethral entrance, voluntary
c. at the urethral exit, involuntary
d. at the urethral exit, voluntary

____ 5. The ureters collect urine from the ____ and transport it to the ____.

a. collecting duct, urethra
b. renal pelvis, urethra
c. renal pelvis, urinary bladder
d. renal pyramids, urinary bladder

6. Identify the labeled structures in Figure 20–5.

Figure 20–5 The Male Urinary Bladder

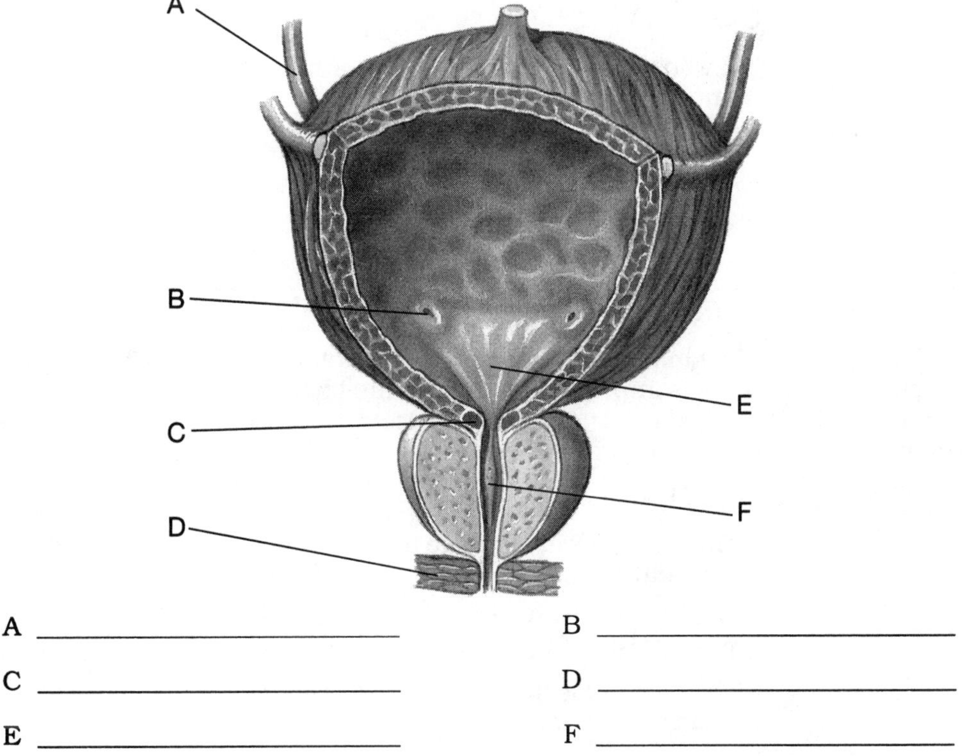

A ____________________ B ____________________

C ____________________ D ____________________

E ____________________ F ____________________

OBJECTIVE 8: Discuss the process of urination and how it is controlled (textbook pp. 497–498).

After studying pp. 497–498 in the text, you should be able to describe how the brain and specialized stretch receptors in the urinary bladder are involved in urination and how the two sphincters help store urine, one under voluntary control and the other under involuntary control.

_____ 1. The first signal that you need to go to the bathroom occurs when the urinary bladder begins to stretch with about _____ mL of urine in it.

a. 10
b. 100
c. 200
d. 500

_____ 2. The involuntary internal sphincter opens when the urinary bladder stretches with about _____ mL of urine in it.

a. 10
b. 100
c. 200
d. 500

_____ 3. The voluntary external sphincter opens when the urinary bladder stretches with about _____ mL of urine in it.

a. 100
b. 200
c. 500
d. It won't open until we're ready for it to open because it's under voluntary control.

4. How is the function of the urethra different in the two sexes?

OBJECTIVE 9: Explain the basic concepts involved in the control of fluid and electrolyte regulation (textbook pp. 498–500).

After studying pp. 498–500 in the text, you should be able to describe why the control of fluid and electrolyte balance is important for maintaining homeostasis in the human body. If the water content of the body changes and the electrolyte balance is disturbed, cellular activities are jeopardized and cells may die.

_____ 1. Where is the ECF located in reference to the ICF?

a. The ECF is located inside the cell, and the ICF is located in the interstitial area.
b. The ECF is located inside the cell, and the ICF is outside the cell.
c. The ECF is located outside the cell, and the ICF is inside the cell.
d. The ECF is located outside the cell, and the ICF is located in the interstitial area.

_____ 2. If the ECF solute concentration rises, water will _____.

a. enter the ICF
b. leave the ICF
c. move into the ICF and back out again to achieve homeostasis
d. not change

_____ 3. As we consume salt, water will _____.

a. cause a rise in the sodium concentration, which will in turn raise blood pressure
b. enter the urinary system and raise the blood pressure
c. leave the ECF and enter the ICF, thus raising blood pressure
d. leave the ICF and enter the ECF, thus raising blood pressure

OBJECTIVE 10: Explain the buffering systems that balance the pH of the intracellular and extracellular fluids (textbook pp. 500–501).

After studying pp. 500–501 in the text, you should be able to describe how buffers resist drastic changes in pH and maintain homeostasis. The body would be out of homeostasis if the pH were to fluctuate beyond the normal range.

_____ 1. Buffers stabilize the pH of body fluids by _____.

a. adding bicarbonate ions to or removing them from the fluid
b. adding hydrogen ions to or removing them from the fluid
c. adding phosphate ions to or removing them from the fluid
d. adding protein to or removing it from the fluid

_____ 2. Hemoglobin acts as a buffer because it _____.

a. adds hydrogen ions to or removes them from the blood
b. adds protein to or removes it from the blood
c. transports oxygen and carbon dioxide
d. all the above

_____ 3. When we exhale, we decrease our blood levels of _____.

a. carbon dioxide and oxygen
b. carbon dioxide and ultimately hydrogen ions
c. oxygen and ultimately hydrogen ions
d. carbon dioxide only

_____ 4. If we were to exhale rapidly and excessively, our blood pH would _____.

a. drop
b. fluctuate
c. not change
d. rise

_____ 5. A rise in carbon dioxide in our blood causes our blood pH to _____.

a. drop
b. fluctuate
c. not change
d. rise

_____ 6. A rise in carbon dioxide in our blood will ultimately cause our breathing rate to _____.

a. decrease
b. fluctuate
c. increase
d. not change

_____ 7. Which part of the kidneys is involved in removing hydrogen ions from or adding them to the blood?

a. the cortex
b. the medulla
c. the nephrons
d. the renal pyramids

_____ 8. Which of the following happens if we hold our breath?

a. Our blood carbon dioxide drops, and the blood pH also drops.
b. Our blood carbon dioxide drops, and the blood pH rises.
c. Our blood carbon dioxide rises, and the blood pH also rises.
d. Our blood carbon dioxide rises, and the blood pH drops.

OBJECTIVE 11: Describe the effects of aging on the urinary system to (textbook p. 501).

After studying p. 501 in the text, you should be able to explain why the need to urinate more frequently and incontinence (the inability to control urination) are quite common in the elderly.

_____ 1. Which of the following affect our kidneys as we age?

a. About 40% of the nephrons quit working.
b. The nephrons become less responsive to ADH.
c. The filtration rate slows down.
d. all the above

_____ 2. As we age, we begin to lose control of the _____.

a. internal urethral sphincter
b. external urethral sphincter
c. prostate gland
d. both a and b

_____ 3. As we age, we have to visit the bathroom more frequently. This is because _____.

a. our nephrons are not responding to ADH
b. we are experiencing incontinence
c. we are lacking ADH
d. we have too much ADH

4. As males age, they may develop prostate swelling. How does this affect their ability to urinate? ____________________________

OBJECTIVE 12: Describe the general symptoms and signs of disorders of the urinary system (textbook pp. 501–503).

The analysis and interpretation of the various signs and symptoms associated with the urinary system helps a physician pinpoint the type of urinary problem the patient is experiencing and determine corrective measures. After studying pp. 501–503 in the text, you should be able to describe the general symptoms and signs of disorders of the urinary system.

_____ 1. Having red-colored urine _____.

a. is a definite sign of hematuria
b. is a definite sign of hemoglobinuria
c. may be an indication of polyuria
d. may be due to certain foods

____ 2. The condition in which a patient urinates very little is called ____.

a. anuria
b. dysuria
c. incontinence
d. oliguria

____ 3. Which of the following is considered a symptom of a possible urinary disorder?

a. cloudy urine
b. dysuria
c. hematuria
d. edema in peripheral tissues

____ 4. Which of the following is a sign of a possible urinary disorder?

a. altered urine color
b. incontinence
c. polyuria
d. pain in the pubic region

____ 5. Incontinence is usually due to ____.

a. problems with the kidney and ureters
b. problems with the kidneys
c. problems with the kidneys, urethra, and urinary bladder
d. problems with the urethra or urinary bladder

OBJECTIVE 13: Give examples of the various disorders of the urinary system (textbook pp. 503–508).

After studying pp. 503–508 in the text, you should be able to describe numerous urinary **system disorders.**

____ 1. Which of the following is not true in the case of inflammation of the kidneys?

a. The blood vessels in the kidneys decrease in size.
b. The glomerular filtration rate decreases.
c. The kidneys swell in size.
d. The nephrons decrease in size.

____ 2. The reflux of urine into the ureter is called ____.

a. chronic pyelonephritis
b. glomerulonephritis
c. nephritis
d. ureteritis

____ 3. Glycosuria is a condition in which ____.

a. blood glucose levels are high but urinary glucose levels are normal
b. blood glucose levels are low but urinary glucose levels are normal
c. blood glucose levels are low but urinary glucose levels are high
d. blood glucose levels are normal but urinary glucose levels are high

_____ 4. An infection that originates elsewhere in the body may produce antigen-antibody complexes. These antigen-antibody complexes may block the filtering apparatus of the nephron, thus reducing filtration. This condition is called _____.

a. anuria
b. glomerulonephritis
c. nephrolithiasis
d. polycystic disease

_____ 5. Patients with chronic renal failure may need to use hemodialysis. In this situation, blood is removed from _____ and is returned to the body via _____.

a. a vein, an artery
b. an artery, a vein
c. the renal vein, a shunt
d. the renal artery, a shunt

OBJECTIVE 14: Give examples of conditions resulting from abnormal renal function (textbook pp. 508–512).

Renal failure can result in other body systems coming out of homeostasis. After studying pp. 508–512 in the text, you should be able to describe some of the conditions that may result from abnormal renal function.

_____ 1. If, due to a urinary disorder, the patient loses an abnormally high amount of water, the body fluids will become _____.

a. hypertonic
b. hypoosmotic
c. hypotonic
d. isotonic

_____ 2. Which of the following may cause an imbalance of electrolytes?

a. excessive water intake
b. excessive water loss
c. excessive loss of ions
d. All of the above

_____ 3. Excessive loss of water may result in _____.

a. hypernatremia, hypokalemia
b. hyponatremia, acidosis
c. hyponatremia, hyperkalemia
d. hyponatremia, hypokalemia

_____ 4. Metabolic acidosis may occur if _____.

a. the kidneys cannot excrete adequate amounts of hydrogen ions
b. we cannot exhale adequate amounts of carbon dioxide
c. the kidneys excrete too many hydrogen ions
d. we exhale too much carbon dioxide

_____ 5. Which of the following is least likely to cause metabolic alkalosis?

a. a large meal
b. elevation of bicarbonate ions
c. excess ketone production
d. frequent vomiting

PART II: CHAPTER-COMPREHENSIVE EXERCISES

Match the terms in column A with the descriptions or statements in column B.

MATCHING I

	(A)	(B)
_____	1. Bowman's capsule	A. Tubes exiting the kidneys and leading to the urinary bladder
_____	2. hilus	B. A single tube exiting the urinary bladder
_____	3. juxtaglomerular apparatus	C. The location of the kidney where blood vessels enter and exit and the ureter exits
_____	4. metabolic acidosis	D. The area of the kidney where the collecting ducts pass through
_____	5. nephrons	E. The main functional unit of the kidneys
_____	6. renal pyramids	F. The first part of the nephron
_____	7. ureters	G. A special area containing cells that release EPO and renin
_____	8. urethra	H. A condition that may occur when excess H^+ are needed by the stomach after a large meal

MATCHING II

	(A)	(B)
_____	1. 10 mL	A. The amount of fluid in the urinary bladder that causes the sphincters to open without our control
_____	2. 180 L	B. The amount of fluid in the urinary bladder that causes us to feel the need to go to the bathroom
_____	3. 200 mL	C. The amount of fluid that enters into the kidneys per day
_____	4. 500 mL	D. The amount of fluid that leaves the nephrons and reenters the circulation
_____	5. 1%	E. The amount of fluid that leaves the nephrons and is on its way out of the body
_____	6. 99%	F. The amount of urine typically left in the urinary bladder after urination
_____	7. cystinuria	G. Excessive loss of the amino acid cystine
_____	8. glomerulonephritis	H. A disorder of the kidneys associated with the accumulation of antigen-antibody complexes
_____	9. urethra	I. Gonorrhea typically causes inflammation of the _____.

CONCEPT MAP

This concept map summarizes and organizes the material in Chapter 20. Use the following terms to complete the mab by filling in the boxes identified by the circled numbers, 1–10.

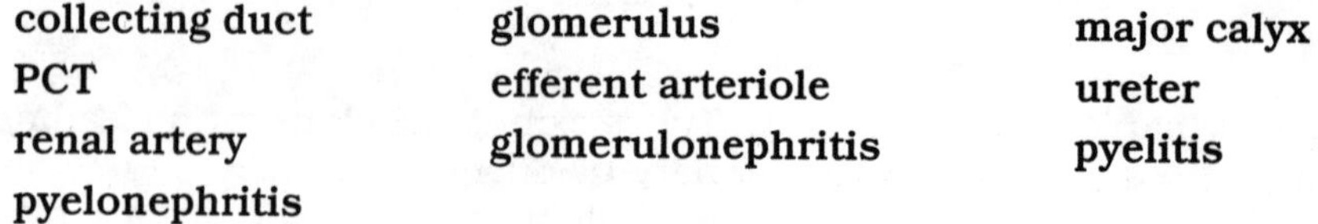

collecting duct
PCT
renal artery
pyelonephritis
glomerulus
efferent arteriole
glomerulonephritis
major calyx
ureter
pyelitis

blood
RENAL CIRCULATION AND FILTRATION
fluid back to the body
in the
①
enters
afferent arteriole
enters
②
blood to
③
blood to
vasa recta
enters the
enters
④
affects
Bowman's capsule
filtrate
⑤ enters
water into
enters
loop of Henle
water into
(descending loop)
ADH and aldosterone
enters
DCT
water into
enters
affect the
⑥
puts water into
urine enters
minor calyx
⑦ enters
renal pelvis
⑧
urinary bladder
urethra
inflammation
⑨
condition
⑩
is a reflux into

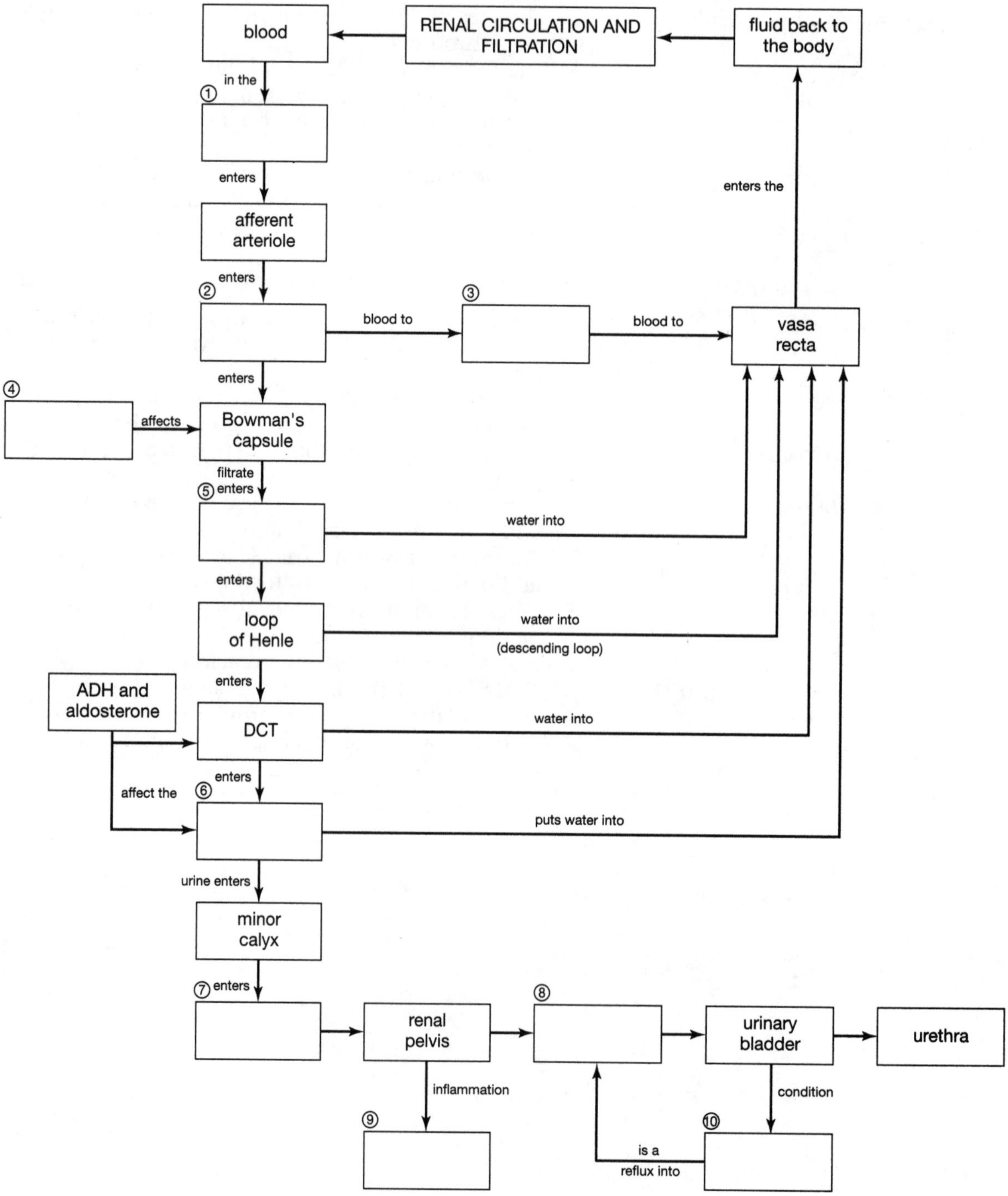

CROSSWORD PUZZLE

The following crossword puzzle reviews the material in Chapter 20. To complete the map, you have to know the answers to the clues given, and you must be able to spell the terms correctly.

ACROSS

4. The area where blood vessels enter and exit the kidneys
5. A tube that leads from the kidneys to the urinary bladder
6. The capillaries inside the Bowman's capsule
9. Elevated levels of sodium ions in the blood
10. This chemical activates angiotensin II
13. This arteriole contains "cleaner" blood because it has been filtered by the glomerulus.
14. At about 500 mL, the _____ sphincter automatically opens.
15. When the prostate gland swells, it constricts the _____, making urination difficult.

DOWN

1. Blood in the urine
2. The _____ ducts pass through the renal pyramid area.
3. Glomerulonephritis is an _____ that occurs at the glomerular capsule (Bowman's capsule).
4. An increase in carbon dioxide will increase the _____ ions in the blood.
7. The arteriole that takes material to the glomerulus
8. An inflammation of the tissues of the urinary bladder
11. The urethral sphincter that is under voluntary control
12. The main functioning unit of the kidney

FILL-IN-THE-BLANK NARRATIVE

Use the following terms to fill in the blanks of this summary of Chapter 20.

minor calyx	**Bowman's**	**DCT**
efferent	**loop of Henle**	**glomerular**
PCT	**afferent**	**vasa recta**
collecting	**blood**	**glomerulonephritis**
filtration		

The kidneys perform many functions, two of which are the elimination of waste and the retention of water to prevent dehydration. Let's follow a drop of water and waste and see what happens to it inside the kidney.

Water and waste in the renal artery enter the kidney and travel through several arterioles leading to the (1) ___________ arteriole. Water and waste enter the (2) ___________ capillaries. Pressure begins to build in these capillaries to the point that some of the water and waste are forced into the first part of the nephron, called the (3) ___________ capsule. The material that does not enter the capsule continues from the glomerular capillaries to the (4) ___________ arteriole. The blood in the efferent arteriole is cleaner than the blood that was in the afferent arteriole. This cleaner blood eventually enters the (5) ___________ and then the renal vein.

The water and waste that have entered the Bowman's capsule continue on through the (6) ___________. Some of the water in the PCT can be reabsorbed into surrounding blood vessels. The rest of the water and waste continue through the descending (7) ___________. As the water descends some of it is reabsorbed into the surrounding blood vessels. As the filtrate travels through the ascending portion sodium ions and chloride ions are reabsorbed.

The filtrate now enters the (8) ___________. Aldosterone targets the DCT, causing it to be permeable to sodium ions. Sodium ions leave the DCT and enter the vasa recta, which becomes hypertonic, causing water to leave the DCT and enter the bloodstream.

ADH also targets the DCT, which puts even more water back into the bloodstream. There is a relatively small amount of water left in the nephron. The remaining filtrate now enters the (9) ___________ duct, which collects urine from several other DCTs. The filtrate, which has been converted into urine, flows out of the collecting duct and into a (10) ___________. The reabsorption of water is complete.

The nephrons are extremely efficient. They remove waste material while recycling about 99 percent of the water into the bloodstream.

One major problem of the urinary system that prevents the nephrons from functioning as they should is an affliction of the glomerular capsule (Bowman's capsule). One such disorder is (11) ___________, in which antigen-antibody complexes may block the glomerular capsule, thus reducing (12) ___________. Another affliction may be a bacterial infection of the glomerular capsule. This infection may create a situation in which erythrocytes may leave the glomerualr capillaries and enter the nephron. Once the erythrocytes enter the nephron, they will appear in the urine. The patient then has (13) ___________ in the urine.

CLINICAL CONCEPT

The following clinical concept applies the information presented in Chapter 20. Following the application is a set of questions to help you understand the concept.

If you want to determine how the pH in the body is affected by the food you eat, ask your instructor for some pH paper. A few minutes before a meal, collect a urine sample and check the pH of the urine by dipping the pH paper into the sample. The normal pH prior to the meal should be about 6.0.

Approximately 30 minutes to 1 hour after you have eaten, check the pH of a new sample of your urine. You probably will find that the pH has changed to 6.5–7.0. How would you explain what has happened?

The answer to this question relates to the chapter on digestion. As food enters your stomach, the stomach stretches. This stretching action causes the release of the hormone gastrin. Gastrin targets specific stomach cells to begin the formation of HCl. In order to digest food, the stomach needs to make HCl. Hydrochloric acid is made from hydrogen ions and chloride ions. Therefore, the more food you have in your stomach, the more HCl your stomach is going to manufacture. To make more HCl, your stomach cells will remove more hydrogen ions from the blood, which will leave fewer hydrogen ions to arrive at the kidneys for excretion. A lower concentration of hydrogen ions in the urine will cause the pH to rise.

1. An increase in hydrogen ion concentration in a solution will do what to the pH?

2. Why does the stomach need to make more HCl?

3. Which is more acidic, a pH of 6.0 or a pH of 6.5?

4. What causes the stomach to begin making HCl?

5. In this scenario, what is happening to the pH of the blood?

THIS CONCLUDES CHAPTER 20 EXERCISES

CHAPTER 21

The Reproductive System

The reproductive system is the only system that is not essential to the life of the individual. Most of the other body systems function to support and maintain the individual, but the reproductive system is specialized to ensure survival, not of the individual, but of the species. The structures and functions of the reproductive system are notably different from those of any other organ system in the human body. The other systems of the body are functional at birth or shortly thereafter; however, the reproductive system does not become functional until it is acted on by hormones during puberty.

Even though major differences exist between the reproductive organs of the male and female, both are primarily involved with propagation of the species and passing genetic material from one generation to another. The reproductive system also produces hormones that control development of secondary sex characteristics. Disorders of the male and female reproductive systems can affect not only the maternal and paternal figures but also the fetus.

The exercises in this chapter will help you review and reinforce your understanding of the structures and mechanics of the remarkable reproductive process, the effects of male and female hormones, the changes that occur during the maturation and aging processes, and disorders of the reproductive system.

Chapter Objectives:

1. Summarize the functions of the human reproductive system and its principal components.
2. Describe the components of the male reproductive system.
3. Describe the process of spermatogenesis.
4. Describe the roles the male reproductive tract and accessory glands play in the maturation and transport of spermatozoa.
5. Describe the hormones involved in sperm production.
6. Describe the components of the female reproductive system.
7. Describe the events of the ovarian and uterine cycles.
8. Describe the hormonal regulation of the ovarian and uterine cycles.
9. Describe the effects of aging on the reproductive system.
10. Describe the general symptoms and signs of male and female reproductive disorders.
11. Give examples of male and female reproductive disorders.

OBJECTIVE 1: Summarize the functions of the human reproductive system and its principal components (textbook pp. 519–520).

After studying pp. 519–520 in the text, you should be able to describe the main function of the male reproductive system, that is, to produce and successfully transport sperm cells out of the male's body, and the main function of the female reproductive system, which is to produce eggs and successfully accept the sperm cell for the purpose of fertilizing the egg, and provide a proper environment for the development of the embryo.

_____ 1. Which of the following is a function of the reproductive system?

a. the perpetuation of the species
b. the survival of the species
c. the production of gametes with 23 chromosomes
d. All the above are functions of the reproductive system.

_____ 2. The reproductive system produces gametes via a process called _____.

a. ejaculation
b. fertilization
c. meiosis
d. none of the above

_____ 3. A fertilized egg contains _____ chromosomes and is called a _____.

a. 23, zygote
b. 23, gamete
c. 46, zygote
d. 46, gamete

_____ 4. The testes produce which of the following?

a. hormones
b. semen
c. sperm
d. all the above

_____ 5. An embryo develops in the _____.

a. ovaries
b. uterine tubes
c. uterus
d. none of the above

OBJECTIVE 2: Describe the components of the male reproductive system (textbook pp. 520–525).

After studying pp. 520–525 in the text, you should be able to trace the pathway of sperm cells through the male reproductive system and identify the reproductive structures along the way.

_____ 1. Which structures inside the testes are responsible for sperm production?

a. epididymis
b. interstitial cells
c. seminiferous tubules
d. sustentacular cells

_____ 2. Which of the following correctly describes the sequence of travel of sperm through the male's body?

a. ductus deferens, epididymis, ejaculatory duct
b. ejaculatory duct, ductus deferens, epididymis
c. epididymis, ductus deferens, ejaculatory duct
d. epididymis, ejaculatory duct, ductus deferens

_____ 3. Which of the following produce(s) fructose for sperm cell survival?

a. bulbourethral gland
b. prostate gland
c. seminal vesicle
d. testes

_____ 4. Which of the following statements about the flow of sperm cells is true?

a. Sperm cells enter the seminal vesicle gland.
b. Sperm cells pass through the prostate gland.
c. Sperm cells enter the bulbourethral gland.
d. All the above are true.

_____ 5. Sperm cells exit the penis by traveling through the _____.

a. corpora cavernosa
b. ductus deferens
c. ejaculatory duct
d. urethra

6. Identify the male structures in Figure 21–1.

Figure 21–1 The Male Reproductive Structures

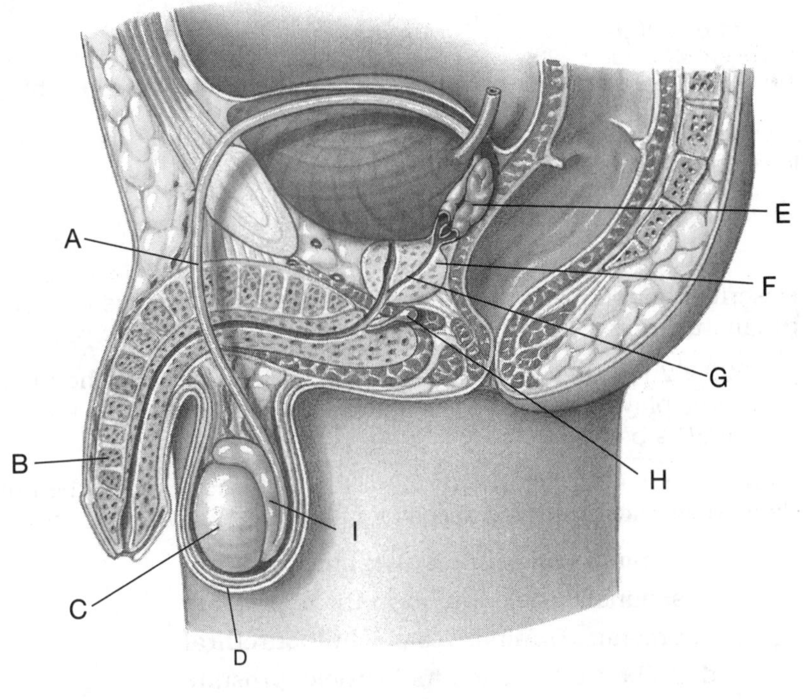

A ______________________ B ______________________

C ______________________ D ______________________

E ______________________ F ______________________

G ______________________ H ______________________

I ______________________

OBJECTIVE 3: Describe the process of spermatogenesis (textbook pp. 521–522).

After studying pp. 521–522 in the text, you should be able to explain the process of spermatogenesis, the formation of sperm.

_____ 1. Spermatogenesis begins in the _____.

a. accessory glands
b. scrotum
c. seminiferous tubules
d. sperm cells

_____ 2. Spermatogenesis ultimately produces _____.

a. male gametes
b. sperm cells
c. sustentacular cells
d. both a and b

_____ 3. Spermatogenesis requires temperatures around _____.

a. 98.6°F
b. 96.6°F
c. 94.6°F
d. 100.6°F

4. Spermatogenesis creates sperm cells with how many chromosomes each? _____

5. Why are the testes located on the outside of the body instead of inside the abdominopelvic region?

OBJECTIVE 4: Describe the roles the male reproductive tract and accessory glands play in the maturation and transport of spermatozoa (textbook pp. 520–525).

After studying pp. 520–525 in the text, you should be able to discuss the major role that the accessory structures play in ensuring the survival of the sperm cells as they travel toward the exit of the male's body.

_____ 1. Which of the following correctly describes the sequence of the flow of sperm cells as they encounter the accessory glands?

a. seminal vesicle, prostate, bulbourethral
b. seminal vesicle, bulbourethral, prostate
c. prostate, seminal vesicle, bulbourethral
d. bulbourethral, seminal vesicle, prostate

_____ 2. Which of the following glands does not produce alkaline semen?

a. seminal vesicle
b. prostate
c. bulbourethral
d. all the above

_____ 3. The failure of which gland may lead to increased urinary tract infection?

a. seminal vesicle
b. prostate
c. bulbourethral
d. These glands are not involved with the urinary system.

_____ 4. Which of the following is involved in the production of sperm?

a. seminal vesicle
b. prostate
c. bulbourethral
d. These glands are not involved in the production of sperm.

_____ 5. Which of the following is involved in the production of semen?

a. seminal vesicle
b. prostate
c. bulbourethral
d. all the above

_____ 6. The major component of semen is (are) _____.

a. sperm cells
b. fructose and alkaline material
c. all the above
d. none of the above

OBJECTIVE 5: Describe the hormones involved in sperm production (textbook p. 525).

After studying p. 525 in the text, you should be able to list the numerous hormones involved with the male reproductive system.

_____ 1. Which of the following hormones initiates sperm production?

a. luteinizing hormone
b. interstitial cell-stimulating hormone
c. gonadotropin-releasing hormone
d. follicle-stimulating hormone

_____ 2. Which of the following hormones is involved in the release of testosterone?

a. luteinizing hormone
b. interstitial cell-stimulating hormone
c. gonadotropin-releasing hormone
d. follicle-stimulating hormone

_____ 3. Testosterone plays a partial role in sperm production because it _____.

a. directly stimulates spermatogenesis
b. causes the production of FSH
c. (at high concentrations) stimulates GnRH to activate the release of FSH
d. (at low concentrations) stimulates GnRH to activate the release of FSH

_____ 4. Which of the following hormones is responsible for the secondary sex characteristics of males?

a. testosterone
b. luteinizing hormone
c. gonadotropin-releasing hormone
d. follicle-stimulating hormone

_____ 5. Which of the following hormones is responsible for the development of some of the male reproductive organs in early development?

a. luteinizing hormone
b. interstitial cell-stimulating hormone
c. gonadotropin-releasing hormone
d. testosterone

OBJECTIVE 6: Describe the components of the female reproductive system (textbook pp. 525–531).

After studying pp. 525–531 and Figure 21-6 of the text, you should be able to list the features of the female reproductive anatomy.

_____ 1. Oogenesis occurs in the _____.

a. uterus
b. uterine tubes
c. ovaries
d. endometrium

_____ 2. The uterine tubes join the _____ to the _____.

a. uterus, vagina
b. ovaries, uterus
c. endometrium, myometrium
d. body of the uterus, cervix

_____ 3. In which part of the uterus does the zygote begin to develop?

a. in the myometrium
b. in the endometrium
c. in the cervix
d. in the body of the uterus

_____ 4. The urethra in the male passes through the penis. The urethra in the female _____.

a. passes posteriorly to the vagina
b. passes through the uterus
c. passes through the vagina
d. passes anteriorly to the vagina

_____ 5. At ovulation, the egg enters the _____.

a. uterus
b. uterine tube
c. ovary
d. cervix of the uterus

6. Identify the female structures in Figure 21–2.

Figure 21–2 The Female Reproductive Structures

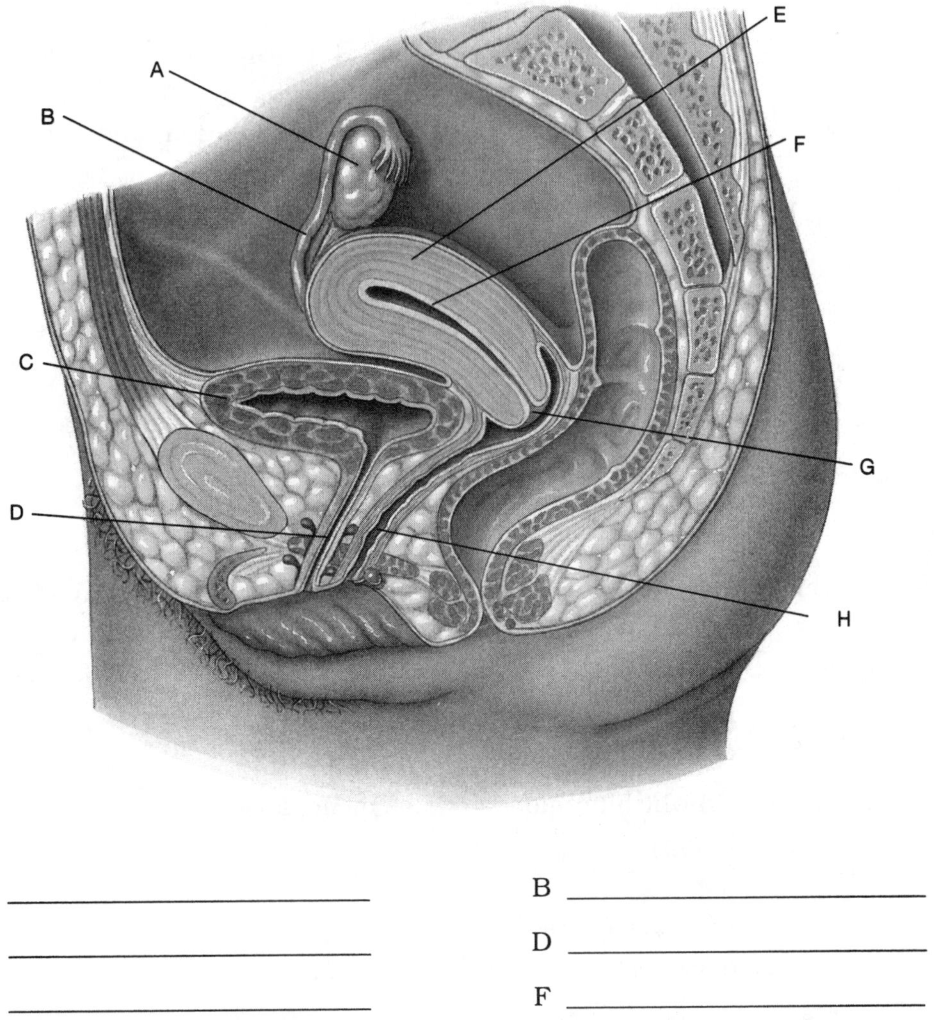

A ______________________ B ______________________

C ______________________ D ______________________

E ______________________ F ______________________

G ______________________ H ______________________

OBJECTIVE 7: Describe the events of the ovarian and uterine cycles (textbook pp. 526–529 and 531–533).

After studying pp. 526–529 and 531–533 in the text, you should be able to explain the menstrual cycle.

_____ 1. A decline in _____ initiates the menstrual cycle.

a. progesterone

b. LH

c. FSH

d. all the above

_____ 2. Which of the following occurs due to a decline in corpus luteum activity?

a. ovulation
b. menstruation
c. menopause
d. Any of the above can occur. It depends on the age of the individual.

_____ 3. Which of the following hormones initiates ovulation, and after ovulation, which hormone is released by the corpus luteum?

a. progesterone, more progesterone
b. LH, progesterone
c. FSH, LH
d. FSH, progesterone

_____ 4. During menstruation, which cells are shed?

a. uterine tube cells
b. myometrial cells
c. endometrial cells
d. cervical cells

_____ 5. Which hormone is involved in the production of a thickened endometrium?

a. progesterone
b. ovarian hormones
c. LH
d. FSH

_____ 6. A decline in which hormone initiates menopause?

a. FSH
b. LH
c. progesterone
d. estrogen

_____ 7. A lack of which hormone is responsible for cessation of egg production?

a. FSH
b. LH
c. progesterone
d. GnRH

OBJECTIVE 8: Describe the hormonal regulation of the ovarian and uterine cycles (textbook pp. 531–533).

After studying pp. 531–533 in the text, you should be able to discuss the hormones of the female reproductive system and their actions that are involved in the survival and development of the embryo.

_____ 1. Which of the following hormones initiates oogenesis?

a. FSH
b. LH
c. progesterone
d. estradiol

_____ 2. Which of the following hormones initiates ovulation?

a. FSH
b. LH
c. progesterone
d. estradiol

_____ 3. After ovulation, the ovarian follicle becomes a corpus luteum. The corpus luteum secretes _____.

a. FSH
b. LH
c. progesterone
d. ovarian hormones

_____ 4. Which of the following hormones causes a thickening of the endometrium to prepare the body for pregnancy?

a. FSH
b. LH
c. progesterone
d. GnRH

_____ 5. During pregnancy, the levels of estrogen and progesterone remain quite high to inhibit the release of _____.

a. FSH
b. LH
c. estradiol
d. progestins

OBJECTIVE 9: Describe the effects of age on the reproductive system (textbook pp. 533–534).

As females age, changes occur in their reproductive system. Their reproductive cycle slows down and may even cease by the age of 45-55. As males age, changes also occur. After studying pp. 533–534 in the text, you should be able to describe the changes that occur in the reproductive systems of males and females as they age.

_____ 1. The uterus and the breasts begin to decrease in size due to the decrease in _____.

a. FSH
b. LH
c. progesterone and estrogen
d. GnRH

_____ 2. A decline in which hormone as females age is linked to the development of osteoporosis?

a. FSH
b. LH
c. estrogen
d. GnRH

_____ 3. A decrease in which hormone may also play a role in causing "hot flashes?"

a. FSH
b. LH
c. estrogen
d. GnRH

4. FSH levels and estrogen levels begin to change in a female's body with age. Which hormone levels, if any, begin to change in men? _____________________

OBJECTIVE 10: Describe the general symptoms and signs of male and female reproductive disorders (textbook pp. 536–538).

The analysis and interpretation of the various signs and symptoms associated with the reproductive systems help a physician pinpoint the type of reproductive problem the patient is experiencing and to determine corrective measures. After studying pp. 536–538 in the text, you should be able to describe some common reproductive disorders.

_____ 1. Which of the following terms is associated with swelling of the scrotum?

a. chylocele
b. hematocele
c. hydrocele
d. all the above

_____ 2. Impotence is a condition characterized by _____.

a. abnormal urination
b. the inability to develop sperm cells
c. the inability to have an ejaculation
d. the inability to have or to maintain an erection

_____ 3. _____ is a technique used to screen for prostatitis.

a. Colposcopy
b. Laparoscopy
c. PAP smear
d. Rectal examination

_____ 4. Amenorrhea is a condition in which _____.

a. the female cannot ovulate
b. the female does not menstruate
c. the male cannot ejaculate
d. the male does not produce sperm

_____ 5. The PAP smear is a technique used to detect _____ and was developed by _____.

a. breast cancer, George Papanicolaou
b. cervical cancer, George Papanicolaou
c. cervical cancer, George Papanosome
d. pelvic aberration, George Papanosome

OBJECTIVE 11: Give examples of male and female reproductive disorders (textbook pp. 538–546).

After studying pp. 538–546 in the text, you should be able to describe the common disorders of the male and female reproductive systems.

____ 1. Orchitis is a disorder characterized by ____.

a. ovarian inflammation
b. testicular inflammation
c. uterus inflammation
d. vaginal inflammation

____ 2. Excessively tight foreskin that results in a narrowed urethral opening is called ____.

a. chancroid
b. oophritis
c. phimosis
d. salpingitis

____ 3. Candidiasis is a ____ infection of the vagina.

a. bacterial
b. protozoan
c. viral
d. yeast

____ 4. Toxic shock syndrome is caused by ____ and infects the ____.

a. bacteria, prostate
b. bacteria, vagina
c. toxic chemical, uterus
d. virus, ovaries

____ 5. Which of the following disorders is not caused by bacteria?

a. chlamydia
b. genital herpes
c. gonorrhea
d. syphilis

____ 6. Ovarian cancer is the most dangerous cancer for women because ____.

a. it cannot be treated
b. it is very difficult to diagnose in early stages
c. it spreads rapidly
d. it is always malignant

____ 7. Fluid retention, pelvic pain, and bloated feelings are ____ of ____.

a. signs, endometriosis
b. signs, PMS
c. symptoms, dysmenorrhea
d. symptoms, PMS

PART II: CHAPTER-COMPREHENSIVE EXERCISES

Match the terms in column A with the descriptions or statements in column B.

MATCHING I

(A)	(B)
_____ 1. acrosome	A. Sex cells that contain only half the original number of chromosomes
_____ 2. chylocele	B. A fertilized egg containing 46 chromosomes
_____ 3. corpora cavernosa	C. An accessory gland that produces acidic semen
_____ 4. gametes	D. The site of sperm production
_____ 5. prostate	E. A cap on the head of a sperm cell that contains enzymes to assist in fertilization
_____ 6. seminiferous tubules	F. Sperm cells travel through this tube to exit the male's body.
_____ 7. urethra	G. Blood flows into this tissue, resulting in an erection.
_____ 8. zygote	H. A swelling of the scrotal cavity due to the accumulation of lymph fluid

MATCHING II

(A)	(B)
_____ 1. candidiasis	A. A ruptured follicle that releases progesterone
_____ 2. corpus luteum	B. Fertilization occurs here.
_____ 3. endometrium	C. The muscular portion of the uterus that is involved in uterine contractions during birth
_____ 4. fimbriae	D. The thickened portion of the uterus that accepts the zygote
_____ 5. fungicide	E. The shedding of excess endometrial tissue if fertilization does not occur
_____ 6. menstruation	F. Structures that help draw the egg into the uterine tube at ovulation
_____ 7. myometrium	G. A condition caused by a fungus that is typically known as a yeast infection.
_____ 8. uterine tube	H. A _____ would be used to treat a yeast infection.

CONCEPT MAP

This concept map summarizes and organizes the ideas discussed in Chapter 21. Use the following terms to complete the map by filling in the boxes identified by the circled numbers, 1–10.

seminiferous tubules	**epididymis**	**prostate**
uterine tube	**seminal vesicle**	**urethra**
progesterone	**orchitis**	**candidiasis**
acidic semen		

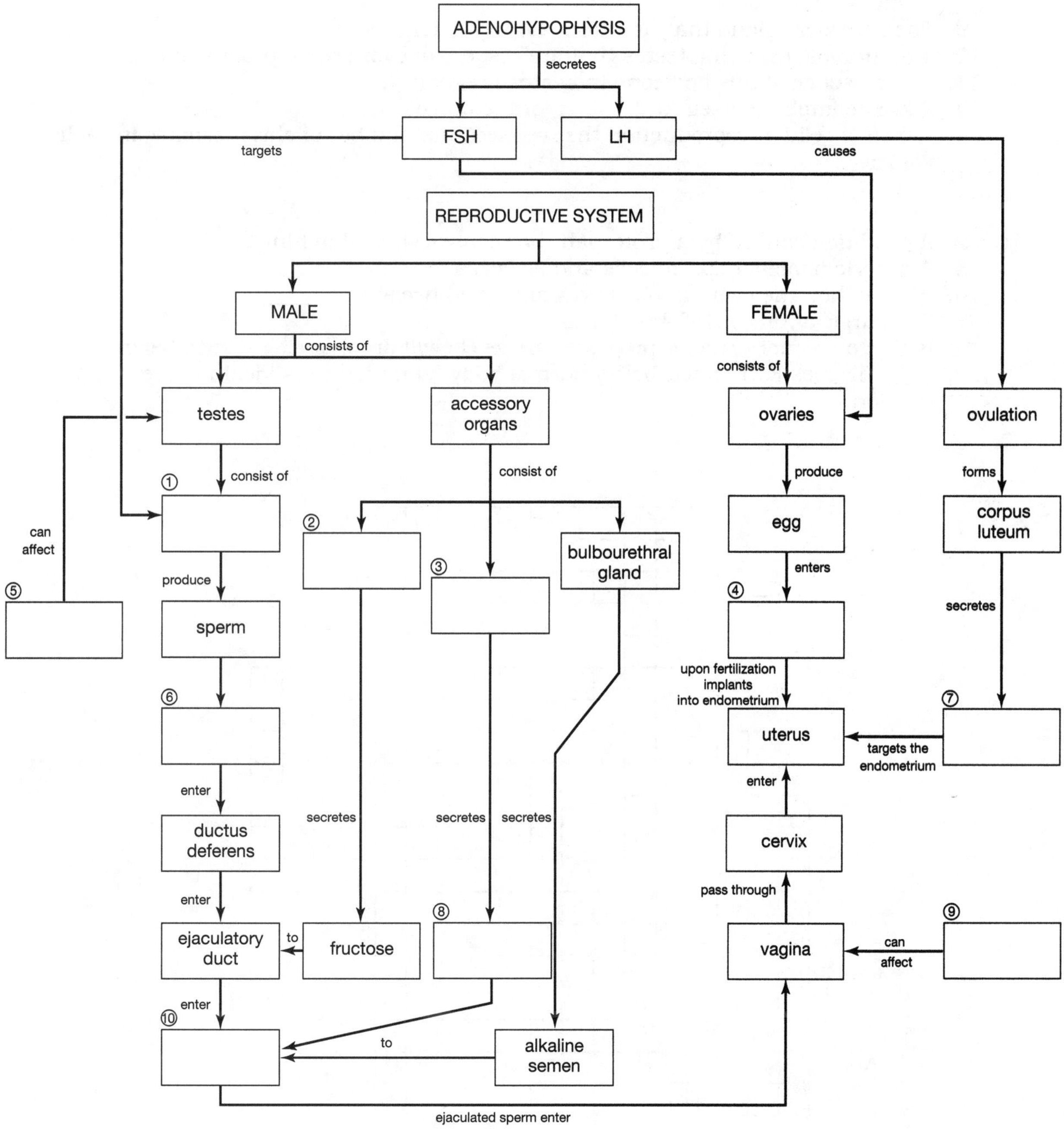

CROSSWORD PUZZLE

The following crossword puzzle reviews the material in Chapter 21. To complete the puzzle, you have to know the answers to the clues given, and you must be able to spell the terms correctly.

ACROSS

1. The luteinizing hormone in males will target the testes and cause them to release _____.
3. The hormone that causes ovulation
4. A condition in which endometrial cells grow in places other than inside the uterus
6. The condition in which the testicles have not descended from the abdominal cavity
9. The accessory gland that produces slightly acidic semen
10. The tubules inside the testes that are responsible for sperm production
11. The presence of this hormone indicates pregnancy.
13. A zygote implants itself in the _____ of the uterus.
14. A type of cellular reproduction that reduces the number of chromosomes in each gamete

DOWN

2. A condition caused by a protozoan that causes vaginal itching
5. A generic name for sperm cells and egg cells
6. At ovulation the ovarian follicle ruptures and becomes a _____.
7. The surgical removal of the uterus
8. If a zygote is not present, progesterone levels will drop, and _____ will begin.
12. _____ degrees Fahrenheit below normal body temperature is ideal for sperm production.

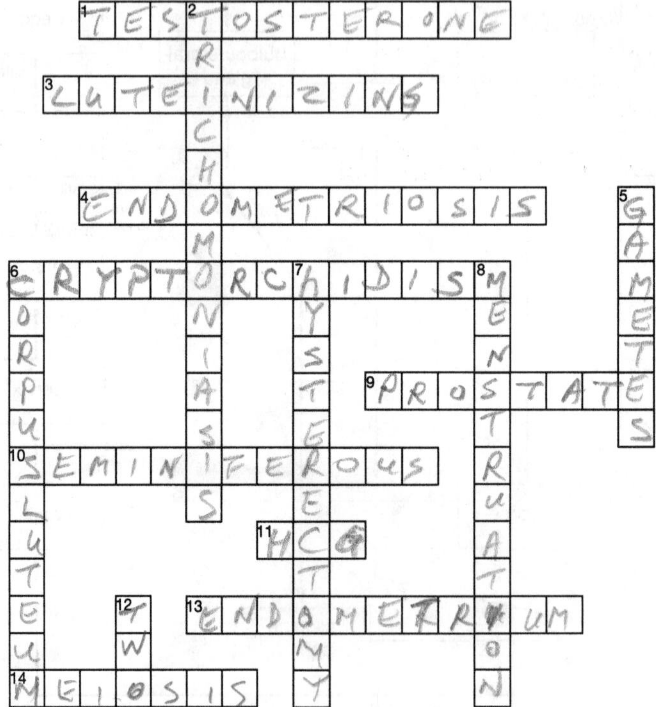

FILL-IN-THE-BLANK NARRATIVE

Use the following terms to fill in the blanks of this summary of Chapter 21.

bulbourethral gland	**seminiferous tubules**	**endometrium**
zygote	**ductus deferens**	**uterine tube**
spermatogenesis	**meiosis**	**ejaculatory duct**
prostate gland	**epididymis**	**corpus luteum**
seminal vesicle gland	**progesterone**	**uterus**
ectopic		

Triggered by the release of FSH from the adenohypophysis, the stem cells in the (1) ___________ undergo (2) ___________, which involves a process called (3) ___________. During this process, four cells, are produced, each of which contains 23 chromosomes. The developing sperm cells continue to mature in the (4) ___________. From there they swim through the (5) ___________. This long tube takes the sperm cells past the (6) ___________, which secretes fructose, a source of energy for the sperm cells. The sperm cells then enter the (7) ___________, which transports the sperm cells through the (8) ___________. This gland releases its slightly acidic semen into the urethra. The sperm cells continue past the (9) ___________, which also secretes alkaline material into the urethra. The sperm cells then exit the male's body and are deposited in the female's body during coitus. The sperm swim through the cervix and enter the uterus.

Triggered by the release of FSH from the adenohypophysis, an egg begins to mature in an ovarian follicle. Luteinizing hormone targets the follicle and causes ovulation. The follicle ruptures, releasing the egg, which then enters the (10) ___________. The ruptured follicle becomes a(n) (11) ___________. This structure releases the hormone (12) ___________, which targets the (13) ___________, causing it to thicken in preparation for the implantation of a fertilized egg.

Some of the sperm cells travel through the uterus and into the uterine tube that contains the egg. Several sperm cells will reach the egg, and one sperm cell will fertilize it. The fertilized egg then becomes a(n) (14) ___________. The cilia in the uterine tube cause the egg to travel down the tube to the uterus. Once inside the uterus, the zygote implants itself in the thickened endometrium, and the process of embryological development begins.

Sometimes a patient develops scar tissue in the uterine tubes. When this happens, the egg cannot travel through the uterine tube to the (15) ___________. The egg implants itself in the uterine tube and begins to develop. This is a "pregnancy outside of the uterus," which is a type of (16) ___________ pregnancy. For successful development, the zygote needs to be implanted in the endometrium.

CLINICAL CONCEPT

The following clinical concept applies the information presented in Chapter 21. Following the application is a set of questions to help you understand the concept.

A couple had been trying for a long time to have a successful pregnancy. The man's sperm count showed that he had about 300 million sperm cells per milliliter of ejaculate. He would be considered sterile if he had fewer than 20 million sperm cells per milliliter of ejaculate. Hormone tests on the female checked her levels of progesterone, FSH, and LH. Results showed that the patient's corpus luteum was failing to produce adequate amounts of progesterone. Without progesterone, the uterus cannot prepare itself for the zygote. One way to correct this situation is to give the patient regular injections of progesterone to raise the level high enough to maintain pregnancy.

1. What do we mean when we say the uterus cannot prepare itself for the zygote?

2. What part of the uterus is involved in preparing itself for the zygote?

3. Where does the corpus luteum originate?

4. How are FSH and LH involved in reproduction?

5. Why does it take so many sperm cells to fertilize one egg?

THIS CONCLUDES CHAPTER 21 EXERCISES

CHAPTER

22

Development and Inheritance

hapter Objectives:

- Describe the process of fertilization.
- List the three prenatal periods and describe the major events of each period.
- List the three primary germ layers and their roles in forming major body systems.
- Describe the roles of the different membranes of the embryo.
- Describe the adjustments of the mother's organ systems in response to the developing embryo.
- Discuss the events that occur during labor and delivery.
- Describe the major stages of life after delivery.
- Describe the basic patterns of inheritance of human traits.
- Describe examples of disruptive factors and genetic disorders in development.
- Describe examples of problems associated with pregnancy.

There are numerous changes that occur in a woman's body during pregnancy. A partial list includes the following:

1. Hormone levels change.
2. The size of the uterus changes.
3. The size of the urinary bladder changes.
4. Body fluid volume changes.

All these changes are designed to support and maintain the development of the embryo.

In previous chapters we learned that the various organ systems function together to help the body maintain homeostasis. During pregnancy, the organ systems not only function together to maintain homeostasis for the pregnant woman but they must now also function together to maintain homeostasis of the embryo.

After reading and studying Chapter 22, you will have a better understanding of the development of a single cell that, if fertilized, will develop into a complex organism consisting of several trillion cells organized into tissues, organs, and organ systems all working together in the body to maintain homeostasis.

The human body does everything it can to maintain homeostasis not only for itself but also for the developing fetus. Unfortunately, there are numerous disorders that can upset the delicate balance of homeostasis.

The latter part of Chapter 22 discusses some disorders and the effects they have on the homeostasis of the developing child.

PART I: OBJECTIVE-BASED QUESTIONS

OBJECTIVE 1: Describe the process of fertilization (textbook pp. 551–552).

After studying pp. 551–552 in the text, you should be able to explain the requirements for successful fertilization, the fusion of a sperm cell and an egg cell.

_____ 1. The fusion of a sperm cell with an egg cell produces a(n) _____ containing _____ chromosomes.

a. embryo, 46
b. gamete, 23
c. zygote, 23
d. zygote, 46

_____ 2. Fertilization typically occurs in the _____.

a. uterus
b. vagina
c. distal 1/3 of the uterine tube in reference to the egg pathway
d. proximal 1/3 of the uterine tube in reference to the egg pathway

_____ 3. Only one sperm cell is needed to fertilize an egg, but sterility may result if the sperm count is less than _____.

a. 10,000/mL
b. 100/mL
c. 20 million/mL
d. 200 million/mL

4. Why are large numbers of sperm cells needed to fertilize an egg if only one will actually fertilize it?____

OBJECTIVE 2: List the three prenatal periods and describe the major events of each period (textbook pp. 552–557).

After studying pp. 552–557 in the text, you should be able to describe the events that occur during each of the three prenatal periods of pregnancy.

_____ 1. The gestation period for humans is _____.

a. 3 months
b. 9 months
c. one trimester
d. none of the above

_____ 2. The components for all the major organ systems appear during _____.

a. the first trimester
b. implantation
c. the second trimester
d. the third trimester

_____ 3. The development of most of the major organ systems is completed by _____.

a. 6 months of gestation
b. the second trimester
c. the first trimester
d. both a and b

_____ 4. Implantation refers to the implantation of the _____ into the _____.

a. embryo, endometrium
b. fetus, myometrium
c. zygote, endometrium
d. zygote, myometrium

_____ 5. Which of the following events occurs during the first trimester?

a. formation of embryonic membranes
b. formation of the placenta
c. implantation
d. all the above

_____ 6. If the corpus luteum stops producing progesterone, the endometrial lining will be lost. To prevent this from happening during pregnancy (especially during the first trimester), _____ targets the corpus luteum to maintain its function.

a. FSH
b. hCG
c. LH
d. progesterone

OBJECTIVE 3: List the three primary germ layers and their roles in forming major body systems (textbook p. 553).

After studying p. 553 in the text, you should be able to describe the three primary germ layers of embryonic tissue that develop into the various organs and tissues of the body.

_____ 1. Germ layer formation occurs during _____.

a. the first trimester
b. the second trimester
c. the third trimester
d. fertilization

_____ 2. Which germ layer is involved in the formation of skin?

a. the ectoderm
b. the endoderm
c. the mesoderm
d. all the above

_____ 3. Which germ layer is involved in the formation of muscle tissue?

a. the ectoderm
b. the endoderm
c. the mesoderm
d. all the above

_____ 4. Which germ layer is involved in the formation of the digestive system?

a. the ectoderm
b. the endoderm
c. the mesoderm
d. all the above

OBJECTIVE 4: Describe the roles of the different membranes of the embryo (textbook p. 554).

After studying p. 554 in the text, you should be able to describe the four major membranous structures associated with the developing embryo and the special functions of each.

_____ 1. Which of the following extraembryonic membranes is involved in blood formation?

a. allantois
b. amnion
c. chorion
d. yolk sac

_____ 2. The fluid within this membranous structure surrounds and cushions the developing embryo.

a. allantois
b. amnion
c. chorion
d. yolk sac

_____ 3. Which of the following membranous structures is involved with the embryo's early urinary system?

a. allantois
b. amnion
c. chorion
d. yolk sac

_____ 4. Which of the following structures will begin to develop the placenta?

a. allantois
b. amnion
c. chorion
d. yolk sac

_____ 5. The chorion eventually forms the placenta. The placenta secretes which of the following hormones?

a. hCG
b. progesterone
c. relaxin
d. all the above

6. In Figure 22–1, identify the developing embryo, amnion, allantois, and yolk sac.

Figure 22–1 Embryonic Membranes

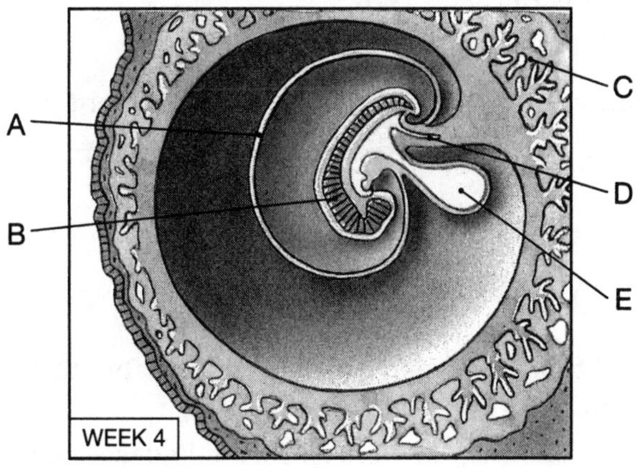

A ____________________ B ____________________

C ____________________ D ____________________

E ____________________

OBJECTIVE 5: Describe the adjustments of the mother's organ systems in response to the developing embryo (textbook p. 557).

After studying p. 557 in the text, you should be able to describe the numerous changes that occur in the mother's organ systems during pregnancy.

_____ 1. By the end of gestation, the uterus will have increased to _____ its original size.

a. two times
b. three times
c. four times
d. five times

_____ 2. What hormone prevents uterine contraction during the early stages of pregnancy?

a. FSH
b. hCG
c. LH
d. progesterone

_____ 3. As the uterus expands anteriorly and inferiorly, it causes the _____ to decrease in size. As the uterus expands laterally, it constricts the size of the _____.

a. small intestine, large intestine
b. stomach, kidneys
c. stomach, large intestine
d. urinary bladder, large intestine

_____ 4. Breathing is difficult during pregnancy because the uterus expands _____.

a. anteriorly
b. posteriorly
c. superiorly
d. inferiorly

5. Explain why pregnant women need to urinate more frequently and experience bouts of constipation. ______________________________

OBJECTIVE 6: Discuss the events that occur during labor and delivery (textbook pp. 557–559).

After studying pp. 557–559 in the text, you should be able to describe the numerous changes that occur during labor and delivery.

_____ 1. Which of the following sequences correctly describes labor and delivery?

a. dilation, expulsion, placental
b. dilation, placental, expulsion
c. expulsion, dilation, placental
d. placental, dilation, expulsion

_____ 2. The dilation stage involves dilation of the _____.

a. cervix
b. uterus
c. vagina
d. none of the above

_____ 3. What happens when the "water breaks?"

a. The amnion sac bursts and releases amniotic fluid.
b. The hymen membrane bursts and releases amniotic fluid.
c. The membranes involved in water retention burst.
d. none of the above

_____ 4. What is "afterbirth?"

a. the expulsion of the placenta
b. the expulsion of the endometrium
c. the expulsion of the amniotic sac
d. the expulsion of fetal waste

OBJECTIVE 7: Describe the major stages of life after delivery (textbook pp. 559–561).

After studying pp. 559–561 in the text, you should be able to describe the stages of development of the child after delivery.

_____ 1. During pregnancy, which of the following systems was (were) controlled by the mother for the baby?

a. immune
b. endocrine
c. respiratory
d. all the above

_____ 2. After the child is born, the mother typically nurses the child. The hormone _____ is required to release the milk from the mammary glands.

a. LH
b. oxytocin
c. progesterone
d. prolactin

_____ 3. Which of the following occurs at the onset of puberty?

a. a decrease in FSH and LH
b. an increase in FSH and LH
c. a decrease in GnRH
d. none of the above

4. Why is colostrum important? ____

OBJECTIVE 8: Describe the basic patterns of inheritance of human traits (textbook pp. 561–564).

After studying pp. 561–564 in the text, you should be able to describe how union of the male gamete (sperm cell) with the female gamete (egg cell) produces a zygote with 46 chromosomes and how the offspring receive their genetic traits from the parents.

_____ 1. If one of the chromosomes in a set has a specific gene that is expressed, that gene is considered to be a _____ gene.

a. dominant
b. recessive
c. sex-linked
d. none of the above

_____ 2. If both genes located on both chromosomes of a pair are required to express a genetic trait, that gene is considered to be a _____ gene.

a. dominant
b. polygenic
c. recessive
d. sex-linked

____ 3. The 46 human chromosomes form 23 pairs. Twenty-two of the pairs are identical, or nearly so, with each other. However, the twenty-third pair of chromosomes are not necessary identical. These chromosomes are known as the ____.

a. gametes
b. polygenic
c. sex chromosomes
d. sex-linked

____ 4. Half of all the sperm cells produced contain ____ [sex chromosome(s)], and the other half contain ____ [sex chromosome(s)]. Each egg cell contains numerous chromosomes and ____ [sex chromosome(s)].

a. two Xs, two Ys, one X
b. one Y, one X, two Xs (XX)
c. two Xs (XX), two Ys (YY), two Xs (XX)
d. one X, one Y, one X

____ 5. The sex of the child is determined by ____ from the mother and ____ from the father.

a. an X or a Y chromosome, a Y chromosome
b. the X chromosome, another X chromosome
c. the X chromosome, a Y chromosome
d. the X chromosome, an X or a Y chromosome

____ 6. Which of the following statements about color blindness is true?

a. The father cannot be a carrier for the color-blind trait.
b. Women can be carriers for the color-blind trait.
c. Women can be color-blind.
d. all the above

____ 7. If the father is color-blind, and the mother is a carrier, which of the following is true?

a. Only the male offspring will be color-blind.
b. Only the female offspring will be color-blind.
c. There is a chance that both male and female offspring will be color- blind.
d. None of the above is true.

____ 8. Suppose the father is not color-blind but the mother is a carrier, which of the following is true?

a. Only the male offspring have the possibility of being color-blind.
b. Only the female offspring have the possibility of being color-blind.
c. Both the female offspring and the male offspring have the possibility of being color-blind.
d. None of the above is true.

____ 9. Which of the following is/are genetic traits carried on only the X chromosomes?

a. color blindness
b. hemophilia
c. muscular dystrophy
d. all the above

OBJECTIVE 9: Describe examples of disruptive factors and genetic disorders in development (textbook pp. 564–567).

Fortunately, the majority of the time, the development of the fetus occurs normally. But occasionally, genetic disorders do occur. After studying pp. 564–567 in the text, you should be able to describe some of the common developmental errors.

____ 1. Which of the following is a teratogen?

a. alcohol
b. syphilis
c. thalidomide
d. all the above

____ 2. Down syndrome is a condition in which the patient ____.

a. has an extra 21st chromosome
b. is missing one of the chromosomes of pair number 21
c. is missing chromosome set number 21 completely
d. None of the above is correct.

____ 3. Which of the following genetic disorders occurs most frequently?

a. Down syndrome
b. Klinefelter syndrome
c. Turner syndrome
d. All the above occur with about the same frequency.

____ 4. Which of the following syndromes is characterized by the lack of one of the two sex chromosomes?

a. Down syndrome
b. fetal alcohol syndrome
c. Klinefelter syndrome
d. Turner syndrome

____ 5. Implantation that occurs someplace other than the endometrial lining is called ____.

a. eclampsia
b. ectopic pregnancy
c. endometriosis
d. in-vivo pregnancy

____ 6. Which of the following is the most common type of ectopic pregnancy?

a. abdominal
b. external uterus
c. peritoneal
d. tubal

_____ 7. In some pregnancies, the placenta forms over the opening of the cervix. This condition is called _____.

a. abruptio placentae
b. eclampsia
c. hydatidiform formation
d. placenta previa

OBJECTIVE 10: Describe examples of problems associated with pregnancy (textbook pp. 567–571).

After studying pp. 567–571 in the text, you should be able to describe some of the problems that may develop during the course of pregnancy.

_____ 1. Preeclampsia is a condition in which _____.

a. fetal blood pressure is decreased
b. fetal blood pressure is increased
c. maternal blood pressure reaches high levels
d. maternal blood pressure reaches low levels

_____ 2. Which of the following describes the normal positioning of the fetus prior to birth?

a. The fetus's face is turned toward the pubis, and the head is closest to the cervix.
b. The fetus's face is turned toward the pubis, and the head is closest to the buttocks.
c. The fetus's face is turned toward the sacrum, and the head is closest to the cervix.
d. The fetus's face is turned toward the sacrum, and the head is closest to the buttocks.

PART II: CHAPTER-COMPREHENSIVE EXERCISES

Match the terms in column A with the descriptions or statements in column B.

MATCHING I

	(A)	(B)
_____	1. Apgar	A. A developing child up to 2 months after fertilization is called a _____.
_____	2. allantois	B. This occurs during the first trimester.
_____	3. Down syndrome	C. The hormone that keeps the corpus luteum functioning
_____	4. embryo	D. The membranous structure that becomes a part of the umbilical cord
_____	5. fraternal	E. The hormone that causes uterine contractions milk release
_____	6. human chorionic gonadotropin	F. The hormone that causes milk production
_____	7. identical	G. When one sperm cell fertilizes one egg cell, and the egg "splits," _____ twins form.
_____	8. implantation	H. When two eggs are released, and a separate sperm fertilizes each egg, _____ twins form.
_____	9. Klinefelter syndrome	I. A rating scale that predicts the survivability of the newborn child and also assesses any neurological disorder
_____	10. oxytocin	J. A condition having 47 chromosomes, specifically an extra 21st chromosome
_____	11. prolactin	K. A condition having 45 chromosomes, specifically a lack of one chromosome at the 23rd set
_____	12. Turner syndrome	L. A condition of having 47 chromosomes, specifically an extra X chromosome at the 23rd set

CONCEPT MAP

This concept map summarizes the information in Chapter 22. Use the following terms to complete the map by filling in the boxes identified by the circled numbers, 1–8.

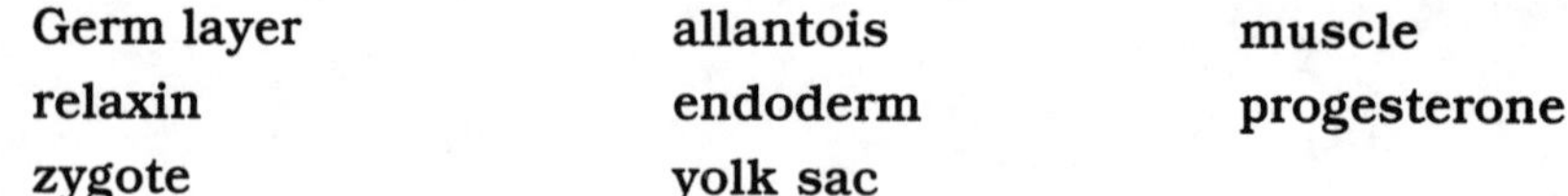

Germ layer	**allantois**	**muscle**
relaxin	**endoderm**	**progesterone**
zygote	**yolk sac**	

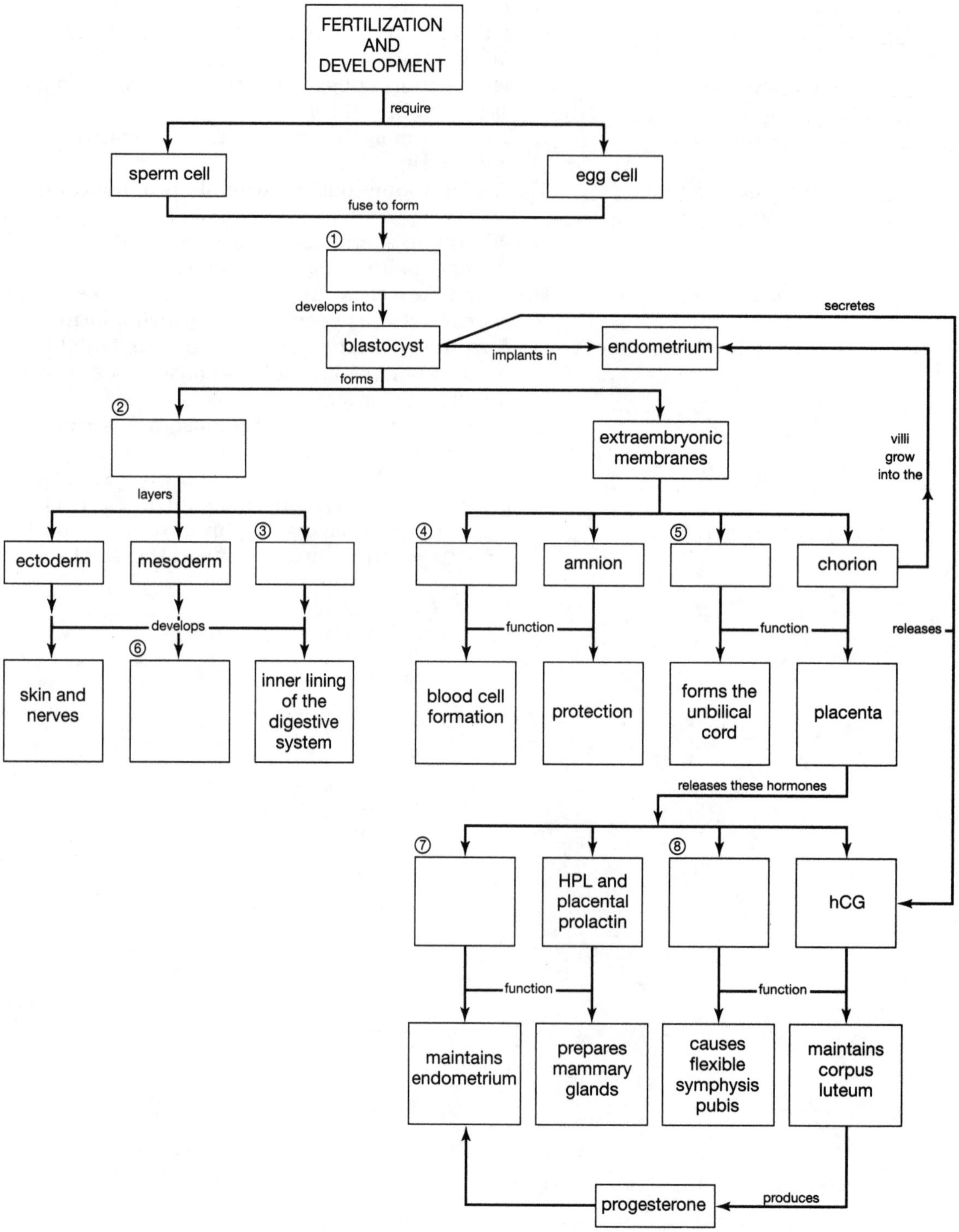

FERTILIZATION AND DEVELOPMENT

require

sperm cell

egg cell

fuse to form

① [ ]

develops into

blastocyst

implants in

endometrium

secretes

forms

② [ ]

extraembryonic membranes

villi grow into the

layers

ectoderm

mesoderm

③ [ ]

④ [ ]

amnion

⑤ [ ]

chorion

develops

function

function

releases

skin and nerves

⑥ [ ]

inner lining of the digestive system

blood cell formation

protection

forms the unbilical cord

placenta

releases these hormones

⑦ [ ]

HPL and placental prolactin

⑧ [ ]

hCG

function

function

maintains endometrium

prepares mammary glands

causes flexible symphysis pubis

maintains corpus luteum

progesterone

produces

CROSSWORD PUZZLE

The following crossword puzzle reviews the material in Chapter 22. To complete the puzzle, you have to know the answers to the clues given, and you must be able to spell the terms correctly.

ACROSS

1. Anything that can alter chromosomes or DNA and result in abnormal fetal development
6. The gestation period is divided into _____.
9. The expulsion of the _____ is referred to as the "afterbirth."
10. The membranous structure that eventually becomes a part of the umbilical cord
12. The membranous structure that forms the placenta
13. The hormone that causes milk release

DOWN

2. The process of removing amnion fluid to examine for chromosome aberrations
3. A female who has a genetic trait that is not expressed is called a _____ of that genetic trait.
4. A fertilized egg
5. The hormone that causes milk production
7. Development of excessively high blood pressure during pregnancy leading to possible convulsions
8. Human _____ gonadotropin hormone targets the corpus luteum.
11. An abnormal positioning of the child at the time of birth

FILL-IN-THE-BLANK-NARRATIVE

Use the following terms to fill in the blanks of this summary of Chapter 22.

amnion	**dilation**	**ectoderm**
endoderm	**expulsion**	**mesoderm**
oxytocin	**placental**	**yolk**
gestation	**oxygen**	**teratogen**
fetal	**gonorrhea**	**syphilis**

Before and during implantation, the zygote releases human chorionic gonadotropin. This hormone targets the corpus luteum and causes it to continue releasing progesterone. Progesterone maintains a thick endometrium for continued development of the zygote. The zygote differentiates and begins to form germ layers that further differentiate to form organs. Muscle tissues are formed by the (1) ___________, the digestive organs are formed by the (2) ___________, and the skin is formed by the (3) ___________.

The embryo is protected by being enclosed in a(n) (4) ___________ filled with amniotic fluid. Blood formation for the embryo begins in the (5) ___________ sac.

As the embryo grows, the mother's uterus increases in size, and her body fluids increase in volume. At the end of the (6) ___________ period, the birthing process begins. The fetus puts internal pressure on the uterus, which signals the neurohypophysis to release (7) ___________. This hormone targets the myometrium of the uterus, thereby causing uterine contractions. The amniotic sac breaks, releasing amniotic fluid, and (8) ___________ of the cervix begins. The cervix eventually dilates to approximately 10 cm, the baby's head appears, (9) ___________ occurs, and a newborn child enters the world.

The final stage of birth is the expulsion of the placenta, which is called the (10) ___________ stage. If all its organ systems and homeostatic mechanisms function properly, the newborn child will grow and develop into a healthy individual.

Unfortunately, numerous substances can cross the placenta barrier to enter the developing fetus. If the substance is considered to be a (11) ___________, the DNA of the fetus may be altered, thus resulting in a developmental abnormality. Chemicals such as carbon monoxide from cigarettes will reduce the maternal (12) ___________ supply. If the maternal oxygen supply is reduced, the (13) ___________ oxygen supply will also be reduced. Bacterial infections that cause (14) ___________ can be passed to the child as it passes through the birth canal. Bacterial infections such as (15) ___________ can pass through the placental barrier and infect the fetus.

CLINCIAL CONCEPTS

The following clinical concept applies the information in Chapter 22. Following the application is a set of questions to help you understand the concept.

Many changes occur in a woman's body during and after pregnancy. Sometimes, new expectant mothers are not aware of what these changes might be.

Mai Ling is pregnant. She is aware that changes will occur, but she is not sure what those changes will be. One of the first things she notices is that she has to urinate more frequently than before. Before pregnancy, she would feel the need go to the bathroom every 2 1/2 to 3 hours. Now it seems she has to go to the bathroom every hour, but she urinates very little each time. Mai Ling also notices that she seems to be constipated quite often.

As the developing embryo gets larger, Mai Ling notices she has difficulty breathing. Her breaths are quite shallow, and sometimes it even hurts a little when she has to take a deep breath.

After the birth of her child, Mai Ling begins to nurse it. While she is nursing she experiences uterine contractions.

1. Why does Mai Ling have to urinate so often?

2. Mai Ling gets the urge to go to the bathroom but when she does, she urinates very little. Why is this so?

3. What accounts for the frequent constipation?

4. Uterine contractions are necessary for the delivery of the child. What hormone is involved in causing uterine contractions?

5. While nursing the child, Mai Ling notices more uterine contractions. Why are there still uterine contractions after birth?

6. Why is it difficult for Mai Ling to breathe normally?

THIS CONCLUDES CHAPTER 22 EXERCISES

CHAPTER 1: An Introduction to Anatomy and Physiology

Part I: Objective-Based Questions

OBJECTIVE 1—p. 2

1. a
2. d
3. a
4. d
5. a
6. b

OBJECTIVE 2—pp. 3–4

1. b
2. b
3. a
4. a
5. b
6. **Figure 1-1**
A. organism
B. organ system
C. organ
D. cellular
E. tissue
F. molecular

OBJECTIVE 3—p. 5

1. b
2. a
3. b
4. d
5. stimulus

OBJECTIVE 4—pp. 5–6

1. b
2. a
3. a
4. The temperature rises in an effort to kill bacteria. The temperature will continue to rise until the bacteria have have died, then will return to normal.
5. Negative feedback
6. **Figure 1-2**
A. A and C
B. B and D
C. Negative feedback

OBJECTIVE 5—pp. 7–10

1. d
2. b
3. d
4 distal
5. standing with the palms facing anteriorly
6. frontal
7. **Figure 1-3**
A. axillary
B. brachial
C. antebrachial
D. crural
E. antecubital
F. inguinal
G. lumbar
H. popliteal
I. olecranal
8. **Figure 1-4**
A. left upper quadrant
B. right upper quadrant
C. left upper quadrant
D. right lower quadrant
9. **Figure 1-5**
A. right hypochondriac
B. right lumbar
C. right iliac
D. epigastric
E. umbilical
F. hypogastric
G. left hypochondriac
H. left lumbar
I. left iliac
10. **Figure 1-6**
A. frontal plane
B. transverse plane
C. sagittal plane

OBJECTIVE 6—pp. 11–12

1. a
2. d
3. c
4. a
5. The diaphragm muscle
6. **Figure 1-7**
A. cranial cavity
B. spinal cavity
C. thoracic cavity
D. abdominal cavity
E. pelvic cavity
F. abdominopelvic cavity
G. pleural cavity
H. pericardial cavity
I. abdominal cavity

OBJECTIVE 7—p. 13

1. b
2. c
3. visceral
4. parietal
5. visceral
6. The visceral membranes surround an organ or several organs, and the parietal membranes line the cavity containing the organs.

OBJECTIVE 8—pp. 13–14

1. c
2. d
3. a
4. c
5. c

Part II: Chapter-Comprehensive Exercises

MATCHING: PART I

1. j
2. a
3. k
4. b
5. i
6. h
7. g
8. d
9. e
10. c
11. f

MATCHING: PART II

1. k
2. j
3. a
4. d
5. e
6. c
7. g
8. b
9. f
10. i
11. h

MATCHING: PART III

1. d
2. b
3. h
4. a
5. e
6. g
7. j
8. i
9. c
10. k
11. f

CONCEPT MAP I

1. cranial cavity
2. spinal cavity
3. abdominopelvic cavity
4. pleural cavity
5. heart
6. abdominal cavity
7. pelvic cavity

CONCEPT MAP II

1. molecules
2. unicellular organism
3. tissues
4. multicellular organisms

CROSSWORD PUZZLE

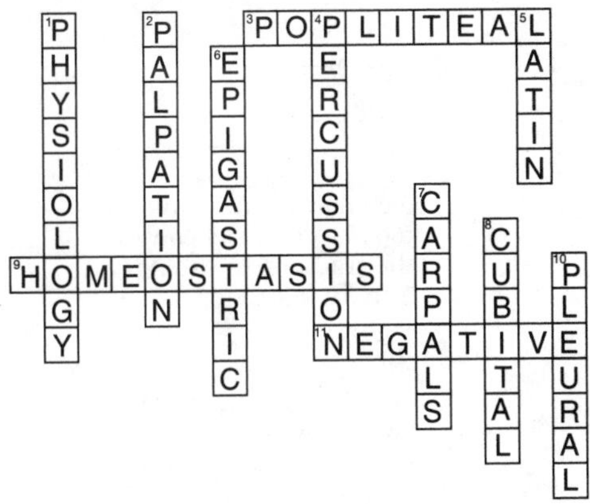

FILL-IN-THE BLANK NARRATIVE

1. Latin
2. confusion
3. popliteal
4. hypogastric
5. urinary bladder
6. auscultation
7. percussion
8. density
9. organ
10. tissues
11. homeostasis
12. positive
13. negative
14. stimuli

CLINICAL CONCEPTS

1. The anatomical position is standing with the palms facing forward (anteriorly).
2. When we are referring to the right side of the body, we are referring to the patient's right side.
3. The right inguinal region (right iliac region) contains the appendix. This patient may have appendicitis.
4. Palpate means to feel a body region.
5. Nurse Abby was not using correct anatomical terminology. Confusion can result between two parties if correct terminology is not used.

CHAPTER 2: Chemistry and the Human Body

Part I: Objective-Based Questions

OBJECTIVE 1—pp. 22–23

1. a
2. b
3. c
4. b
5. b
6. **Figure 2-1**

A. 2, nucleus
B. 2, nucleus
C. 2, electron shell

7.

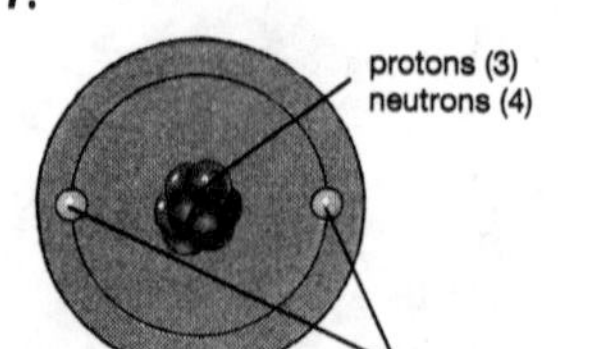

8. **Figure 2–2**

A. Hydrogen-1 does not have any neutrons.
B. Hydrogen-2 has one neutron.
C. Hydrogen-3 has two neutrons.

8.(continued)

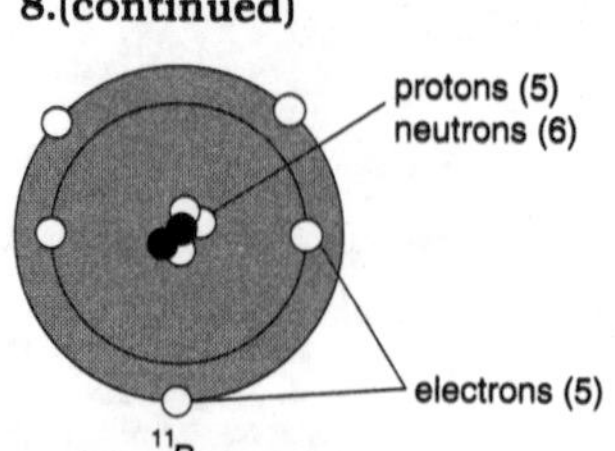

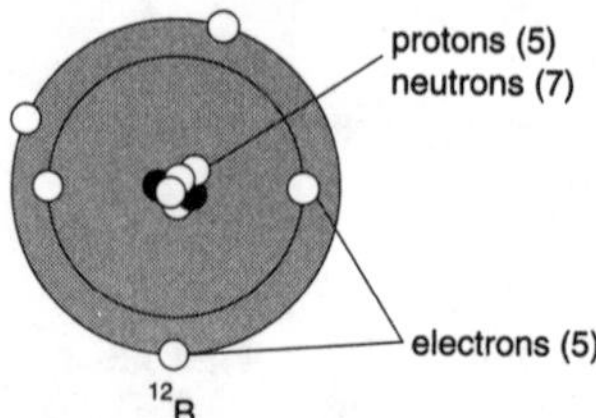

OBJECTIVE 2—pp. 24–25

1. b
2. c
3. c
4. b
5. c
6. **Table 2–2**

A. 20
B. 13
C. 8
D. 7
E. 13
F. 2
G. Ca^{2+}
H. ^{8}Li
I. 8
J. 22
K. 29

OBJECTIVE 3—pp. 25–26

1. c
2. b
3. d
4. c
5. equilibrium

OBJECTIVE 4—p. 26

1. b
2. A. inorganic
   B. inorganic
   C. organic
   D. organic
3. carbon
4. Early chemists long thought that carbon dioxide was inorganic, so today's chemists continue the tradition of calling it inorganic.

OBJECTIVE 5—p. 27

1. a
2. b
3. d
4. c
5. two-thirds of the total body weight

OBJECTIVE 6—pp. 27–29

1. a
2. a
3. d
4. b
5. c
6. a
7. pH
8. **Figure 2–3**

A. acidic
B. alkaline
C. pH of stomach acid (1)
D. pH of urine (6)
E. pH of water (6.8)
F. pH of blood (7.35–7.45)

OBJECTIVE 7—pp. 29–30

1. a
2. d
3. c
4. a
5. b

OBJECTIVE 8—pp. 30-32

1. c
2. d
3. c
4. d
5. a
6. a
7. **Figure 2–4**

A. thymine
B. thymine
C. guanine
D. adenine
E. cytosine

8.

Table 2-3 Classification of Various Organic Molecules

	Carbohydrate	Lipid	Protein	Nucleic acid	High-energy compound
Amino Acid			X		
ATP					X
Cholesterol		X			
Cytosine				X	
Disaccharide	X				
DNA				X	
Fatty acid		X			
Glucose	X				
Glycerol		X			
Glycogen	X				
Guanine				X	
Monosaccharide	X				
Nucleotide				X	
Phospholipid		X			
Polysaccharide	X				
RNA				X	
Starch	X				

OBJECTIVE 9—pp. 32–33

1. d
2. a
3. d
4. d
5. denature

OBJECTIVE 10—p. 33

1. d
2. a
3. c
4. b
5. d

Part II: Chapter-Comprehensive Exercises

MATCHING I

1. a
2. d
3. i
4. b
5. e
6. g
7. j
8. h
9. c
10. k
11. f
12. l

MATCHING II

1. c
2. j
3. d
4. h
5. b
6. i
7. l
8. a
9. g
10. f
11. k
12. e

CONCEPT MAP

1. atoms
2. protons
3. ion
4. isotope
5. radioactive
6. carbohydrates
7. proteins
8. ATP
9. steroids
10. DNA
11. sucrose

CROSSWORD PUZZLE

```
    H Y D R O G E N
    A
  G L U C O S E
    F
N   L       P
E   I S O T O P E S
G   F       S
A D E N O S I N E
T     E     T
I     U   L I P I D S
V     T     V # O
E   P R O T E I N
      O
    A N T I G E N S
      S
```

FILL-IN-THE-BLANK NARRATIVE

1. protons
2. chemical
3. covalent
4. carbohydrate
5. polysaccharide
6. glycogen
7. glucose
8. ATP
9. metabolism
10. enzymes
11. hydrogen ions
12. drop
13. Buffers
14. pH
15. homeostasis
16. hormones
17. tracer

CLINICAL CONCEPTS

1. The use of ^{99}Tc is considered to be a noninvasive test. The gamma ray pictures revealed that Patricia had an ulcer rather than a bad appendix. Without the use of radioactive technetium, the doctor would have ordered exploratory surgery. The surgeon would have found a perfectly normal appendix.
2. Technetium-99 has 56 neutrons (99 - 43=56)
3. The number 99 represents the mass number for that specific isotope.
4. Patricia was using terms that were too vague and not descriptive. The term "belly" could be interpreted as the stomach area rather than the intestine area.

CHAPTER 3: Cells: Their Structure and Function

Part I: Objective-Based Questions

OBJECTIVE 1—pp. 40–41

1. a
2. d
3. c
4. phospholipid bilayer
5. cell membrane
6. **Figure 3–1**
A. phospholipid
B. protein
C. carbohydrate portion of a glycolipid
D. cholesterol

OBJECTIVE 2—pp. 41–43

1. d
2. a
3. b
4. c
5. b
6. **Figure 3–2**
A. B
B. C
7. **Figure 3–3**
A. 2
B. 3
C. 1

OBJECTIVE 3—pp. 43–44

1. d
2. b
3. a
4. a
5. a
6. **Figure 3–4**
A. centriole
B. Golgi body
C. mitochondrion
D. rough endoplasmic reticulum
E. nucleus
F. lysosome
G. smooth endoplasmic reticulum
H. nucleolus

OBJECTIVE 4—pp. 45–46

1. b
2. a
3. c
4. a
5. b
6. **Figure 3–5**
A. nuclear pore
B. fixed ribosome
C. free ribosome
D. nucleolus

OBJECTIVE 5—pp. 46–48

1. a
2. c
3. a
4. d
5. a
6. d
7. a
8. amino acids
9. **Figure 3–6**
A. process of transcription
B. DNA
C. tRNA
D. process of translation
E. ribosome
F. mRNA
G. nuclear pore

OBJECTIVE 6—pp. 49–50

1. a
2. d
3. a
4. c
5. c
6. **Figure 3–7**
A. prophase
B. metaphase
C. anaphase
D. telophase
E. centriole
F. chromatids
G. spindle fibers
H. centromere

OBJECTIVE 7—pp. 50–51

1. a
2. b
3. b
4. differentiate
5. specialization

OBJECTIVE 8—p. 51

1. a
2. d
3. d
4. d
5. b

OBJECTIVE 9—p. 52

1. d
2. c
3. a
4. d
5. a

Part II: Chapter-Comprehensive Exercises

MATCHING I

1. c
2. h
3. d
4. e
5. f
6. g
7. a
8. b
9. i

MATCHING II

1. c
2. g
3. h
4. d
5. f
6. i
7. a
8. b
9. e

CONCEPT MAP

1. lipid bilayer
2. proteins
3. organelles
4. cytosol
5. ribosomes
6. nucleolus
7. lysosomes
8. mitochondria
9. Tay-Sachs disease
10. Parkinson's disease

CROSSWORD PUZZLE

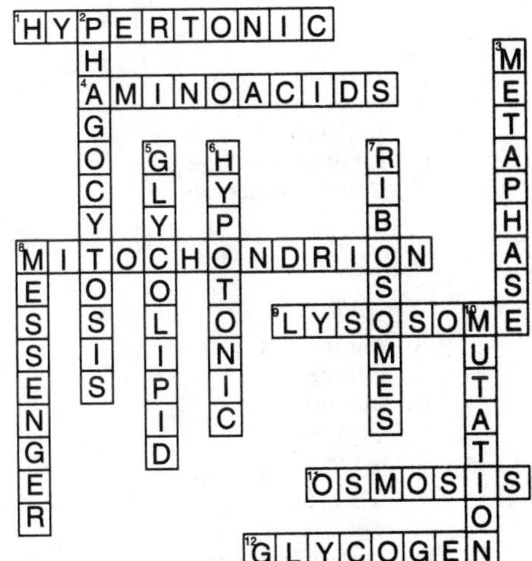

FILL-IN-THE BLANK NARRATIVE

1. homeostasis
2. organelles
3. phagocytosis
4. lysosomes
5. ribosome
6. DNA
7. nucleus
8. codons
9. mRNA
10. amino acids
11. solute
12. hypertonic
13. lose
14. hypotonic
15. mitosis
16. chromosomes
17. mutation
18. tumor
19. genes
20. carcinogenic

CLINICAL CONCEPTS

1. In glycogen storage disease, there is an accumulation of glycogen in the cytosol of the cell due to malfunctioning lysosomes, which disrupts cellular activity. In Tay-Sach's disease, the nerve cells begin to accumulate lipids due to malfunctioning lysosomes, which disrupts cellular activity. Both are diseases caused by the accumulation of cellular products due to malfunctioning lysosomes.
2. In glycogen storage disease the lysosomes fail to break down glycogen. In Tay-Sachs disease the lysosomes fail to break down lipids.
3. Streptomycin disrupts the function of bacterial ribosomes, so they cannot produce protein. Without protein, the bacteria die.
4. The structure of human ribosomes is different from that of bacterial ribosomes. Streptomycin affects only bacterial ribosomes and not human ribosomes.
5. If human ribosomes were structurally the same as bacterial ribosomes, then streptomycin would disrupt the human ribosomes as well. The human ribosomes would not be able to produce protein, and the human would die just as the bacteria do when exposed to streptomycin.

CHAPTER 4: Tissues and Body Membranes

Part I: Objective-Based Questions

OBJECTIVE 1—p. 59

1. b
2. d
3. a
4. b
5. b
6. protection

OBJECTIVE 2—pp. 60–61

1. d
2. b
3. a
4. d
5. c
6. Epithelial tissues cover both external and internal body surfaces.

Figure 4–1

7. B
8. D
9. C
10. E
11. A
12. F
13. H
14. L
15. I
16. J
17. G
18. K

OBJECTIVE 3—pp. 62–64

1. a
2. b
3. b
4. c
5. d
6. a
7. d
8. a
9. plasma
10. **Figure 4–2**

A. adipose
B. loose connective
C. collagen fibers
D. central canal
E. bone
F. canaliculi
G. matrix
H. hyaline cartilage
I. dense connective

OBJECTIVE 4—pp. 65–66

1. d
2. d
3. a
4. d
5. d
6. b
7. b
8. **Figure 4–3**

A. cutaneous
B. mucous
C. serous
D. synovial

OBJECTIVE 5—pp. 66–67

1. c
2. d
3. b
4. muscle fibers
5. **Figure 4–4**

A. intercalated disc
B. cardiac muscle
C. striations
D. skeletal muscle
E. smooth muscle
F. nucleus

OBJECTIVE 6—pp. 68–69

1. d
2. d
3. c
4. a
5. b
6. **Figure 4–5**

A. axon
B. nucleus
C. dendrite
D. soma

OBJECTIVE 7—p. 69

1. d
2. d
3. a
4. inflammation

OBJECTIVE 8—p. 70

1. d
2. b
3. d

OBJECTIVE 9—pp. 70–71

1. b
2. c
3. b
4. b
5. a

OBJECTIVE 10—p. 71

1. c
2. e
3. d
4. c
5. a

OBJECTIVE 11—p. 72

1. c
2. b
3. c
4. a
5. because there are so many types of cancer and so many different mechanisms, a single cure is probably not possible.

Part II: Chapter-Comprehensive Exercises

MATCHING : Part I

1. g
2. e
3. d
4. f
5. i
6. c
7. a
8. b
9. h
10. j

MATCHING: Part II

1. j
2. b
3. a
4. d
5. h
6. e
7. f
8. c
9. i
10. g

CONCEPT MAP

1. connective
2. columnar
3. skeletal
4. neuron
5. loose
6. adipose
7. ligaments
8. cartilage
9. blood
10. immunity
11. carcinoma
12. sarcoma
13. glioma

CROSSWORD PUZZLE

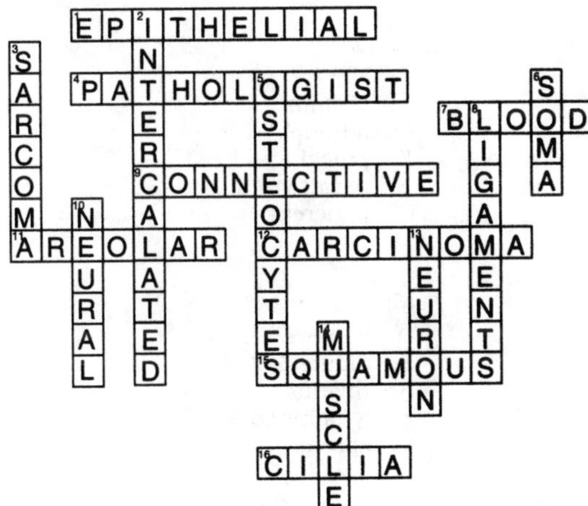

FILL-IN-THE-BLANK NARRATIVE

1. four
2. homeostasis
3. Epithelial
4. squamous
5. lining
6. blood
7. matrix
8. areolar
9. Bone
10. plasma
11. contract
12. neurons
13. skeletal
14. voluntary
15. tendons
16. ligaments
17. connective
18. glioma
19. benign

CLINICAL CONCEPTS

1. The cilia become paralyzed and are no longer be able to move mucus and debris away from the lungs.
2. The mucus traps foreign debris that might be inhaled.
3. Stratified epithelial, columnar
4. The cells in the trachea of a smoker will undergo abnormal mitosis.
5. The cell types change in order to provide a greater resistance to the drying effects and chemical irritation the smoker is experiencing.

CHAPTER 5: Organ Systems: An Overview

Part I: Objective-Based Questions

OBJECTIVE 1—p. 79

1. d
2. b
3. d

OBJECTIVE 2—pp. 79–83

1. b
2. d
3. b
4. d
5. c
6. d
7. c
8. c
9. As skeletal muscles contract, they generate heat.
10. skin
11. **Figure 5–1**
A. nails
B. hair
C. skin
12. **Figure 5–2**
A. axial
B. axial
C. axial
D. appendicular
E. axial
F. appendicular
G. appendicular
H. appendicular
I. appendicular
J. appendicular
K. appendicular
13. **Figure 5–3**
A. axial
B. axial
C. appendicular
D. appendicular

OBJECTIVE 3—pp. 84–86

1. a
2. c
3. a
4. bloodstream
5. hormones
6. **Figure 5–4**
A. central nervous system
B. peripheral nervous system
7. **Figure 5–5**
A. pituitary
B. thyroid
C. thymus
D. adrenal
E. pineal
F. parathyroid
G. pancreas

OBJECTIVE 4—pp. 87–89

1. d
2. b
3. d
4. c
5. c
6. lymphatic system
7. **Figure 5–6**
A. heart
B. liver
C. lungs
D. kidneys
8. **Figure 5–7**
A. thymus
B. lymph nodes
C. spleen

OBJECTIVE 5—pp. 90–93

1. c
2. d
3. c
4. d
5. alveoli
6. **Figure 5–8**
A. trachea
B. diaphragm
C. bronchi
D. lungs
7. **Figure 5–9**
A. esophagus
B. liver
C. small intestine
D. salivary glands
E. stomach
F. large intestine
8. **Figure 5–10**
A. urinary bladder
B. urethra
C. kidneys
D. ureter

OBJECTIVE 6—pp. 94–95

1. c
2. b
3. a
4. c
5. **Figure 5–11**
A. seminal vesicle
B. prostate gland
C. ductus deferens
D. urethra
E testis
F. penis
G. scrotum
H. mammary gland
I. uterine tube
J. ovary
K. uterus
L. vagina

Part II: Chapter-Comprehensive Exercises

MATCHING

1. b
2. d
3. e
4. g
5. k
6. i
7. h
8. c
9. j
10. f
11. a
12. i
13. e
14. h
15. d
16. j
17. c
18. b
19. g
20. k
21. a
22. f

CONCEPT MAP

1. integumentary
2. endocrine
3. lymphatic
4. respiratory
5. skeletal
6. thyroid
7. ureter
8. tonsils
9. pancreas
10. uterus

CROSSWORD PUZZLE

FILL-IN-THE-BLANK NARRATIVE

1. homeostasis
2. species
3. nutrients
4. cardiovascular
5. digestive
6. respiratory
7. lymphatic
8. plasma
9. antibodies
10. calcium
11. skeletal
12. urinary
13. muscular
14. nervous
15. integumentary
16. endocrine
17. oxytocin
18. uterine
19. positive

CLINICAL CONCEPT

1. The normal pH range for blood is 7.35–7.45.
2. High pH makes it difficult for the erythrocytes to release oxygen to the body tissues.
3. A low pH represents too high a hydrogen ion concentration.
4. Stress may stimulate the autonomic nerves, which may in turn cause the stomach to begin producing digestive juices, which are acidic. This excess acid may lead to an ulcer.
5. The answers will vary, but a good place to start is with the circulatory system. Check the blood pH. If the blood pH is abnormal, you can then run tests to determine why it is abnormal.

CHAPTER 6: Mechanisms of Disease

Part I: Objective-Based Questions

OBJECTIVE 1—p. 102

1. d
2. disease
3. homeostasis

OBJECTIVE 2—pp. 102–103

1. c
2. d
3. c
4. c
5. a

OBJECTIVE 3—p. 103

1. a
2. b
3. c
4. b
5. a

OBJECTIVE 4—p. 104

1. a
2. d
3. c
4. c
5. a

OBJECTIVE 5—p. 105

1. d
2. d
3. d
4. d
5. c
6. c

OBJECTIVE 6—p. 106

1. d
2. d

OBJECTIVE 7—pp. 106–107

1. c
2. c
3. a
4. b

OBJECTIVE 8—p. 107

1. b
2. c
3. a

OBJECTIVE 9—p. 107

1. d
2. d

OBJECTIVE 10—p. 108

1. a
2. d
3. a

OBJECTIVE 11—pp. 108–109

1. a
2. c
3. c
4. a
5. b

Part II: Chapter-Comprehensive Exercises

MATCHING

1. h
2. d
3. b
4. a
5. f
6. i
7. g
8. c
9. e
10. j

CONCEPT MAP

1. viruses
2. fungi
3. protozoans
4. urethra
5. semen
6. vectors
7. sneezing
8. utensil
9. biological
10. foodborne

CROSSWORD PUZZLE

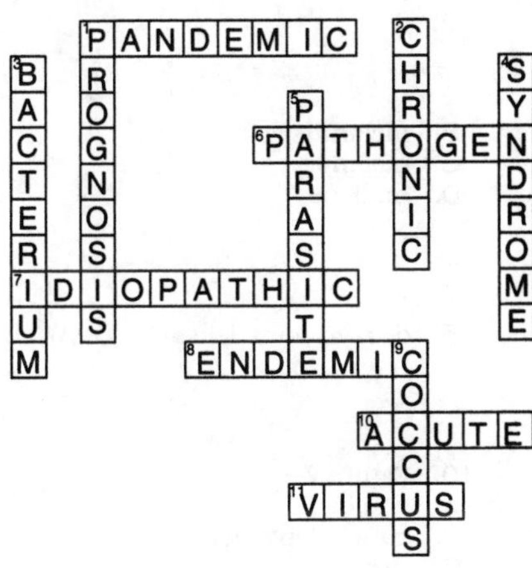

FILL-IN-THE-BLANK NARRATIVE

1. skin
2. Ringworm
3. neoplasm
4. virus
5. bacterial
6. arthropod
7. antibiotic
8. diagnosis
9. homeostasis

CLINICAL CONCEPT

1. Streptococci are spherical-shaped bacteria that typically occur in chains as opposed to clusters, as in the case of staphylococcus.
2. An antiviral medication would not be effective against bacteria. The bacteria would continue to thrive and Nadine would continue to have strep throat.
3. The uvula is the fleshy lobe that hangs down at the back of your throat. When it is stimulated by the passage of food, it will initiate the swallowing reflex.
4. The body often responds to a bacterial infection by creating inflammation. Inflammation is the dilation of blood vessels in an effort to bring white blood cells to the infected site. Due to the dilation of the blood vessels, the area appears red.
5. Streptococci can be spread by sneezing, coughing, and using fomites (utensils, drinking glasses, and towels, for example).

Chapter 7: The Integumentary System

Part I: Objective-Based Questions

OBJECTIVE 1—pp. 115–116

1. d
2. c
3. a
4. b
5. **Figure 7–1**
A. epidermis
B. dermis
C. hypodermis
D. sebaceous gland
E. arrector pili muscle
F. touch and pressure receptors
G. hair follicles
H. sweat glands

OBJECTIVE 2—pp. 117–118

1. a
2. a
3. d
4. c
5. d
6. d
7. a
8. c
9. d

OBJECTIVE 3—pp. 118–119

1. d
2. c
3. a
4. a
5. a
6. melanin
7. **Figure 7–2**
A. stratum germinativum
B. melanin
C. melanocyte
D. nucleus

OBJECTIVE 4—pp. 119–120

1. d
2. d
3. d
4. sunlight
5. deoxyribonucleic acid (DNA)

OBJECTIVE 5—pp. 120–122

1. b
2. a
3. c
4. b
5. The nails protect the tips of the fingers and toes.
6. The nose hairs prevent the entry of foreign particles.
7. The eyelashes prevent the entry of foreign particles into the eye.
8. arrector pili muscles
9. **Figure 7–3**
A. sebaceous gland
B. arrector pili muscle
C. hair follicle
D. sweat gland
10. **Figure 7–4**
A. nail bed
B. cuticle (eponychium)
C. free edge
D. lunula

OBJECTIVE 6—p. 122

1. b
2. d
3. b
4. a protein substance called keratin

OBJECTIVE 7—p. 123

1. c
2. d
3. a
4. d

OBJECTIVE 8—pp. 123-124

1. d
2. c
3. a
4. b
5. b

OBJECTIVE 9—pp. 124-125

1. b
2. b
3. a
4. a
5. a
6. a

OBJECTIVE 10—p. 125

1. d
2. a
3. a
4. c
5. c

Part II: Chapter-Comprehensive Exercises

MATCHING

1. h
2. l
3. i
5. b
4. e
6. k
7. a
8. m
9. n
10. g
11. d
12. c
13. j
14. k

CONCEPT MAP

1. dermis
2. glands
3. synthesizes vitamin D
4. lubrication
5. production of secretions
6. first-degree
7. second-degree
8. third-degree
9. sensory reception

CROSSWORD PUZZLE

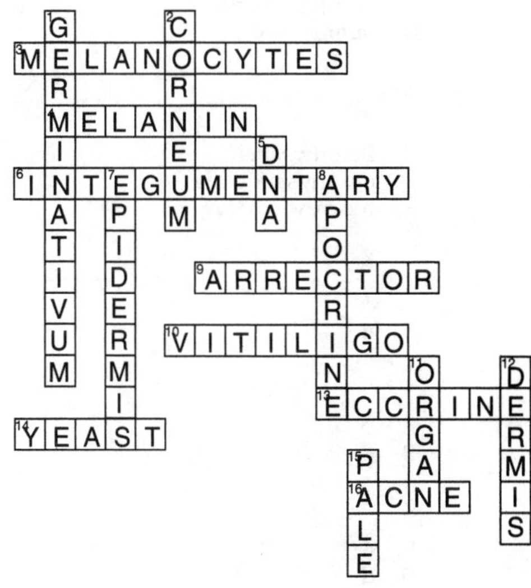

FILL-IN-THE-BLANK NARRATIVE

1. organ
2. homeostasis
3. eccrine
4. cooling
5. sebaceous
6. lubricates
7. Melanocytes
8. melanin
9. DNA
10. mutate
11. cancer
12. pathogens
13. stratum germinativum
14. fibroblasts
15. epidermis
16. scar
17. organ systems

CLINICAL CONCEPT

1. The skin begins to bulge because of the buildup of sebum.
2. Pimples are typically red due to the increased blood flow to the site, which brings white bloods cells to fight the bacteria that have entered the body.
3. The whitehead is pus, which consists of dead white blood cells and dead bacteria.
4. Popping the pimple might force some of the live bacteria into deeper tissue.
5. One possible reason is excess autonomic nerve activity and hormone production. The autonomic nerves activate the sebaceous glands. Overactive sebaceous glands produce excess sebum.

CHAPTER 8: The Skeletal System

Part I: Objective-Based Questions

OBJECTIVE 1—p. 132

1. a
2. b
3. c
4. c

OBJECTIVE 2—pp. 132–134

1. a
2. a
3. a
4. a
5. d
6. a
7. b
8. **Figure 8–1**
   A. proximal epiphysis
   B. marrow cavity
   C. compact bone
   D. periosteum
9. **Figure 8–2**
   A. marrow cavity
   B. osteon
   C. compact bone
   D. spongy bone

OBJECTIVE 3—pp. 135–136

1. a
2. b
3. a
4. d
5. **Figure 8–3**
   A. epiphysis
   B. diaphysis
   C. blood vessel
   D. epiphyseal plate
   E. marrow cavity

OBJECTIVE 4—pp. 136–137

1. d
2. d
3. b
4. b
5. a

OBJECTIVE 5—p. 137

1. c
2. c
3. d

OBJECTIVE 6—p. 138

1. d
2. d
3. d
4. a
5. d
6. d

OBJECTIVE 7—pp. 138–142

1. b
2. b
3. d
4. d
5. b
6. a
7. b
8. d
9. a
10. c
11. a
12. **Figure 8–4**
   A. parietal
   B. temporal
   C. occipital
   D. mastoid process
   E. coronal suture
   F. frontal
   G. sphenoid
   H. nasal
   I. zygomatic
   J. maxilla
   K. mandible
13. **Figure 8–5**
   A. sphenoid
   B. temporal
   C. ethmoid
   D. lacrimal
   E. zygomatic
   F. maxilla
   G. mandible
   H. frontal
   I. coronal suture
   J. nasal
   K. mastoid process
   L. perpendicular plate of the ethmoid
   M. vomer
   N. occipital
   O. maxilla
   P. palatine
   Q. zygomatic arch
   R. occipital condyle
   S. foramen magnum
14. **Figure 8–6**
   A. frontal
   B. cribiform plate of the ethmoid
   C. sphenoid
   D. temporal
   E. occipital
   F. crista galli
   G. depression for the pituitary gland
   H. frontal sinus
   I. ethmoid
   J. vomer
   K. maxilla
   L. mandible

OBJECTIVE 8—pp. 143–147

1. c
2. c
3. b
4. a
5. b
6. a
7. a
8. b
9. c
10. manubrium, body, and xiphoid
11. **Figure 8–7**
   A. cervical
   B. thoracic
   C. lumbar
   D. sacrum
   E. coccygeal
12. **Figure 8–8**
   A. lamina
   B. pedicle
   C. spinous process
   D. transverse process
   E. vertebral body
13. **Figure 8–9**
   A. manubrium
   B. body
   C. xiphoid
   D. floating ribs, 2 pairs
   E. true ribs, 7 pairs
   F. false ribs, 5 pairs

OBJECTIVE 9—pp. 147–151

1. b
2. d
3. c
4. c
5. a
6. b
7. c
8. d
9. c
10. c
11. c
12. a
13. b
14 b
15. a
16. b
17. d
18. a
19. **Figure 8–10**
A. acromion
B. coracoid process
C. clavicle
D. manubrium
E. body
F. greater tubercle
G. capitulum
H. medial epicondyle
I. head of the radius
J. olecranon
K. radius
L. scaphoid
M. trapezium
N. trapezoid
O. lunate
P. triquetal
Q. pisiform
R. hamate
S. capitate
20. **Figure 8–11**
A. ilium
B. pubis
C. ischeum
D. greater trochanter
E. lateral epicondyle
F. medial epicondyle
G. lateral malleolus
H. tibial tuberosity
I. medial malleolus
J. cuboid
K. calcaneus
L. talus
M. navicular
N. cuneiform

OBJECTIVE 10—pp. 151–153

1. c
2. d
3. a
4. c
5. a
6. **Figure 8–12**
A. articular cartilage
B. bursa
C. patella
D. meniscus
E. joint cavity

OBJECTIVE 11—pp. 153–154

1. a
2. d
3. a
4. d
5. c
6. **Figure 8–13**
A. flexion
B. extension
C. flexion
D. extension
E. abduction
F. adduction
G. abduction
H. adduction
I. supination
J. pronation

OBJECTIVE 12—pp. 154–156

1. b
2. c
3. b
4. d
5. a

OBJECTIVE 13—pp. 156–157

1. a
2. b
3. a
4. d
5. a
6. a

Part II: Chapter-Comprehensive Exercises

MATCHING I

1. d
2. j
3. c
4. e
5. i
6. g
7. a
8. b
9. h
10. f
11. l
12. k

MATCHING II

1. l
2. c
3. b
4. a
5. d
6. k
7. i
8. f
9. j
10. g
11. e
12. h

CONCEPT MAP

1. support
2. osteons
3. osteocytes
4. osteosarcoma
5. lacunae
6. myeloma
7. axial
8. sternum
9. pectoral girdle
10. lumbar
11. true
12. tibia

CROSSWORD PUZZLE

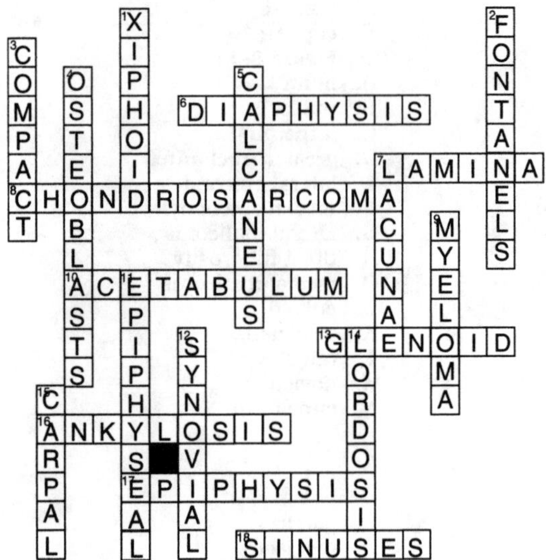

FILL-IN-THE-BLANK NARRATIVE

1. Osteoblast
2. Osteoclast
3. homeostasis
4. osteoporosis
5. storage
6. calcium
7. red marrow
8. red blood
9. movement
10. tendons
11. attachments
12. greater tubercle
13. olecranon process
14. acromion
15. greater trochanter
16. tibial tuberosity
17. transverse processes
18. diarthrosis
19. lambdoidal
20. sagittal
21. synarthrosis
22. Arthritis
23. Rheumatoid
24. cartilage
25. extend
26. pronated
27. supinate

CLINICAL CONCEPT

1. Calcium ions are stored in the compact bone area.
2. Osteoblasts need calcium ions to replace cartilage with bone.
3. During pregnancy, the mother needs calcium ions not only to maintain her bones but also to supply her developing child with enough calcium ions for its bone growth. Without the dietary supplement, the developing child will utilize calcium ions from the mother's bones, thereby depleting them of calcium. The mother's bones will begin to deteriorate.
4. Christine's bones were beginning to deteriorate because her diet contained inadequate amounts of calcium ions. Christine's osteoclasts were removing the calcium ions from her bones in order to supply her developing child needed calcium ions for bone growth and development.
5. osteoporosis

CHAPTER 9: The Muscular System

Part I: Objective-Based Questions

OBJECTIVE 1—p. 164

1. a
2. c
3. b
4. b

OBJECTIVE 2—pp. 164–165

1. a
2. b
3. d
4. b
5. a
6. **Figure 9–1**
A. muscle fascicle
B. muscle fiber
C. myofibril

OBJECTIVE 3—p. 166

1. c
2. a
3. c
4. c
5. **Figure 9–2**
A. myosin
B. actin
C. z line
D. cross-bridge

OBJECTIVE 4—pp. 167–168

1. d
2. b
3. b
4. a
5. d
6. **Figure 9–3**
A. myosin
B. actin
C. cross-bridge
D. relaxed sarcomere
E. contracted sarcomere

OBJECTIVE 5—pp. 168–169

1. b
2. c
3. a
4. d
5. **Figure 9–4**
A. nerve
B. muscle
C. motor unit

OBJECTIVE 6—pp. 169–170

1. d
2. a
3. a
4. b
5. d
6. **Figure 9–5**
A. isotonic
B. isometric

OBJECTIVE 7—pp. 170–171

1. c
2. b
3. d
4. d
5. b
6. d

OBJECTIVE 8—pp. 171–173

1. a
2. d
3. c
4. e
5. b
6. **Figure 9–6**
A. frontalis
B. masseter
C. pectoralis major
D. temporalis
E. sternocleidomastoid
F. serratus anterior
G. external oblique
H. rectus abdominis
I. trapezius
J. infraspinatus
K. latissimus dorsi
L. occipitalis

OBJECTIVE 9—pp. 173–176

1. a
2. d
3. d
4. flexor carpi radialis
5. extensor digitorum
6. a
7. b
8. d
9. a
10. biceps femoris
11. gracilis
12. **Figure 9–7**
A. deltoid
B. biceps brachii
C. flexor carpi radialis
D. flexor digitorum
E. rectus femoris
F. vastus lateralis
G. vastus medialis
H. tibialis anterior
I. brachioradialis
J. flexor carpi ulnaris
K. gracilis
L. sartorius
13. **Figure 9–8**
A. brachioradialis
B. semimembranosus
C. semitendinosus
D. biceps femoris
E. gastrocnemius
F. triceps brachii
G. extensor digitorum
H. extensor carpi ulnaris
I. gluteus medius
J. gluteus maximus

OBJECTIVE 10—p. 177

1. a
2. a
3. c
4. b

OBJECTIVE 11—pp. 177–178

1. a
2. b
3. a
4. d
5. d

OBJECTIVE 12—pp. 178–179

1. c
2. b
3. a
4. c
5. a

Part II: Chapter-Comprehensive Exercises

MATCHING I

1. a
2. f
3. e
4. c
5. h
6. i
7. j
8. b
9. d
10. g
11. k

MATCHING II

1. j
2. c
3. b
4. d
5. e
6. k
7. i
8. h
9. l
10. a
11. f
12. g

CONCEPT MAP

1. maintain body temperature
2. maintain posture
3. actin
4. myosin
5. axial
6. torso
7. latissimus dorsi
8. deltoid
9. triceps brachii
10. gastrocnemius

CROSSWORD PUZZLE

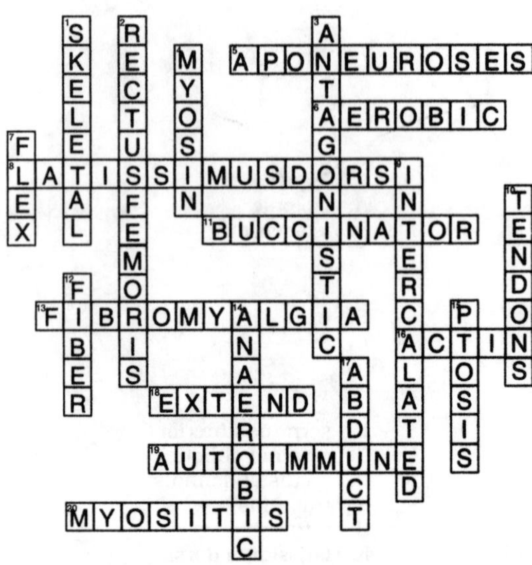

FILL-IN-THE-BLANK NARRATIVE

1. heat
2. arrector pili
3. superficial
4. isometric
5. gluteus maximus
6. flex
7. biceps femoris
8. hamstrings
9. iliopsoas
10. rectus femoris
11. extending
12. deltoid
13. abducts
14. nervous
15. neuromuscular
16. homeostasis
17. myesthenia gravis
18. muscular dystrophy
19. trichinosis
20. hypocalcemia

CLINICAL CONCEPT

1. A nerve impulse causes the release of calcium ions from the sarcoplasmic reticulum of muscles.
2. Adenosine triphosphate (ATP) is needed to create movement of the cross-bridges.
3. Oxygen is necessary to complete the formation of ATP.
4. When the body runs out of oxygen, it will run out of ATP.

Chapter 10: The Nervous System I: Nerve Cells and the Spinal Cord

Part I: Objective-Based Questions

OBJECTIVE 1—p. 185

1. b
2. c
3. a
4. a
5. d

OBJECTIVE 2—pp. 186–187

1. b
2. d
3. b
4. c
5. a
6. **Figure 10–1**
A. nucleus
B. soma
C. axon
D. dendrite
E. neuromuscular junction
7. **Figure 10–2**
A. myelinated axon
B. neuron
C. astrocyte
D. microglial cell
E. oligodendrocyte

OBJECTIVE 3—p. 188

1. a
2. c
3. a
4. d

OBJECTIVE 4—pp. 188–190

1. b
2. c
3. a
4. d
5. d
6. neurotransmitter
7. neurotransmitter
8. **Figure 10–3**
A. axon
B. synapse
C. dendrite
D. soma
E. E to F

OBJECTIVE 5—pp. 190–193

1. c
2. a
3. b
4. a
5. a
6. L1–L5 or S1–S5
7. **Figure 10–4**
A. cervical nerves
B. thoracic nerves
C. lumbar nerves
D. sacral nerves
E. cauda equina
8. **Figure 10–5**
A. gray matter
B. ventral root
C. dorsal root
D. posterior spinal cord
E. anterior spinal cord
F. white matter

OBJECTIVE 6—pp. 193–194

1. a
2. d
3. c
4. a
5. 2
6. **Figure 10–6**
A. sensory nerve
B. site of stimulus
C. effector
D. spinal cord
E. motor nerve

OBJECTIVE 7—pp. 195–196

1. d
2. a
3. a
4. a
5. a
6. a
7. **Figure 10–7**
A. sympathetic chain
B. spinal cord
C. parasympathetic nerve
D. sympathetic nerve

OBJECTIVE 8—p. 197

1. c
2. b
3. c
4. c
5. b

OBJECTIVE 9—pp. 197–198

1. c
2. c
3. a
4. c
5. a

Part II: Chapter-Comprehensive Exercises

MATCHING I

1. d
2. h
3. a
4. g
5. f
6. e
7. b
8. c
9. i

MATCHING II

1. i
2. f
3. h
4. a
5. c
6. b
7. d
8. e
9. g

CONCEPT MAP

1. PNS
2. somatic nerves
3. spinal cord
4. sympathetic
5. glands
6. soma
7. neurotransmitter
8. gray matter

CROSSWORD PUZZLE

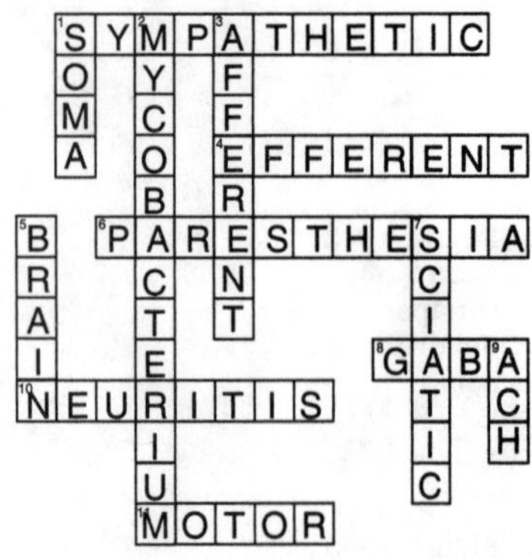

FILL-IN-THE-BLANK NARRATIVE

1. stimulus
2. permeable
3. sensory
4. motor
5. acetylcholine
6. neurotransmitter
7. synapse
8. actin
9. myosin
10. ATP
11. reflex arc
12. autonomic
13. parasympathetic
14. homeostasis
15. negative

CLINICAL CONCEPTS

1. The nerve membrane becomes impermeable to sodium ions. Without sodium ions, there can be no impulse to travel to the brain for interpretation as pain.
2. Novocain causes the nerve membrane to become impermeable to sodium ions.
3. As soon as the Novocain wears off, the membrane becomes permeable to sodium ions. When this happens, impulses arrive at the brain that are interpreted as pain.
4. The myelin sheath provides protection for the nerve.
5. Multiple sclerosis is a degeneration of the myelin sheath around peripheral nerves. As the sheath is destroyed, the nerves malfunction. Paralysis may occur.

CHAPTER 11: The Nervous System II: The Brain and Cranial Nerves

Part I: Objective-Based Questions

OBJECTIVE 1—pp. 205–206

1. c
2. d
3. a
4. d
5. c
6. d
7. **Figure 11–1**
A. thalamus
B. pons
C. medulla oblongata
D. cerebrum
E. cerebellum

OBJECTIVE 2—pp. 206–207

1. c
2. d
3. a
4. d
5. b
6. **Figure 11–2**
A. skull
B. dura mater
C. arachnoid mater
D. cerebrospinal fluid area
E. pia mater
F. brain tissue

OBJECTIVE 3—pp. 207–209

1. c
2. d
3. d
4. d
5. d
6. choroid plexus
7. **Figure 11–3**
A. lateral ventricle
B. third ventricle
C. fourth ventricle
D. pons
E. medulla oblongata
8. **Figure 11–4**
A. choroid plexus
B. third ventricle area
C. pons
D. fourth ventricle area
E. medulla oblongata
F. cerebellum
G. central canal

OBJECTIVE 4—pp. 210–211

1. a
2. b
3. b
4. d
5. a
6. **Figure 11–5**
A. frontal lobe
B. temporal lobe
C. central sulcus
D. parietal lobe
E. occipital lobe

OBJECTIVE 5—pp. 211–212

1. b
2. d
3. c
4. hippocampus
5. **Figure 11–6**
The limbic system is B.

OBJECTIVE 6—pp. 212–213

1. d
2. c
3. c
4. b
5. **Figure 11–7**
A. thalamus
B. hypothalamus

OBJECTIVE 7—pp. 213–214

1. b
2. b
3. c
4. a
5. b
6. reticular activating system

OBJECTIVE 8—pp. 214–215

1. d
2. d
3. c
4. b
5. occipital

OBJECTIVE 9—pp. 215–217

1. a
2. b
3. b
4. b
5. b
6. b
7. b
8. d
9. c
10. d
11. accessory nerve
12. **Figure 11–8**
A. olfactory (CN I)
B. oculomotor (CN III)
C. trigeminal (CN V)
D. vestibulocochlear (CN VIII)
E. glossopharyngeal (CN IX)
F. hypoglossal (CN XII)
G. optic (CN II)
H. abducens (CN VI)
I. trochlear (CN IV)
J. facial (CN VII)
K. vagus (CN X)
L. accessory (CN XI)

OBJECTIVE 10—pp. 217–218

1. c
2. d
3. c
4. a

OBJECTIVE 11—p. 218

1. b
2. a
3. a
4. d

OBJECTIVE 12—p. 219

1. d
2. b
3. a
4. d
5. c

Part II: Chapter-Comprehensive Exercises

MATCHING

1. j
2. o
3. d
4. i
5. c
6. f
7. m
8. b
9. g
10. h
11. e
12. l
13. n
14. a
15. k

CONCEPT MAP

1. meninges
2. diencephalon
3. cerebellum
4. pons
5. thalamus
6. blood pressure
7. body temperature

CROSSWORD PUZZLE

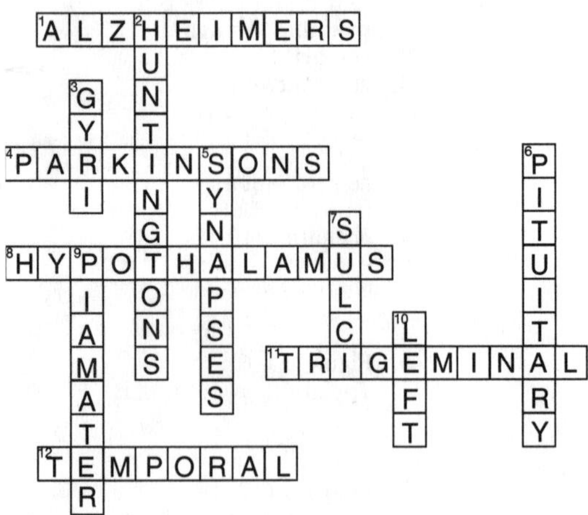

FILL-IN-THE-BLANK NARRATIVE

1. olfactory
2. temporal
3. oculomotor
4. occipital
5. optic
6. cerebellum
7. trigeminal
8. masseter
9. hypoglossal
10. accessory
11. hemisphere
12. alert
13. sleep
14. synaptic
15. Parkinson's disease
16. levodopa
17. dopamine
18. Bell's palsy

CLINICAL CONCEPTS

1. The facial nerve is cranial nerve number 7 (CN VII).
2. Bell's palsy affects cranial nerve 7 (facial nerve), which controls the lacrimal glands. The lacrimal glands produce tears to lubricate the surfaces of the eye. If the facial nerve malfunctions, the patient will not be able to produce tears, and the eyes will become very dry and damaged.
3. The cerebrum initiates the command. The cerebellum carries out the command in a smooth, coordinated manner.
4. The cerebellum carries out the cerebral commands in a smooth, coordinated manner.
5. Cancer of the cerebellum will lead to a malfunction of the cerebellum. If the cerebellum cannot function properly, the patient will not be able to carry out smooth, coordinated processes and may walk with a jerky motion.

CHAPTER 12: The Senses

Part I: Objective-Based Questions

OBJECTIVE 1—p. 226

1. b
2. a
3. b
4. The "special" senses have sensory receptors that are located in specific areas of the body, such as the organ of Corti, which detects hearing and is located in the cochlea of the ear. The "general" senses have sensory receptors that are located throughout the body.

OBJECTIVE 2—pp. 226–227

1. c
2. d
3. d
4. a
5. tactile

OBJECTIVE 3—pp. 227–228

1. d
2. a
3. cribriform plate
4. dissolved
5. Their olfactory receptor surface area is 72 times greater than ours.
6. **Figure 12–1**
   A. olfactory nerve
   B. cribriform plate
   C. olfactory receptor
   D. olfactory cilia

OBJECTIVE 4—pp. 228–229

1. b
2. a
3. b
4. a
5. dissolve
6. **Figure 12–2**
   A. papilla
   B. taste buds
   C. papilla
   D. papilla
   E. papilla
   F. taste bud
   G. taste buds
   H. gustatory cell
   I. taste bud

OBJECTIVE 5—pp. 230–232

1. d
2. b
3. c
4. d
5. d
6. lacrimal
7. six
8. **Figure 12-3**
   A. ciliary bodies
   B. lens
   C. pupil
   D. cornea
   E. vitreous chamber
   F. optic disc
   G. optic nerve
   H. retina
   I. sclera
9. **Figure 12–4**
   A. superior oblique
   B. superior rectus
   C. lateral rectus
   D. inferior rectus
   E inferior oblique
   F. medial rectus

OBJECTIVE 6—pp. 232–234

1. c
2. a
3. a
4. b
5. three
6. cones
7. **Figure 12–5**
   A. rods
   B. cones
   C. inner layer
   D. outer layer
8. **Figure 12–6**
   A. myopia
   B. hyperopia
   C. lens
   D. retina
   E. cornea

OBJECTIVE 7—pp. 234–235

1. a
2. c
3. d
4. b
5. endolymph
6. **Figure 12–7**
   A. outer ear
   B. middle ear
   C. inner ear
   D. semicircular canals
   E. cranial nerve 8 (CN VIII)

OBJECTIVE 8—p. 236

1. a
2. d
3. b
4. c
5. c
6. **Figure 12–8**
   A. malleus
   B. incus
   C. stapes
   D. tympanum
   E. oval window
   F. perilymph area
   G. endolymph area
   H. area of low-pitched sounds
   I. area of high-pitched sounds

OBJECTIVE 9—p. 237

1. a
2. c
3. c
4. a
5. b
6. a

OBJECTIVE 10—p. 238

1. d
2. b
3. d
4. c
5. a
6. d

Part II: Chapter-Comprehensive Exercises

MATCHING I

1. h
2. i
3. e
4. c
5. g
6. d
7. b
8. f
9. a
10. j
11. k

MATCHING II

1. j
2. e
3. a
4. g
5. f
6. h
7. b
8. c
9. d
10. i

CONCEPT MAP

1. pain
2. taste
3. vision
4. proprioceptors
5. tactile receptors
6. gustatory receptors
7. organ of Corti
8. affliction of N I
9. affliction of N VII
10. glaucoma
11. otitis media
12. vertigo

CROSSWORD PUZZLE

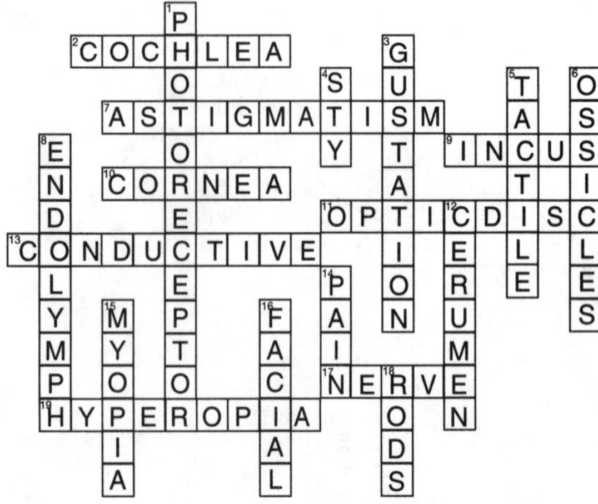

FILL-IN-THE-BLANK NARRATIVE

1. cones
2. rods
3. myopia
4. emmetropia
5. olfactory
6. dissolved
7. gustatory
8. facial
9. Corti
10. cochlea
11. vestibulocochlear
12. semicircular canals
13. vestibular
14. efferent
15. tactile

CLINICAL CONCEPT

1. Motion of the fluid in the semicircular canals causes the sensory hairs to bend, which activates the sensory receptors. The sensory receptors then send signals to the brain, where position is interpreted.
2. When a person stops moving, the fluid in the semicircular canals continues to rock back and forth until it, too, comes to a stop. Each time the fluid moves, it activates the sensory receptors.
3. A change in temperature causes the fluid in the semicircular canals to move, even though you are standing still. The movement of the fluid activates the sensory receptors.
4. Low blood pressure may result in low blood flow to the brain. A decreased flow of blood to the brain may result in dizziness.
5. Bacteria in the inner ear may cause the fluid in the semicircular canal to begin moving, activating the sensory receptors of motion.

CHAPTER 13: The Endocrine System

Part I: Objective-Based Questions

OBJECTIVE 1—p. 245

1. c
2. c
3. endocrine
4. nervous
5. endocrine

OBJECTIVE 2—pp. 245–246

1. c
2. a
3. b
4. c
5. d

OBJECTIVE 3—pp. 246–247

1. c
2. b
3. a
4. c
5. receptor molecules
6. **Figure 13–1**
A. endocrine gland
B. bloodstream
C. receptor site
D. target organ

OBJECTIVE 4—pp. 247–248

1. b
2. b
3. a
4. d
5. d

OBJECTIVE 5—pp. 248–251

1. c
2. a
3. b
4. a
5. a
6. c
7. b
8. a
9. b
10. **Figure 13–2**
A. hypothalamus
B. pituitary
C. thyroid
D. thymus
E. adrenals
F. ovaries
G. pineal
H. parathyroid
I. atria of the heart
J. pancreas
K. testis
Figure 13–3
11. a
12. h
13. e
14. c
15. f
16. g
17. d
18. b
19. l
20. m
21. k
22. i
23. j
24. n
25. q
26. o
27. p

OBJECTIVE 6—pp. 252–253

1. b
2. b
3. d
4. Leptin. It binds to the appetite control centers of the hypothalamus. It suppresses the appetite and increases metabolism.
5. There is less oxygen at a high altitude. A decrease in oxygen causes the kidneys to release erythropoietin to increase red blood cell production in an effort to deliver more oxygen to the body tissues.
6. **Figure 13–4**
A. thymus
B. area of the small intestine
C. heart
D. kidney

OBJECTIVE 7—pp. 253–254

1. a
2. d
3. d
4. d
5. Stress causes a change in hormone production, which may result in a change in metabolic processes. These changes may result in electrolyte imbalance, high blood pressure, and a decrease in lipids, all of which may lead to heart failure.

OBJECTIVE 8—p. 254

1. b
2. b
3. a
4. a
5. a

OBJECTIVE 9—p. 255

1. a
2. a
3. c
4. c
5. b
6. d
7. weakened bones

Part II: Chapter-Comprehensive Exercises

MATCHING I

1. i
2. e
3. b
4. h
5. g
6. j
7. d
8. f
9. a
10. c

MATCHING II

1. b
2. d
3. g
4. f
5. e
6. h
7. a
8. c
9. j
10. i

CONCEPT MAP

1. TSH
2. calcitonin
3. LH
4. mammary glands
5. skeleton
6. ACTH
7. posterior
8. kidneys
9. uterus

CROSSWORD PUZZLE

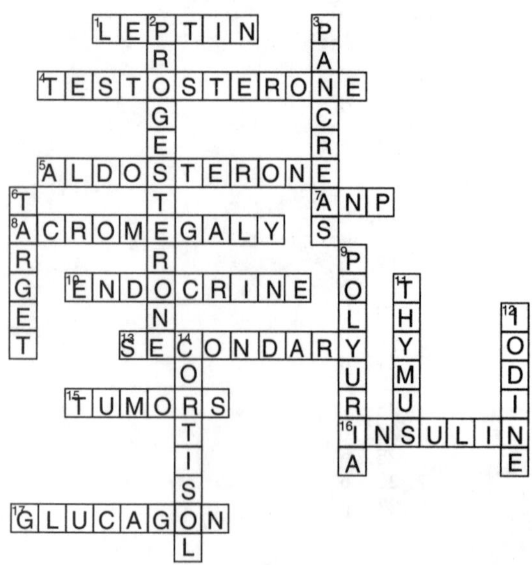

FILL-IN-THE-BLANK NARRATIVE

1. oxytocin
2. hypothalamus
3. target
4. thyroid hormones
5. ACTH
6. glucocorticoids
7. FSH
8. LH
9. thyroxine
10. calcitonin
11. osteoblast
12. parathyroid
13. osteoclasts
14. thymus
15. thymosin
16. ANP
17. sodium
18. blood pressure
19. cortisol
20. epinephrine
21. EPO
22. oxygen
23. bone marrow
24. insulin
25. testosterone
26. progesterone

CLINICAL CONCEPT

1. The hypothalamus produces the antidiuretic hormone.
2. The posterior pituitary gland releases ADH into the bloodstream.
3. ADH targets tubules inside the kidneys.
4. Antidiuretic hormone causes the kidneys to recirculate water into the bloodstream.
5. If 2000 mL of water passed through the kidneys, only about 20 mL would enter the urinary bladder to be excreted. (2000 mL X 0.01 = 20 mL)

CHAPTER 14: The Blood

Part I: Objective-Based Questions

OBJECTIVE 1—p. 262

1. d
2. b

OBJECTIVE 2—pp. 262–263

1. d
2. c
3. c
4. b
5. Blood absorbs the heat generated by muscle activity and redistributes the heat to other body tissues.

OBJECTIVE 3—pp. 263–264

1. a
2. d
3. a
4. b
5. erythrocytes and leukocytes
6. **Figure 14–1**
   A. plasma
   B. white blood cells
   C. red blood cells

OBJECTIVE 4—pp. 264–265

1. d
2. d
3. d
4. c
5. fibrinogen

OBJECTIVE 5—pp. 265–266

1. b
2. b
3. b
4. Anemia results when tissues do not receive adequate oxygen. Oxygen bonds to iron to be delivered to body tissues. Without iron, oxygen cannot be delivered.
5. erythrocytes

Figure 14–2

6. A

OBJECTIVE 6—pp. 266–268

1. a
2. d
3. a
4. a
5. b
6. a
7. The lab report indicated that your sample of blood consisted of higher than normal numbers of lymphocytes. Lymphocytes respond to viral infections.
8. lymphocytes
9. **Figure 14–3**

   Cell A
   i. neutrophil
   ii. 50–70%
   iii. fights bacteria

   Cell B
   i basophil
   ii. less than 1%
   iii. promotes inflammation

   Cell C
   i. lymphocyte
   ii. 20–30%
   iii. fights viruses

   Cell D
   i. monocyte
   ii. 6–8%
   iii. fights fungus

   Cell E
   i. eosinophil
   ii. 2–4%
   iii. fights allergens and pathogens

OBJECTIVE 7—pp. 269–270

1. d
2. d
3. b
4. b
5. d
6. erythropoiesis
7. leukopoiesis
8. thrombopoiesis
9. **Figure 14–4**
   A. platelets
   B. red blood cell
   C. megakaryocyte
   D. white blood cell

OBJECTIVE 8—pp. 270–271

1. d
2. c
3. b
4. a
5. b
6. from platelets or cells of damaged blood vessels
7. blood clotting factors convert prothrombin to thrombin
8. fibrin forms fibers that help in the blood clotting process

OBJECTIVE 9—pp. 271–273

1. b
2. d
3. c
4. a
5. a
6. In blood clumping the plasma antibodies cause the red blood cells to begin sticking to one another. Blood clotting is a series of chemical reactions initiated by the platelets, which release a clotting chemical.
7. **Figure 14–5**
   A. B antigen
   B. A antigen
8. **Figure 14-6**
   A. safe
   B. unsafe
   C. unsafe
   D. safe
   E. unsafe
   F. safe
   G. unsafe
   H. safe
   I. safe
   J. safe
   K. safe
   L. safe
   M. unsafe
   N. unsafe
   O. unsafe
   P. safe

OBJECTIVE 10—p. 274

1. a
2. b
3. d
4. d
5. a

OBJECTIVE 11—p. 275

1. b
2. d
3. a
4. b
5. d

Part II: Chapter-Comprehensive Exercises

MATCHING I

1. f
2. a
3. e
4. h
5. g
6. d
7. b
8. c
9. i
10. j

MATCHING II

1. h
2. d
3. c
4. b
5. a
6. f
7. e
8. g
9. i
10. j

CONCEPT MAP

1. solutes
2. leukocytes
3. oxygen
4. albumin
5. neutrophil
6. eosinophil
7. lymphocytes
8 viruses

CROSSWORD PUZZLE

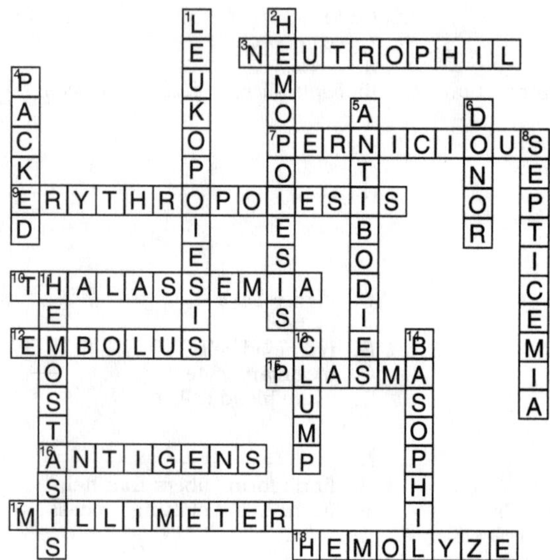

FILL-IN-THE-BLANK NARRATIVE

1. homeostasis
2. median cubital
3. antecubital
4. 6000–9000/mm^3
5. differential
6. lymphocytes
7. eosinophils
8. neutrophils
9. monocytes
10. 20–30%
11. 39%
12. fungicides
13. cubic millimeter
14. carbon monoxide
15. iron
16. erythropoietin
17. erythroblasts
18. erythrocytes

CLINICAL CONCEPT

1. The incidence of sickle cell anemia is higher in the African American population that in the Caucasian population.
2. False. The cells carry the normal amount of oxygen but when they release the oxygen, they become sickle-shaped and may cause blockage in the capillaries.
3. False. Sickle cell anemia can be treated by blood transfusions but not cured. Sickle cell disease is an inherited disorder.
4. Because the sickle-shaped red blood cells have pointed ends, they can get stuck in capillary walls. If enough cells get stuck, they may cause blockage.

CHAPTER 15: The Heart and Circulatory System

Part I: Objective-Based Questions

OBJECTIVE 1—pp. 282–283

1. b
2. d
3. b
4. atrioventricular
5. aortic semilunar valve
6. **Figure 15–1**
A. right side
B. coronary vessels
C. aortic arch
D. pulmonary trunk
E. base
F. left side
G. apex

OBJECTIVE 2—pp. 283–284

1. c
2. b
3. b
4. tricuspid valve
5. left ventricle
6. **Figure 15–2**
A. aortic arch
B. superior vena cava
C. right atrium
D. tricuspid valve
E. chordae tendinae
F. inferior vena cava
G. pulmonary trunk
H. pulmonary semilunar valve
I. pulmonary arteries
J. pulmonary veins
K. left atrium
L. bicuspid valve
M. left ventricle
N. interventricular septum

OBJECTIVE 3—p. 285

1. c
2. d
3. d
4. serous membrane
5. **Figure 15–3**
A. intercalated disc
B. epicardium
C. myocardium
D. endocardium

OBJECTIVE 4—pp. 286–287

1. a
2. c
3. b
4. stethoscope
5. The closing of the pulmonary semilunar valves and the closing of the aortic semilunar valves prevents the backflow of blood into the ventricles.
6. **Figure 15–4**
A. tricuspid valve
B. bicuspid valve
C. aortic semilunar valve
D. pulmonary semilunar valve
E. the atrioventricular valves
F. the semilunar valves

OBJECTIVE 5—pp. 287–289

1. d
2. b
3. a
4. right atrium
5. sinoatrial node
6. **Figure 15–5**
A. sinoatrial node
B. atrioventricular node
C. bundle branches
D. Purkinje fibers
7. **Figure 15–6**
A. 1
B. 2
8. P
9. QRS
10. T

OBJECTIVE 6—p. 289

1. b
2. b
3. c
4. a
5. a
6. medium-sized veins
7. Varicose veins occur when the valves in the veins become weak or distorted, and blood begins to pool in the lower extremities. As the blood pools, the veins further distend, creating discomfort and a cosmetic problem.

OBJECTIVE 7—p. 290

1. b
2. a
3. b
4. a
5. A. decrease blood pressure
B. decrease blood pressure
C. increase blood pressure
D. increase blood pressure

OBJECTIVE 8—pp. 291–292

1. a
2. b
3. a
4. a
5. c
6. **Figure 15–7**
A. vein
B. venules
C. capillaries
D. artery
E. arteriole

OBJECTIVE 9—pp. 293–297

1. c
2. d
3. b
4. a
5. a
6. d
7. c
8. c
9. d
10. a
11. a
12. a
13. b
14. b
15. a
16. b
17. **Figure 15–8**
A. right carotid artery
B. right jugular vein
C. brachiocephalic vein
D. right pulmonary arteries
E. superior vena cava
F. right pulmonary vein
G. inferior vena cava
H. left carotid artery
I. aortic arch
J. left subclavian artery
K. pulmonary trunk
L. descending aorta
18. **Figure 15–9**
A. right common carotid artery
B. right subclavian artery
C. brachiocephalic artery
D. brachial artery
E. radial artery
F. ulnar artery
G. popliteal artery
H. posterior tibial artery
I. anterior tibial artery
J. left common carotid artery
K. aortic arch
L. left subclavian artery
M. axillary artery
N. common iliac artery
O. femoral artery
19. **Figure 15–10**
A. external jugular vein
B. subclavian vein
C. axillary vein
D. cephalic vein
E. brachial vein
F. basilic vein
G. median cubital vein
H. radial vein
I. great saphenous vein
J. popliteal vein
K. brachiocephalic vein
L. superior vena cava
M. inferior vena cava
N. common iliac vein
O. femoral vein
P. posterior tibial vein
Q. anterior tibial vein

OBJECTIVE 10—p. 298

1. b
2. d
3. c
4. d

OBJECTIVE 11—pp. 298–299

1. c
2. a
3. d
4. b
5. b
6. a

OBJECTIVE 12—pp. 299–300

1. a
2. b
3. d
4. d
5. a

Part II: Chapter-Comprehensive Exercises

MATCHING I

1. k
2. a
3. c
4. j
5. b
6. h
7. i
8. f
9. e
10. g
11. d
12. l

MATCHING II

1. k
2. b
3. g
4. l
5. f
6. c
7. e
8. h
9. j
10. i
11. a
12. d

CONCEPT MAP

1. endocardium
2. atria
3. right ventricle
4. left atrium
5. tricuspid
6. lungs
7. subclavian artery
8. superior vena cava

CROSSWORD PUZZLE

	1P										2F	E	3M	O	R	A	L
4C	A	P	I	L	L	A	5R	I	E	S			Y				
	L						E						O				
	P						N						C			6T	
	I		7P			8V	A	S	O	D	I	L	A	T	O	R	
	T		A				L						R			I	
	A		C		9P								D			C	
10A	T	H	E	R	O	S	C	L	E	R	O	S	I	S		U	
	I		M		P								U			S	
	O		A		L								M			P	
	N		K		I											I	
	S		E		T		12M									D	
		13A	R	T	E	R	I	E	S				14S				
	15B				A		T						Y				
	16A	X	I	L	L	A	R	Y					S				
	S						A			17L	E	F	T				
	E			18O	V	19A	L	E					O				
						N							L				
						20P	U	R	K	I	N	J	E				

FILL-IN-THE BLANK NARRATIVE

1. sinoatrial
2. atrioventricular
3. apex
4. Purkinje
5. ventricles
6. pulmonary arteries
7. oxygen
8. pulmonary veins
9. left ventricle
10. aortic arch
11. capillaries
12. carbon dioxide
13. deoxygenated
14. inferior vena cava
15. superior vena cava
16. right atrium
17. murmur
18. left
19. atrium
20. oxygenated
21. enlarged

CLINICAL CONCEPT

1. The body normally has 0 cervical ribs, 12 pairs of thoracic ribs, and 0 lumbar ribs.
2. A cervical rib might put pressure on the aortic arch or perhaps on a subclavian artery.
3. The number 60 represents the diastolic value.
4. The normal range for leukocytes is 6000–9000/mm^3 of blood. The normal range for erythrocytes is 4.8 million–5.4 million/mm^3 of blood.

CHAPTER 16: The Lymphatic System and Immunity

Part I: Objective-Based Questions

OBJECTIVE 1—pp. 307–308

1. d
2. d
3. b
4. d
5. thymosin
6. **Figure 16–1**

A. cervical nodes
B. thymus gland
C. thoracic duct
D. inguinal nodes
E. axillary nodes
F. spleen

OBJECTIVE 2—p. 309

1. c
2. a
3. d
4. bone marrow, natural killers

OBJECTIVE 3—pp. 309–310

1. a
2. d
3. c
4. b
5. The stomach has a mucus coating that is acidic. This acidic mucus will kill many pathogens.

OBJECTIVE 4—pp. 310–311

1. b
2. a
3. c
4. c
5. d

OBJECTIVE 5—p. 311

1. a
2. d
3. The word *humoral* means "liquid." The antibodies produced by the B cells are transported in plasma, which is a liquid.

OBJECTIVE 6—pp. 311–312

1. c
2. b
3. a
4. b

OBJECTIVE 7—p. 312

1. b
2. c
3. c

OBJECTIVE 8—pp. 312–313

1. c
2. d
3. Due to age, the immune surveillance cells decline in number; therefore, tumor cells are not eliminated as effectively.

OBJECTIVE 9—p. 313

1. d
2. b
3. a
4. c
5. d

OBJECTIVE 10—p. 314

1. c
2. d
3. c
4. b

Part II: Chapter-Comprehensive Exercises

MATCHING I

1. f
2. b
3. a
4. c
5. h
6. e
7. d
8. i
9. g

MATCHING II

1. h
2. i
3. a
4. g
5. e
6. f
7. c
8. d
9. b

CONCEPT MAP

1. nonspecific
2. inflammation
3. skin
4. interferon
5. innate
6. NK cell activity
7. helper T cells
8. AIDS
9. plasma B cells
10. antibodies
11. autoantibodies

CROSSWORD PUZZLE

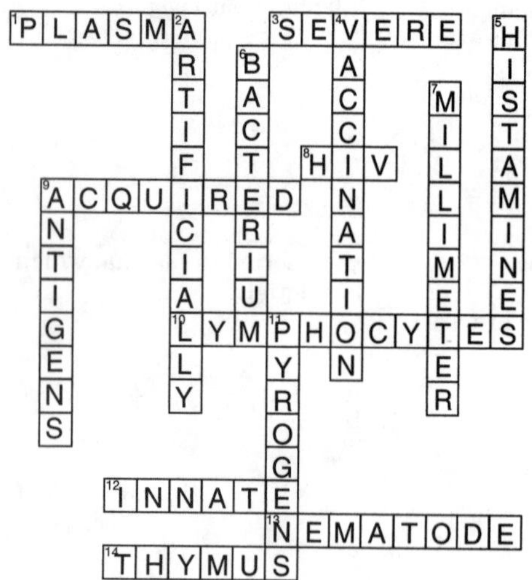

FILL-IN-THE-BLANK NARRATIVE

1. pathogens
2. innate
3. antigen
4. macrophages
5. helper T cells
6. B cells
7. plasma cells
8. antibodies
9. specific
10. immunity
11. secondary response
12. AIDS
13. HIV
14. secondary

CLINICAL CONCEPTS

1. Each individual flu virus has its own identifiable antigens.
2. The vaccine consists of the antigens of a specific virus. These antigens are on either a weakened form of the virus or a killed virus. Sam developed antibodies specific for the flu antigen just as he would have if he had been exposed to the live flu virus only without developing ill symptoms.
3. The flu Sam became sick with had different antigens. Sam was not immune to these particular antigens.
4. When Sam gets a vaccination, he is developing an artificially acquired immunity.
5. When Sam gets the "flu bug," he is developing a naturally acquired immunity.

CHAPTER 17: The Respiratory System

Part I: Objective-Based Questions

OBJECTIVE 1—pp. 321–322

1. b
2. b
3. b
4. lungs
5. cells
6. **Figure 17–1**
A. oxygen going into the bloodstream
B. carbon dioxide going into an alveolus
C. pulmonary circuit
D. systemic circuit
E. alveolus
F. bloodstream
G. oxygen going into a muscle cell
H. carbon dioxide leaving a muscle cell

OBJECTIVE 2—pp. 322–324

1. b
2. a
3. a
4. a
5. The hairs within the nose help prevent the entry of foreign particles.The nasal conchae within the nose warm the air. The mucus that coats the inside lining of the nose prevents the entry of foreign particles.
6. **Figure 17–2**
A. pharynx
B. nasal concha
C. larynx
D. tracheal cartilage
7. **Figure 17–3**
A. mucus layer
B. ciliated columnar cell
C. goblet cell

OBJECTIVE 3—pp. 324–325

1. a
2. d
3. b
4. d
5. d
6. **Figure 17–4**
A. highest concentration of carbon dioxide
B. movement of oxygen into the bloodstream
C. movement of carbon dioxide into the alveolus
D. highest concentration of oxygen

OBJECTIVE 4—pp. 325–327

1. a
2. c
3. c
4. b
5. a
6. **Figure 17–5**
A. a decrease in thoracic cavity size and therefore increased internal pressure
B. increased thoracic cavity size and therefore a decreased internal pressure
7. A. exhalation
B. inhalation
8. A. decreases
B. increases

OBJECTIVE 5—p. 327

1. a
2. a
3. a
4. c
5 d

OBJECTIVE 6—pp. 328–329

1. b
2. c
3. b
4. a
5. transported to the alveoli of the lungs to be exhaled.
6. An increased level of CO_2 lowers the blood pH
7. **Figure 17–6**
A. 7%
B. 23%
C. 70%

OBJECTIVE 7—pp. 329–331

1. d
2. b
3. e
4. stretch, vagus
5. carbon dioxide
6. **Figure 17-7**
A. pons
B. medulla oblongata
7. E
8. C and D
9. C and D

OBJECTIVE 8—p. 331

1. b
2. c
3. c

OBJECTIVE 9—p. 332

1. d
2. a
3. c
4. c
5 c

OBJECTIVE 10—pp. 332–333

1. d
2. b
3. d
4. a
5. c
6. b
7. c
8. c

Part II: Chapter-Comprehensive Exercises

MATCHING I

1. a
2. f
3. g
4. i
5. j
6. d
7. c
8. h
9. b
10. e
11. k

MATCHING II

1. a
2. b
3. d
4. j
5. k
6. c
7. h
8. i
9. e
10. g
11. f

CONCEPT MAP

1. diaphragm
2. decreased internal air pressure
3. glottis
4. increased internal air pressure
5. cystic fibrosis
6. bronchi
7. bronchioles
8. emphysema
9. pneumonia

CROSSWORD PUZZLE

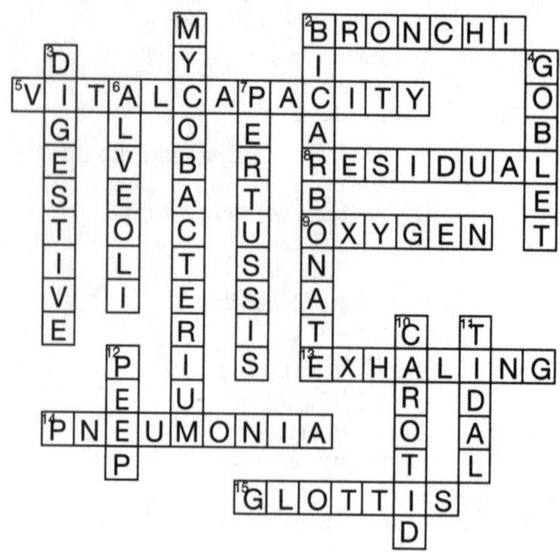

FILL-IN-THE-BLANK NARRATIVE

1. diaphragm
2. nasal conchae
3. epiglottis
4. bronchi
5. bronchioles
6. lobules
7. alveoli
8. pulmonary veins
9. pulmonary arteries
10. cystic fibrosis
11. cilia
12. lumen
13. destruction
14. pneumonia
15. homeostasis

CLINICAL CONCEPT

1. The Heimlich maneuver is performed by pushing up on the diaphragm muscle. Whenever the diaphragm muscle goes up, it decreases the size of the thoracic cavity, which increases internal pressure. This increased pressure forces the lodged material out.
2. If you were to slap someone on the back, you might dislodge the material and force it farther down the tracheal tubes.
3. If you are talking at the same time you are trying to swallow, the epiglottis will be open. If the epiglottis is open, there is a chance for food to enter the trachea.
4. When you prepare to swallow, the epiglottis closes over the glottis, which is the opening into the trachea. With the epiglottis closed, food can go down only the esophagus, not the trachea.
5. The Heimlich maneuver is applied to the diaphragm muscle.

CHAPTER 18: The Digestive System

Part I: Objective-Based Questions

OBJECTIVE 1—p. 340

1. d
2. c
3. a
4. c
5. The chemical breakdown of food into small molecules.

OBJECTIVE 2—pp. 341–342

1. b
2. c
3. b
4. c
5. d
6. **Figure 18–1**

A. liver
B. duodenum of the small intestine
C. large intestine
D. esophagus
E. stomach
F. pancreas
G. small intestine

OBJECTIVE 3—pp. 342–343

1. d
2. d
3. b
4. peristalsis

OBJECTIVE 4—pp. 343–344

1. c
2. a
3. a
4. c
5. a
6. a
7. b
8. **Figure 18–2**

A. cuspid
B. incisors
C. molars
D. bicuspids

OBJECTIVE 5—pp. 344–345

1. a
2. c
3. b
4. c
5. chyme
6. **Figure 18-3**

A. esophagus
B. esophageal sphincter area
C. body of the stomach
D. pyloric sphincter
E. duodenum
F. rugae
G. fundus

OBJECTIVE 6—p. 346

1. d
2. d
3. d
4. c
5. c
6. b

OBJECTIVE 7—pp. 346–348

1. a
2. a
3. b
4. d
5. d
6. a
7. **Figure 18–4**

A. hepatic ducts
B. gallbladder
C. duodenum
D. liver
E. common bile duct
F. pancreas

OBJECTIVE 8—pp. 348–349

1. c
2. c
3. d
4. a
5. d
6. **Figure 18-5**

A. transverse colon
B. ascending colon
C. cecum
D. appendix
E. rectum
F. descending colon
G. sigmoid colon

OBJECTIVE 9—pp. 349–350

1. a
2. c
3. b
4. c
5. d

OBJECTIVE 10—p. 350

1. a
2. c
3. If the diet of the elderly is deficient in calcium, tooth loss will occur, and the bone mass will be reduced. Calcium ions are necessary to maintain the teeth and bone mass.
4. oral
5. constipation

OBJECTIVE 11—p. 351

1. b
2. d
3. b
4. a
5. b

OBJECTIVE 12—pp. 351–352

1. c
2. d
3. a
4. a
5. b
6. a

Part II: Chapter-Comprehensive Exercises

MATCHING

1. f
2. h
3. i
4. a
5. k
6. j
7. c
8. b
9. e
10. l
11. g
12. m
13. d

CONCEPT MAP

1. pancreas
2. duodenum
3. salivary amylase
4. pepsin
5. large intestine
6. lipase
7. CCK
8. carbohydrates
9. proteins
10. fats

CROSSWORD PUZZLE

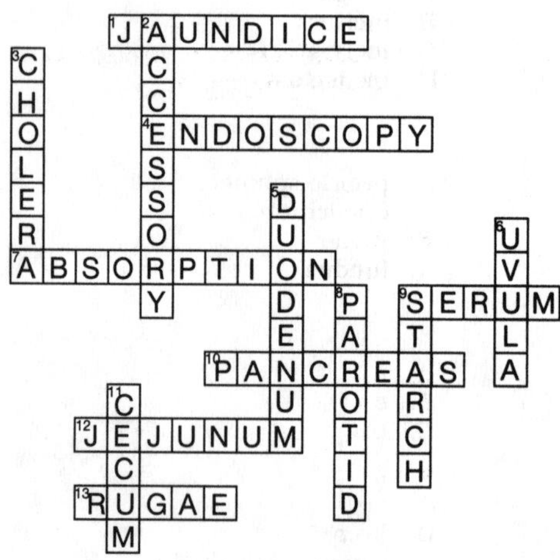

FILL-IN-THE-BLANK NARRATIVE

1. incisors
2. pharynx
3. uvula
4. epiglottis
5. esophagus
6. peristalsis
7. esophageal
8. rugae
9. gastrin
10. pepsin
11. pyloric
12. duodenum
13. cholecystokinin
14. proteinases
15. carboxypeptidase
16. jejunum
17. villi
18. ileum
19. cecum
20. diarrhea
21. dehydration
22. constipation

CLINICAL CONCEPT

1. Bile is produced by the liver once the liver has been targeted by the hormone secretin. Bile leaves the liver and eventually becomes stored in the gallbladder.
2. Bile consists of bile salts and water. To concentrate the bile salts, the gallbladder releases water to the bloodstream. Sometimes, the bile salts become too concentrated and coalesce, thus forming stones.
3. Cholecystokinin causes the gallbladder to release bile into the duodenum. Bile assists lipase by emulsifying fat. This emulsification process makes it easier for lipase to do its joB.
4. Steve needs to reduce his fat intake because with the removal of the gallbladder, he no longer has an adequate supply of bile to emulsify the fat. His liver continues to make bile, so Steve can still consume some fat but not in large quantities.
5. Bile leaves the liver and enters the common bile duct. From the common bile duct, bile enters the duodenum.

CHAPTER 19: Nutrition and Metabolism

Part I: Objective-Based Questions

OBJECTIVE 1—p. 359

1. d
2. a
3. d

OBJECTIVE 2—pp. 359–360

1. b
2. a
3. a
4. a
5. b

OBJECTIVE 3—pp. 360–362

1. c
2. c
3. c
4. b
5. a
6. c
7. **Figure 19-1**

A. fatty acids
B. glycerol
C. glucose
D. amino acids
E. mitochondrion

OBJECTIVE 4—pp. 362–363

1. a
2. c
3. a
4. c

OBJECTIVE 5—p. 363

1. d
2. c
3. c
4. Fats, oils, and sweets are placed at the top of the food pyramid to stress that we need to restrict our intake of those food groups.

OBJECTIVE 6—p. 364

1. d
2. d
3. b
4. b
5. d

OBJECTIVE 7—pp. 364–365

1. d
2. a
3. d

OBJECTIVE 8—pp. 365–366

1. d
2. a
3. c
4. a
5. d

OBJECTIVE 9—pp. 366–367

1. b
2. a
3. a
4. d
5. a
6. d

Part II: Chapter-Comprehensive Exercises

MATCHING

1. d
2. b
3. c
4. a
5. g
6. h
7. f
8. j
9. i
10. e
11. k
12. l

CONCEPT MAP

1. anabolism
2. carbohydrates
3. lipids
4. protein
5. amino acids
6. glycerol
7. essential
8. tyrosine
9. lysine
10. PKU
11. phenylalanine
12. linoleic acid
13. ketones

CROSSWORD PUZZLE

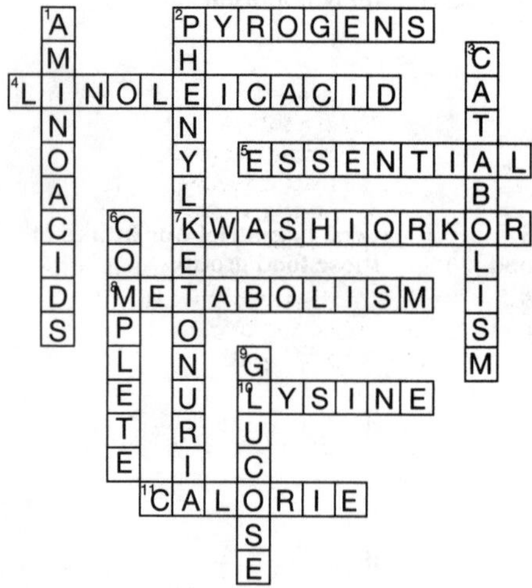

FILL-IN-THE BLANK NARRATIVE

1. catabolism
2. anabolism
3. metabolism
4. anaerobic
5. aerobic
6. ketones
7. pH
8. enzymes

CLINICAL CONCEPTS

1. Corn is considered to be an incomplete protein because it does not contain all the essential amino acids.
2. As the body continues to metabolize fat, a byproduct (ketone) is produced.
3. Sodium ions are the predominant positive ion on the outside of the nerve membrane. When these ions enter the membrane, an impulse is created. A lack of sodium ions can therefore be linked to nerve problems.
4. Some fad diets are high in protein. By consuming excess protein and no carbohydrates, the body begins to metabolize fats. This will cause weight loss, but excessive metabolism of fat will produce ketones.

CHAPTER 20: The Urinary System and Body Fluids

Part I: Objective-Based Questions

OBJECTIVE 1—pp. 373–374

1. c
2. c
3. b
4. The left kidney is slightly higher than the right kidney.
5. **Figure 20–1**
A. kidney
B. ureters
C. urinary bladder
D. urethra

OBJECTIVE 2—pp. 374–375

1. d
2. b
3. c
4. **Figure 20–2**
A. renal cortex
B. renal medulla
C. renal pelvis
D. ureter
E. major calyx
F. minor calyx
G. renal capsule
H. renal pyramids

OBJECTIVE 3—pp. 376–378

1. c
2. d
3. d
4. a
5. d
6. **Figure 20–3**
A. efferent arteriole
B. afferent arteriole
C. Bowman's capsule
D. glomerulus
E. proximal convoluted tubule
F. loop of Henle
G. distal convoluted tubule
H. collecting duct
7. **Figure 20–4**
A. Bowman's capsule
B. efferent arteriole
C. afferent arteriole
D. glomerulus
E. proximal convoluted tubule
F. distal convoluted tubule
G. collecting duct
H. loop of Henle

OBJECTIVE 4—pp. 378–379

1. c
2. b
3. d
4. b
5. b

OBJECTIVE 5—pp. 379–380

1. a
2. b
3. b
4. d

OBJECTIVE 6—p. 380

1. b
2. d
3. d

OBJECTIVE 7—pp. 380–381

1. b
2. d
3. a
4. d
5. c
6. **Figure 20–5**
A. ureter
B. ureteral openings
C. internal urethral sphincter
D. external urethral sphincter
E. trigone
F. urethra

OBJECTIVE 8—p. 382

1. c
2. d
3. c
4. The male urethra is involved in the transportation of urine as well as sperm cells. The female urethra transports only urine.

OBJECTIVE 9—pp. 382–383

2. c
2. b
3. d

OBJECTIVE 10—pp. 383–384

1. b
2. a
3. b
4. d
5. a
6. c
7. c
8. d

OBJECTIVE 11—p. 384

1. d
2. b
3. a
4. As the prostate gland swells it constricts the urethra, thus restricting the flow of urine.

OBJECTIVE 12—pp. 384–385

1. d
2. a
3. b
4. a
5. d

OBJECTIVE 13—pp. 385–386

1. c
2. a
3. d
4. b
5. b

OBJECTIVE 14—p. 386

1. a
2. d
3. a
4. a
5. c

Part II: Chapter-Comprehensive Exercises

MATCHING I

1. f
2. c
3. g
4. h
5. e
6. d
7. a
8. b

MATCHING II

1. f
2. c
3. b
4. a
5. e
6. d
7. g
8. h
9. i

CONCEPT MAP

1. renal artery
2. glomerulus
3. efferent arteriole
4. glomerulonephritis
5. PCT
6. collecting duct
7. major calyx
8. ureter
9. pyelitis
10. pyelonephritis

CROSSWORD PUZZLE

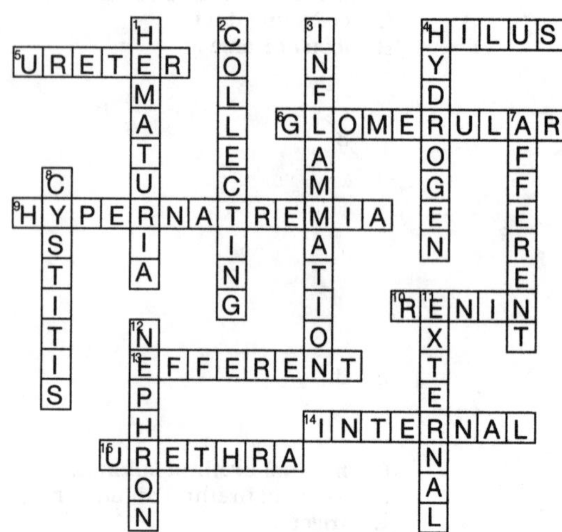

FILL-IN-THE-BLANK NARRATIVE

1. afferent
2. glomerular
3. Bowman's
4. efferent
5. vasa recta
6. PCT
7. loop of Henle
8. DCT
9. collecting
10. minor calyx
11. glomerulonephritis
12. filtration
13. blood

CLINICAL CONCEPT

1. An increase in hydrogen ions will cause a decrease in pH.
2. The stomach needs to make more HCl in this scenario because the consumption of more food requires more HCl to assist in the formation of pepsin, which is a digestive enzyme.
3. A pH value of 6.0 is more acidic than a pH value of 6.5.
4. The presence of more food in the stomach causes the stomach to release gastrin. Gastrin in turn will cause the stomach to begin making HCl.
5. The pH of the blood is rising, since the stomach cells are removing the hydrogen ions from the blood to make more HCl.

CHAPTER 21: The Reproductive System

Part I: Objective-Based Questions

OBJECTIVE 1—p. 393

1. d
2. c
3. c
4. d
5. c

OBJECTIVE 2—pp. 394–395

1. c
2. c
3. c
4. b
5. d
6. **Figure 21-1**

A. ductus deferens
B. penis
C. testis
D. scrotum
E. seminal vesicle gland
F. prostate gland
G. ejaculatory duct
H. bulbourethral gland
I. epididymis

OBJECTIVE 3—pp. 395–396

1. c
2. d
3. b
4. 23
5. In order for the sperm cells to develop, the temperature must be about 2°F cooler than body temperature. If the testes were inside the abdominopelvic region, the temperature would be too high and the sperm cells would not develop properly.

OBJECTIVE 4—pp. 396–397

1. a
2. b
3. d
4. d
5. d
6. b

OBJECTIVE 5—p. 397

1. d
2. a
3. d
4. a
5. d

OBJECTIVE 6—pp. 398–399

1. c
2. b
3. b
4. d
5. b
6. **Figure 21-2**

A. ovary
B. uterine tube
C. urinary bladder
D. urethra
E. myometrium
F. endometrium
G. cervix
H. vagina

OBJECTIVE 7—pp. 399–400

1. a
2. b
3. b
4. c
5. a
6. a
7. a

OBJECTIVE 8—pp. 400–401

1. a
2. b
3. c
4. c
5. a

OBJECTIVE 9—pp. 401–402

1. c
2. c
3. c
4. testosterone

OBJECTIVE 10—p. 402

1. d
2. d
3. d
4. b
5. b

OBJECTIVE 11—p. 403

1. b
2. c
3. d
4. b
5. b
6. b
7. d

Part II: Chapter-Comprehensive Exercises

MATCHING I

1. e
2. h
3. g
4. a
5. c
6. d
7. f
8. b

MATCHING II

1. g
2. a
3. d
4. f
5. h
6. e
7. c
8. b

CONCEPT MAP

1. seminiferous tubules
2. seminal vesicle
3. prostate gland
4. uterine tube
5. orchitis
6. epididymis
7. progesterone
8. acidic semen
9. candidiasis
10. urethra

CROSSWORD PUZZLE

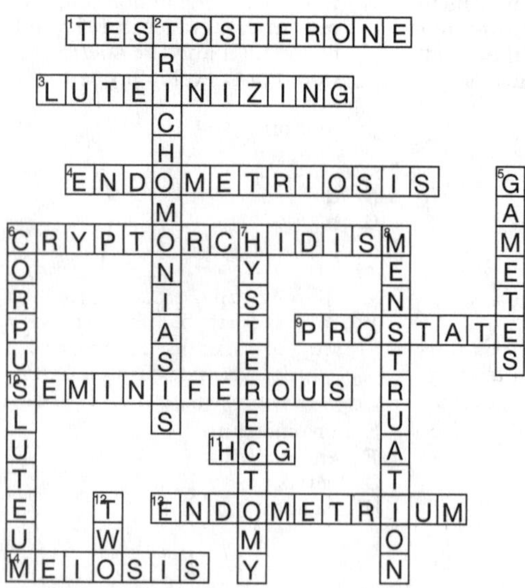

FILL-IN-THE-BLANK NARRATIVE

1. seminiferous tubules
2. spermatogenesis
3. meiosis
4. epididymis
5. ductus deferens
6. seminal vesicle gland
7. ejaculatory duct
8. prostate gland
9. bulbourethral gland
10. uterine tube
11. corpus luteum
12. progesterone
13. endometrium
14. zygote
15. uterus
16. ectopic

CLINICAL CONCEPT

1. By saying that the uterus cannot prepare itself for the zygote, we mean that the endometrium cannot thicken enough to prepare for the implantation of the fertilized egg. The fertilized egg will not implant and will therefore degenerate and die.
2. The endometrium is the part of the uterus that is involved in preparing for the zygote.
3. The corpus luteum is a follicle that has ruptured to release the egg.
4. The hormone FSH stimulates egg production, and the hormone LH causes ovulation.
5. An individual sperm cell does not have enough hyaluronidase to break down the egg barrier, but several sperm cells together will accumulate enough hyaluronidase to begin the breakdown of the egg barrier. As soon as the barrier is broken, a sperm cell can enter and fertilize the egg.

CHAPTER 22: Development and Inheritance

Part I: Objective-Based Questions

OBJECTIVE 1—p. 410

1. d
2. d
3. c
4. An individual sperm cell does not have enough hyaluronidase to break down the egg barrier. It requires several sperm cells to accumulate enough hyaluronidase to do the job. As soon as the egg barrier begins to decompose, a sperm will enter and fertilize the egg.

OBJECTIVE 2—pp. 410–411

1. b
2. a
3. b
4. c
5. d
6. b

OBJECTIVE 3—p. 411–412

1. a
2. a
3. c
4. b

OBJECTIVE 4—pp. 412–413

1. d
2. b
3. a
4. c
5. d
6. **Figure 22–1**

A. amnion
B. developing embryo
C. chorion
D. allantois
E. yolk sac

OBJECTIVE 5—pp. 413–414

1. c
2. d
3. d
4. c
5. The need to urinate is quite frequent during pregnancy because the expansion of the uterus causes a constriction of the urinary bladder. Therefore, the urinary bladder cannot hold as much urine as it could before. Also, increased body fluids during pregnancy cause the urinary bladder to fill more rapidly than before. Several bouts of constipation seem to occur during pregnancy because the expanding uterus causes constriction of the large intestine.

OBJECTIVE 6—p. 414

1. a
2. a
3. a
4. a

OBJECTIVE 7—p. 415

1. d
2. b
3. b
4. Colostrum consists of proteins. Many of these proteins are antibodies that help to fight infections.

OBJECTIVE 8—pp. 415–417

1. a
2. c
3. d
4. d
5. d
6. d
7. c
8. a
9. d

OBJECTIVE 9—pp. 417–418

1. d
2. a
3. a
4. d
5. b
6. d
7. d

OBJECTIVE 10—p. 418

1. c
2. c

Part II: Chapter-Comprehensive Exercises

MATCHING

1. i
2. d
3. j
4. a
5. h
6. c
7. g
8. b
9. l
10. e
11. f
12. k

CONCEPT MAP

1. zygote
2. germ layer
3. endoderm
4. muscle
5. yolk sac
6. allantois
7. progesterone
8. relaxin

CROSSWORD PUZZLE

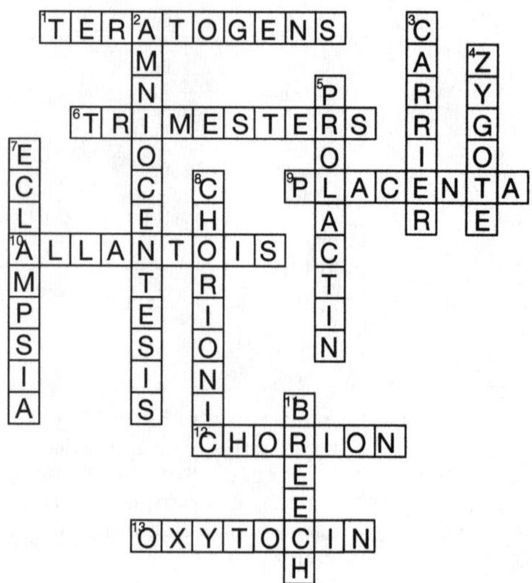

FILL-IN-THE BLANK-NARRATIVE

1. mesoderm
2. endoderm
3. ectoderm
4. amnion
5. yolk
6. gestation
7. oxytocin
8. dilation
9. expulsion
10. placental
11. teratogen
12. oxygen
13. fetal
14. gonorrhea
15. syphilis

CLINICAL CONCEPT

1. Mai Ling has to urinate often because the expanded uterus causes a constriction of the urinary bladder, so it cannot hold as much urine as before. Also, during pregnancy, the fluid volume of the female's body increases tremendously, so a smaller urinary bladder has to contend with a larger volume of fluid.
2. The urinary bladder decreases in size due to the expanded uterus. The urinary bladder now holds very little urine when full.
3. Constipation is rather frequent during pregnancy because the expanded uterus also causes constriction of the large intestine.
4. Oxytocin is produce by the hypothalamus and is released by the pituitary gland. Oxytocin is responsible for causing uterine contractions.
5. Oxytocin is released to create uterine contractions in order for the child to be born. Oxytocin also triggers milk release for the nursing child. As the child nurses, pressure signals to the brain tell it to release more oxytocin so the child can receive more milk. Oxytocin released to target the mammary glands, also targets the uterus, thereby causing contractions.
6. Mai Ling has difficulty breathing because the expanded uterus is putting pressure on the inferior side of the diaphragm muscle. This does not allow the diaphragm muscle to lower adequately during inhalation.